CAMBRIDGE LIBRARY COLLECTION

Books of enduring scholarly value

Darwin

Two hundred years after his birth and 150 years after the publication of 'On the Origin of Species', Charles Darwin and his theories are still the focus of worldwide attention. This series offers not only works by Darwin, but also the writings of his mentors in Cambridge and elsewhere, and a survey of the impassioned scientific, philosophical and theological debates sparked by his 'dangerous idea'.

Principles of Geology

In 1830, Charles Lyell laid the foundations of evolutionary biology with Principles of Geology, a pioneering book that Charles Darwin took with him on the Beagle. Lyell championed the ideas of geologist James Hutton, who formulated one of the fundamental principles of modern geology – uniformitarianism. This proposed that natural processes always operate according to the same laws, and that this allows us to understand how features of the Earth's surface were produced by physical, chemical, and biological processes over long periods of time. Volume 1 consists of 26 chapters, a comprehensive index and woodcut illustrations of various mechanisms of geological change. Lyell begins with a definition of geology and then reviews ancient theories of the successive destruction and renovation of the world. He mentions James Hutton's ideas in chapter four, and goes on to discuss the effects of climate change, running water, volcanic eruptions and earthquakes on the Earth's crust.

Cambridge University Press has long been a pioneer in the reissuing of out-of-print titles from its own backlist, producing digital reprints of books that are still sought after by scholars and students but could not be reprinted economically using traditional technology. The Cambridge Library Collection extends this activity to a wider range of books which are still of importance to researchers and professionals, either for the source material they contain, or as landmarks in the history of their academic discipline.

Drawing from the world-renowned collections in the Cambridge University Library, and guided by the advice of experts in each subject area, Cambridge University Press is using state-of-the-art scanning machines in its own Printing House to capture the content of each book selected for inclusion. The files are processed to give a consistently clear, crisp image, and the books finished to the high quality standard for which the Press is recognised around the world. The latest print-on-demand technology ensures that the books will remain available indefinitely, and that orders for single or multiple copies can quickly be supplied.

The Cambridge Library Collection will bring back to life books of enduring scholarly value across a wide range of disciplines in the humanities and social sciences and in science and technology.

Principles of Geology

An Attempt to Explain the Former Changes of the Earth's Surface, by Reference to Causes now in Operation

VOLUME 1

CHARLES LYELL

CAMBRIDGE UNIVERSITY PRESS

Cambridge New York Melbourne Madrid Cape Town Singapore São Paolo Delhi

Published in the United States of America by Cambridge University Press, New York

www.cambridge.org
Information on this title: www.cambridge.org/9781108001359

This edition first published 1830
This digitally printed version 2009

ISBN 978-1-108-00135-9

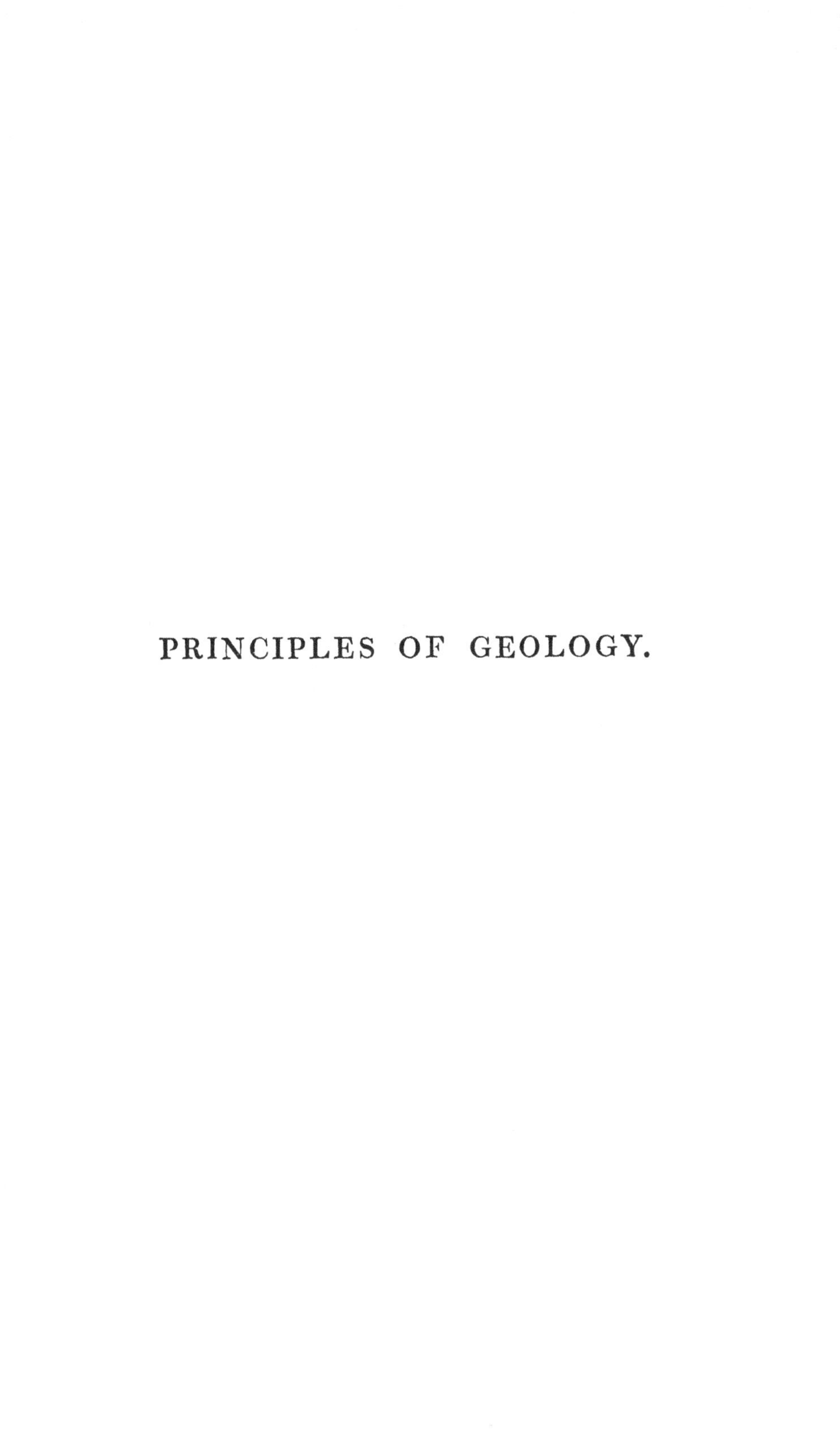

PRINCIPLES OF GEOLOGY.

"Amid all the revolutions of the globe the economy of Nature has been uniform, and her laws are the only things that have resisted the general movement. The rivers and the rocks, the seas and the continents have been changed in all their parts; but the laws which direct those changes, and the rules to which they are subject, have remained invariably the same."—PLAYFAIR, *Illustrations of the Huttonian Theory*, § 374.

T. Bradley, Sc.

Present state of the Temple of Serapis at Puzzuoli.

London, Published by John Murray, Albemarle St. June, 1830.

PRINCIPLES

OF

GEOLOGY,

BEING

AN ATTEMPT TO EXPLAIN THE FORMER CHANGES
OF THE EARTH'S SURFACE,

BY REFERENCE TO CAUSES NOW IN OPERATION.

BY

CHARLES LYELL, Esq., F.R.S.
FOR. SEC. TO THE GEOL. SOC., &c.

IN TWO VOLUMES.

VOL. I.

LONDON:
JOHN MURRAY, ALBEMARLE-STREET.

MDCCCXXX.

CONTENTS.

PAGE

CHAPTER I.

CHAPTER II.

CHAPTER III.

PAGE

CHAPTER VIII.

CHAPTER IX.

CHAPTER X.

CHAPTER XI.

PAGE

CHAPTER XV.

CHAPTER XVI.

CHAPTER XVII.

CHAPTER XVIII.

CHAPTER XIX.

CHAPTER XX.

CHAPTER XXI.

CHAPTER XXII.

CHAPTER XXIII.

CHAPTER XXIV.

CHAPTER XXV.

CHAPTER XXVI.

ERRATA.

Page	35	*line*	2	*from*	top,	*for*	or	*read*	and
—	35	—	13	—	bottom,	—	Alp	—	Alps
—	150	—	12	—	top,	—	law	—	laws
—	154	—	11	—	do.,	—	Of	—	In
—	190	—			heading	—	LOUISLAND	—	LOUISIANA
—	226	—			do.	—	DELTA	—	DELTAS
—	326	—	6	—	bottom	—	Pl. 3	—	Pl. 2.

LIST OF PLATES AND WOOD-CUTS

IN

THE FIRST VOLUME.

PLATES.

WOOD-CUTS.

PRINCIPLES OF GEOLOGY.

CHAPTER I.

Geology defined—Compared to History—Its relation to other Physical Sciences—Its distinctness from all—Not to be confounded with Cosmogony.

Geology is the science which investigates the successive changes that have taken place in the organic and inorganic kingdoms of nature; it enquires into the causes of these changes, and the influence which they have exerted in modifying the surface and external structure of our planet.

By these researches into the state of the earth and its inhabitants at former periods, we acquire a more perfect knowledge of its *present* condition, and more comprehensive views concerning the laws *now* governing its animate and inanimate productions. When we study history, we obtain a more profound insight into human nature, by instituting a comparison between the present and former states of society. We trace the long series of events which have gradually led to the actual posture of affairs; and by connecting effects with their causes, we are enabled to classify and retain in the memory a multitude of complicated relations—the various peculiarities of national character—the different degrees of moral and intellectual refinement, and numerous other circumstances, which, without historical associations, would be uninteresting or imperfectly understood. As the present condition of nations is the result of many antecedent changes, some extremely remote and others recent, some gradual, others sudden and violent, so the state of the natural world is the result of a long succession of events, and if we would enlarge our experience of the present economy of nature, we must investigate the effects of her operations in former epochs.

We often discover with surprise, on looking back into the chronicles of nations, how the fortune of some battle has influenced the fate of millions of our contemporaries, when it has long been forgotten by the mass of the population. With this remote event we may find inseparably connected the geographical boundaries of a great state, the language now spoken by the inhabitants, their peculiar manners, laws, and religious opinions. But far more astonishing and unexpected are the connexions brought to light, when we carry back our researches into the history of nature. The form of a coast, the configuration of the interior of a country, the existence and extent of lakes, valleys, and mountains, can often be traced to the former prevalence of earthquakes and volcanoes, in regions which have long been undisturbed. To these remote convulsions the present fertility of some districts, the sterile character of others, the elevation of land above the sea, the climate, and various peculiarities, may be distinctly referred. On the other hand, many distinguishing features of the surface may often be ascribed to the operation at a remote era of slow and tranquil causes—to the gradual deposition of sediment in a lake or in the ocean, or to the prolific growth in the same of corals and testacea. To select another example, we find in certain localities subterranean deposits of coal, consisting of vegetable matter, formerly drifted into seas and lakes. These seas and lakes have since been filled up, the lands whereon the forests grew have disappeared or changed their form, the rivers and currents which floated the vegetable masses can no longer be traced, and the plants belonged to species which for ages have passed away from the surface of our planet. Yet the commercial prosperity, and numerical strength of a nation, may now be mainly dependant on the local distribution of fuel determined by that ancient state of things.

Geology is intimately related to almost all the physical sciences, as is history to the moral. An historian should, if possible, be at once profoundly acquainted with ethics, politics, jurisprudence, the military art, theology; in a word, with all branches of knowledge, whereby any insight into human affairs, or into the moral and intellectual nature of man, can be obtained. It would be no less desirable that a geologist should be well versed in chemistry, natural philosophy, mineralogy, zoology, comparative anatomy, botany; in short, in every science relating to organic and inorganic nature. With these accomplishments the historian and geologist would rarely fail to draw correct and philosophical conclusions from the various monuments transmitted to them of former occurrences. They would know to what combination of causes analogous effects were referrible, and they would often be enabled to supply by inference, information concerning many events unrecorded in the defective archives of former ages. But the brief duration of human life, and our limited powers, are so far from permitting us to aspire to such extensive acquisitions, that excellence even in one department is within the reach of few, and those individuals most effectually promote the general progress, who concentrate their thoughts on a limited portion of the field of inquiry. As it is necessary that the historian and the cultivators of moral or political science should reciprocally aid each other, so the geologist and those who study natural history or physics stand in equal need of mutual assistance. A comparative anatomist may derive some accession of knowledge from the bare inspection of the remains of an extinct quadruped, but the relic throws much greater light upon his own science, when he is informed to what relative era it belonged, what plants and animals were its contemporaries, in what degree of latitude it once existed, and other historical details. A fossil shell may interest a conchologist, though he be ignorant of the locality from which it came; but it will be of more value when he learns with what other species it was associated, whether they were marine or fresh-water, whether the strata containing them were at a certain elevation above the sea, and what relative position they held in regard to other groups of strata, with many other particulars determinable by an experienced geologist alone. On the other hand, the skill of the comparative anatomist and conchologist are often indispensable to those engaged in geological research, although it will rarely happen that the geologist will himself combine these different qualifications in his own person.

Some remains of former organic beings, like the ancient temple, statue, or picture, may have both their intrinsic and their historical value, while there are others which can never be expected to attract attention for their own sake. A

B 2

painter, sculptor, or architect, would often neglect many curious relics of antiquity, as devoid of beauty and uninstructive with relation to their own art, however illustrative of the progress of refinement in some ancient nation. It has therefore been found desirable that the antiquary should unite his labours to those of the historian, and similar co-operation has become necessary in geology. The field of inquiry in living nature being inexhaustible, the zoologist and botanist can rarely be induced to sacrifice time in exploring the imperfect remains of lost species of animals and plants, while those still existing afford constant matter of novelty. They must entertain a desire of promoting *geology* by such investigations, and some knowledge of its objects must guide and direct their studies. According to the different opportunities, tastes, and talents of individuals, they may employ themselves in collecting particular kinds of minerals, rocks, or organic remains, and these, when well examined and explained, afford data to the geologist, as do coins, medals, and inscriptions to the historian.

It was long ere the distinct nature and legitimate objects of geology were fully recognized, and it was at first confounded with many other branches of inquiry, just as the limits of history, poetry, and mythology were ill-defined in the infancy of civilization. Werner appears to have regarded geology as little other than a subordinate department of mineralogy, and Desmarest included it under the head of Physical Geography. But the identification of its objects with those of Cosmogony has been the most common and serious source of confusion. The first who endeavoured to draw a clear line of demarcation between these distinct departments, was Hutton, who declared that geology was in no ways concerned "with questions as to the origin of things." But his doctrine on this head was vehemently opposed at first, and although it has gradually gained ground, and will ultimately prevail, it is yet far from being established. We shall attempt in the sequel of this work to demonstrate that geology differs as widely from cosmogony, as speculations concerning the creation of man differ from history. But before we enter more at large on this controverted question, we shall endeavour to trace the progress of opinion on this topic, from the earliest ages, to the commencement of the present century.

CHAPTER II.

Oriental Cosmogony—Doctrine of the successive destruction and renovation of the world—Origin of this doctrine—Common to the Egyptians—Adopted by the Greeks—System of Pythagoras—Of Aristotle—Dogmas concerning the extinction and reproduction of genera and species—Strabo's theory of elevation by earthquakes—Pliny—Concluding Remarks on the knowledge of the Ancients.

THE earliest doctrines of the Indian and Egyptian schools of philosophy, agreed in ascribing the first creation of the world to an omnipotent and infinite Being. They concurred also in representing this Being, who had existed from all eternity, as having repeatedly destroyed and reproduced the world and all its inhabitants. In the "Institutes of Menù," the sacred volume of the Hindoos, to which, in its present form, Sir William Jones ascribes an antiquity of at least eight hundred and eighty years before Christ, we find this system of the alternate destruction and renovation of the world, proposed in the following remarkable verses.

"The Being, whose powers are incomprehensible, having created me (Menù) and this universe, again became absorbed in the supreme spirit, changing the time of energy for the hour of repose.

"When that power awakes, then has this world its full expansion; but when he slumbers with a tranquil spirit, then the whole system fades away. . . . . For while he reposes as it were, embodied spirits endowed with principles of action depart from their several acts, and the mind itself becomes inert."

Menù then describes the absorption of all beings into the Supreme essence, and the Divine soul itself is said to slumber, and to remain for a time immersed in "the first idea, or in darkness." He then proceeds, (verse fifty-seven,) "Thus that immutable power, by waking and reposing alternately, revivifies and destroys, in eternal succession, this whole assemblage of locomotive and immoveable creatures."

It is then declared that there has been a long succession of *manwantaras*, or periods, each of the duration of many thousand ages, and—

"There are creations also, and destructions of worlds innumerable: the Being, supremely exalted, performs all this with as much ease as if in sport, again and again for the sake of conferring happiness *."

The compilation of the ordinances of Menù was not all the work of one author nor of one period, and to this circumstance some of the remarkable inequalities of style and matter are probably attributable. There are many passages, however, wherein the attributes and acts of the "Infinite and Incomprehensible Being" are spoken of with much grandeur of conception and sublimity of diction, as some of the passages above cited, though sufficiently mysterious, may serve to exemplify. There are at the same time such puerile conceits and monstrous absurdities in the same cosmogony, that some may impute to mere accident any slight approximation to truth, or apparent coincidence between the oriental dogmas and observed facts. This pretended revelation, however, was not purely an effort of the unassisted imagination, nor invented without regard to the opinions and observations of naturalists. There are introduced into the same chapter, certain astronomical theories, evidently derived from observation and reasoning. Thus for instance, it is declared that, at the North Pole, the year was divided into a long day and night, and that their long day was the northern, and their night the southern course of the sun;* and to the inhabitants of the moon it is said, one day is equal in length to one month of mortals †. If such statements cannot be resolved into mere conjectures, we have no right to refer, to mere chance, the prevailing notion, that the earth and its inhabitants had formerly undergone a succession of revolutions and catastrophes, interrupted by long intervals of tranquillity.

Now there are two sources in which such a theory may

* Institutes of Hindoo Law, or the Ordinances of Menù, from the Sanscrit, translated by Sir William Jones, 1796.

† Menù Instit. c. i. 66 and 67.

have originated. The marks of former convulsions on every part of the surface of our planet are obvious and striking. The remains of marine animals imbedded in the solid strata are so abundant, that they may be expected to force themselves on the observation of every people who have made some progress in refinement; and especially where one class of men are expressly set apart from the rest for study and contemplation. If these appearances are once recognized, it seems natural that the mind should come to the conclusion, not only of mighty changes in past ages, but of alternate periods of repose and disorder—of repose when the fossil animals lived, grew, and multiplied—of disorder, when the strata wherein they were buried became transferred from the sea to the interior of continents, and entered into high mountain chains. Those modern writers, who are disposed to disparage the former intellectual advancement and civilization of eastern nations, might concede some foundation of observed facts for the curious theories now under consideration, without indulging in exaggerated opinions of the progress of science; especially as universal catastrophes of the world, and exterminations of organic beings, in the sense in which they were understood by the Brahmin, are untenable doctrines. We know that the Egyptian priests were aware, not only that the soil beneath the plains of the Nile, but that also the hills bounding the great valley, contained marine shells; and it could hardly have escaped the observation of Eastern philosophers, that some soils were filled with fossil remains, since so many national works were executed on a magnificent scale by oriental monarchs in very remote eras. Great canals and tanks required extensive excavations; and we know that in more recent times (the fourteenth century of our era) the removal of soil necessary for such undertakings, brought to light geological phenomena, which attracted the attention of a people less civilized than were many of the older nations of the East *.

* This circumstance is mentioned in a Persian MS. copy of the historian Ferishta, in the library of the East India Company, relating to the rise and progress of the Mahomedan Empire in India, and procured from the library of Tippoo Sultaun in 1799; and has been recently referred to at some length by Dr. Buckland.—(Geol. Trans. 2d Series, vol. ii. part iii. p. 389.)—It is stated that, in the year 762, (or 1360 of our era) the king employed fifty thousand labourers in cutting through a

But although we believe the Brahmins, like the priests of Egypt, to have been acquainted with the existence of fossil remains in the strata, it is probable that the doctrine of successive destructions and renovations of the world merely received corroboration from such proofs; and that it was originally handed down, like the religious dogmas of most nations, from a ruder state of society. The true source of the system must be sought for in the exaggerated traditions of those partial, but often dreadful catastrophes, which are sometimes occasioned by various combinations of natural causes. Floods and volcanic eruptions, the agency of water and fire, are the chief instruments of devastation on our globe. We shall point out in the sequel the extent of these calamities, recurring at distant intervals of time, in the present course of nature; and shall only observe here, that they are so peculiarly calculated to inspire a lasting terror, and are so often fatal in their consequences to great multitudes of people, that it scarcely requires the passion for the marvellous, so characteristic of rude and half-civilized nations, still less the exuberant imagination of eastern writers, to augment them into general cataclysms and conflagrations.

Humboldt relates the interesting fact, that after the annihilation of a large part of the inhabitants of Cumana, by an earthquake in 1766, a season of extraordinary fertility ensued, in consequence of the great rains which accompanied the subterranean convulsions. "The Indians," he says, "celebrated, after the ideas of an antique superstition, by festivals and dancing, the destruction of the world and the approaching epoch of its regeneration*."

The existence of such rites among the rude nations of South America is most important, for it shews what effects may be produced by great catastrophes of this nature, recurring at distant intervals of time, on the minds of a barbarous and uncultivated race. The superstitions of a savage tribe are transmitted through all the progressive stages of society, till they exert a powerful influence on the mind of the philosopher. He

mound, so as to form a junction between the rivers Selima and Sutluj, and in this mound were found the bones of elephants and men, some of them petrified, and some of them resembling bone. The gigantic dimensions attributed to the human bones shew them to have belonged to some of the larger pachydermata.

* Humboldt et Bonpland, Voy. Relat. Hist. vol. i. p. 30

may find, in the monuments of former changes on the earth's surface, an apparent confirmation of tenets handed down through successive generations, from the rude hunter, whose terrified imagination drew a false picture of those awful visitations of floods and earthquakes, whereby the whole earth as known to him was simultaneously devastated.

Respecting the cosmogony of the Egyptian priests, we gather much information from writers of the Grecian sects, who borrowed almost all their tenets from Egypt, and amongst others that of the former successive destruction and renovation of the world *. We learn from Plutarch, that this was the theme of one of the hymns of Orpheus, so celebrated in the fabulous ages of Greece. It was brought by him from the banks of the Nile; and we even find in his verses, as in the Indian systems, a definite period assigned for the duration of each successive world †. The returns of great catastrophes were determined by the period of the Annus Magnus, or great year, a cycle composed of the revolutions of the sun, moon, and planets, and terminating when these return together to the same sign whence they were supposed at some remote epoch to have set out. The duration of this great cycle was variously estimated. According to Orpheus, it was 120,000 years; according to others, 300,000; and by Cassander it was taken to be 360,000 years ‡. We learn particularly from the Timœus of Plato, that the Egyptians believed the world to be subject to occasional conflagrations and deluges, whereby the gods arrested the career of human wickedness, and purified the earth from guilt. After each regeneration, mankind were in a state of virtue and happiness, from which they gradually degenerated again into vice and immorality. From this Egyptian doctrine, the poets derived the fable of the decline from the golden to the iron age. The sect of Stoics adopted most fully the system of catastrophes destined at certain intervals to destroy the world. These they taught were of two kinds—the Cataclysm, or destruction by deluge, which sweeps away the

* Prichard's Egypt. Mythol. p. 177.

† Plut. de Defectu Oraculorum, cap. 12. Censorinus de die Nat. See also Prichard's Egypt. Mythol. p. 183.

‡ Prichard's Egypt. Mythol. p. 182.

whole human race, and annihilates all the animal and vegetable productions of nature; and the Ecpyrosis, or conflagration, which dissolves the globe itself. From the Egyptians also they derived the doctrine of the gradual debasement of man from a state of innocence. Towards the termination of each era the gods could no longer bear with the wickedness of men, and a shock of the elements or a deluge overwhelmed them; after which calamity, Astrea again descended on the earth, to renew the golden age *.

The connexion between the doctrine of successive catastrophes and repeated deteriorations in the moral character of the human race, is more intimate and natural than might at first be imagined. For, in a rude state of society, all great calamities are regarded by the people, as judgments of God on the wickedness of man. Thus, in our own time, the priests persuaded a large part of the population of Chili, and perhaps believed themselves, that the great earthquake of 1822 was a sign of the wrath of heaven for the great political revolution just then consummated in South America. In like manner, in the account given to Solon by the Egyptian priests, of the submersion of the island of Atlantis under the waters of the ocean, after repeated shocks of an earthquake, we find that the event happened when Jupiter had seen the moral depravity of the inhabitants †. Now, when the notion had once gained ground, whether from causes before suggested or not, that the earth had been destroyed by several general catastrophes, it would next be inferred that the human race had been as often destroyed and renovated. And, since every extermination was assumed to be *penal*, it could only be reconciled with divine justice, by the supposition that man, at each successive creation, was regenerated in a state of purity and innocence.

A very large portion of Asia, inhabited by the earliest nations whose traditions have come down to us, has been always subject to tremendous earthquakes. Of the geographical boundaries of these, and their effects, we shall, in the proper place, have occasion to speak. Egypt has, for the most part, been exempt from this scourge, and the tra-

* Prichard's Egypt. Mythol. p. 193.

† Plato's Timœus.

dition of catastrophes in that country was perhaps derived from the East.

One extraordinary fiction of the Egyptian mythology was the supposed intervention of a masculo-feminine principle, to which was assigned the development of the embryo world, somewhat in the way of incubation. For the doctrine was, that when the first chaotic mass had been produced, in the form of an egg, by a self-dependent and eternal Being, it required the mysterious functions of this masculo-feminine demi-urgus to reduce the component elements into organized forms. Although it is scarcely possible to recall to mind this conceit without smiling, it does not seem to differ essentially in principle from some cosmological notions of men of great genius and science in modern Europe. The Egyptian philosophers ventured on the perilous task of seeking out some analogy to the mode of operation employed by the Author of Nature in the first creation of organized beings, and they compared it to that which governs the birth of new individuals by generation. To suppose that some general rules might be observed in the first origin of created beings, or the first introduction of new species into our system, was not absurd, nor inconsistent with anything known to us in the economy of the universe. But the hypothesis, that there was any analogy between such laws, and those employed in the continual reproduction of species once created, was purely gratuitous. In like manner, it is not unreasonable or derogatory to the attributes of Omnipotence, to imagine that some general laws may be observed in the creation of new worlds; and if man could witness the birth of such worlds, he might reason by induction upon the origin of his own. But in the absence of such data, an attempt has been made to fancy some analogy between the agents now employed to destroy, renovate, and perpetually vary the earth's surface, and those whereby the first chaotic mass was formed, and brought by supposed nascent energy from the embryo to the habitable state. By how many shades the elaborate systems, constructed on these principles, may differ from the mysteries of the "Mundane Egg" of Egyptian fable, we shall not inquire. It would, perhaps, be dangerous ground, and some of our contemporaries might not sit as patiently as the Athenian audience, when the fiction of the chaotic egg, engrafted by Orpheus upon their own mythology, was turned into ridicule by Aristophanes. That comedian introduced his birds singing, in a solemn hymn, "How sable-plumaged night conceived in the boundless bosom of Erebus, and laid an egg, from which, in the revolution of ages, sprung Love, resplendent with golden pinions. Love fecundated the dark-winged chaos, and gave origin to the race of birds *."

Pythagoras, who resided for more than twenty years in Egypt, and, according to Cicero, had visited the East, and conversed with the Persian philosophers, introduced into his own country, on his return, the doctrine of the gradual deterioration of the human race from an original state of virtue and happiness; but if we are to judge of his theory concerning the destruction and renovation of the earth, from the sketch given by Ovid, we must concede it to have been far more philosophical than any known version of the cosmologies of Oriental or Egyptian sects. Although Pythagoras is introduced by the poet as delivering his doctrine in person, some of the illustrations are derived from natural events which happened after the death of the philosopher. But notwithstanding these anachronisms, we may regard the account as a true picture of the tenets of the Pythagorean school in the Augustan age; and although perhaps partially modified, it must have contained the substance of the original scheme. Thus considered, it is extremely curious and instructive; for we here find a comprehensive and masterly summary of almost all the great causes of change now in activity on the globe, and these adduced in confirmation of a principle of perpetual and gradual revolution inherent in the nature of our terrestrial system. These doctrines, it is true, are not directly applied to the explanation of *geological* phenomena; or, in other words, no attempt is made to estimate what may have been, in past ages, or what may hereafter be, the aggregate amount of change brought about by such never-ending fluctuations. Had this been the case, we might have been called upon to admire so extraordinary an anticipation with no less interest than astronomers, when they endeavour to divine by what means the Samian

* Aristophanes' Birds, 694.

philosopher came to the knowledge of the Copernican theory. Let us now examine the celebrated passages to which we have been adverting *:—

"Nothing perishes in this world; but things merely vary and change their form. To be born, means simply that a thing begins to be something different from what it was before; and dying, is ceasing to be the same thing. Yet, although nothing retains long the same image, the sum of the whole remains constant." These general propositions are then confirmed by a series of examples, all derived from natural appearances, except the first, which refers to the golden age giving place to the age of iron. The illustrations are thus consecutively adduced.

1. Solid land has been converted into sea.

2. Sea has been changed into land. Marine shells lie far distant from the deep, and the anchor has been found on the summit of hills.

3. Valleys have been excavated by running water, and floods have washed down hills into the sea †.

4. Marshes have become dry ground.

5. Dry lands have been changed into stagnant pools.

6. During earthquakes some springs have been closed up, and new ones have broken out. Rivers have deserted their channels, and have been re-born elsewhere; as the Erasinus in Greece, and Mysus in Asia.

7. The waters of some rivers, formerly sweet, have become bitter, as those of the Anigris in Greece, &c. ‡

8. Islands have become connected with the main land by the growth of deltas and new deposits, as in the case of Antissa joined to Lesbos, Pharos to Egypt, &c.

9. Peninsulas have been divided from the main land, and have become islands, as Leucadia; and according to tradition Sicily, the sea having carried away the isthmus.

10. Land has been submerged by earthquakes: the Grecian

* Ovid's Metamor. lib. 15.

† Eluvie mons est deductus in æquor, v. 267. The meaning of this last verse is somewhat obscure, but taken with the context, may be supposed to allude to the abrading power of floods, torrents, and rivers.

‡ The impregnation from new mineral springs, caused by earthquakes in volcanic countries, is, perhaps, here alluded to.

cities of Helice and Buris, for example, are to be seen under the sea, with their walls inclined.

11. Plains have been upheaved into hills by the confined air seeking vent, as at Træzen in the Peloponnese.

12. The temperature of some springs varies at different periods.

13. The waters of others are inflammable *.

14. Extraordinary medicinal and deleterious effects are produced by the water of different lakes and springs †.

15. Some rocks and islands, after floating, and having been subject to violent movements, have at length become stationary and immoveable, as Delos and the Cyanean Isles ‡.

16. Volcanic vents shift their position; there was a time when Etna was not a burning mountain, and the time will come when it will cease to burn. Whether it be that some caverns become closed up by the movements of the earth, and others opened, or whether the fuel is finally exhausted, &c. &c.

The various causes of change in the inanimate world having been thus enumerated, the doctrine of equivocal generation is next propounded, as illustrating a corresponding perpetual flux in the animate creation §.

* This is probably an allusion to the escape of inflammable gas, like that in the district of Baku, west of the Caspian; at Pietra-mala, in the Tuscan Apennines; and several other places.

† Many of those described seem fanciful fictions, like the virtues still so commonly attributed to mineral waters.

‡ Raspe, in a learned and judicious essay (chap. 19, de novis insulis), has made it appear extremely probable that all the traditions of certain islands in the Mediterranean having at some former time frequently shifted their position, and at length become stationary, originated in the great change produced in their form by earthquakes and submarine eruptions, of which there have been modern examples in the new islands raised in the time of history. When the series of convulsions ended, the island was said to become fixed.

§ It is not inconsistent with the Hindoo mythology to suppose, that Pythagoras might have found in the East not only the system of universal and violent catastrophes and periods of repose in endless succession, but also that of periodical revolutions, effected by the continued agency of ordinary causes. For Brahma, Vishnu, and Siva, the first, second, and third persons of the Hindoo triad, severally represented the Creative, the Preserving, and the Destroying powers of the Deity. The co-existence of these three attributes, all in simultaneous operation, might well accord with the notion of perpetual but partial alterations finally bringing about a complete change. But the fiction expressed in the verses before quoted from Menû, of eternal vicissitudes in the vigils and slumbers of the Infinite Being seems accommodated to the system of great general catastrophes followed by new creations and periods of repose.

In the Egyptian and Eastern cosmogonies, and in the Greek version of them, no very definite meaning can, in general, be attached to the term "destruction of the world," for sometimes it would seem almost to imply the annihilation of our planetary system, and at others a mere revolution of the surface of the earth.

From the works now extant of Aristotle, and from the system of Pythagoras, as above exposed, we might certainly infer that these philosophers considered the agents of change now operating in Nature, as capable of bringing about in the lapse of ages a complete revolution; and the Stagyrite even considers occasional catastrophes, happening at distant intervals of time, as part of the regular and ordinary course of Nature. The deluge of Deucalion, he says, affected Greece only, and principally the part called Hellas, and it arose from great inundations of rivers during a rainy winter. But such extraordinary winters, he says, though after a certain period they return, do not always revisit the same places*. Censorinus quotes it as Aristotle's opinion, that there were general inundations of the globe, and that they alternated with conflagrations, and that the flood constituted the winter of the great year, or astronomical cycle, while the conflagration, or destruction by fire, is the summer or period of greatest heat†. If this passage, as Lipsius supposes, be an amplification by Censorinus, of what is written in "the Meteorics," it is a gross misrepresentation of the doctrine of the Stagyrite, for the general bearing of his reasoning in that treatise tends clearly in an opposite direction. He refers to many examples of changes now constantly going on, and insists emphatically on the great results which they must produce in the lapse of ages. He instances particular cases of lakes that had dried up, and deserts that had at length become watered by rivers and fertilized. He points to the growth of the Nilotic delta since the time of Homer, to the shallowing of the Palus Mæotis within sixty years from his own time, and although, in the same chapter, he says nothing of earthquakes, yet in others of the same treatise‡, he shews himself not unacquainted with their effects.

* Meteor. lib. i. cap. xii.

† De Die. Nat.

‡ Lib. ii. cap. 14, 15, and 16.

He alludes, for example, to the upheaving of one of the Eolian islands, previous to a volcanic eruption. "The changes of the earth, he says, are so slow in comparison to the duration of our lives, that they are overlooked (λανθανει); and the migrations of people after great catastrophes, and their removal to other regions, cause the event to be forgotten*." When we consider the acquaintance displayed by Aristotle with the destroying and renovating powers of nature in his various works, the introductory and concluding passages of the twelfth chapter of his "Meteorics" are certainly very remarkable. In the first sentence he says, "the distribution of land and sea in particular regions does not endure throughout all time, but it becomes sea in those parts where it was land, and again it becomes land where it was sea, and there is reason for thinking that these changes take place according to a certain system, and within a certain period." The concluding observation is as follows: "As time never fails, and the universe is eternal, neither the Tanais, nor the Nile, can have flowed for ever. The places where they rise were once dry, and there is a limit to their operations, but there is none to time. So also of all other rivers, they spring up and they perish, and the sea also continually deserts some lands and invades others. The same tracts, therefore, of the earth are not some always sea, and others always continents, but every thing changes in the course of time."

It seems, then, that the Greeks had not only derived from preceding nations, but had also, in some degree, deduced from their own observations, the theory of great periodical revolutions in the inorganic world, but there is no ground for imagining that they contemplated former changes in the races of animals and plants. Even the fact, that marine remains were inclosed in solid rocks, although observed by many, and even made the groundwork of geological speculation, never stimulated the industry or guided the inquiries of naturalists. It is not impossible that the theory of equivocal generation might have engendered some indifference on this subject, and that a belief in the spontaneous production of living beings from the earth, or corrupt matter, might have caused the organic world

* Lib. ii. cap. 14, 15, and 16.

to appear so unstable and fluctuating, that phenomena indicative of former changes would not awaken intense curiosity. The Egyptians, it is true, had taught, and the Stoics had repeated, that the earth had once given birth to some monstrous animals, which existed no longer; but the prevailing opinion seems to have been, that after each great catastrophe the same species of animals were created over again. This tenet is implied in a passage of Seneca, where, speaking of a future deluge, he says, "Every animal shall be generated anew, and men free from guilt shall be given to the earth *." An old Arabian version of the doctrine of the successive revolutions of the globe, translated by Abraham Ecchellensis †, seems to form a singular exception to the general rule, for here we find the idea of different genera and species having been created. The Gerbanites, a sect of astronomers who flourished some centuries before the Christian era, taught as follows:—"That after every period of thirty-six thousand years, there were produced twenty-five pair of *every* species of animals, male and female, from whom animals might be propagated and inhabit this lower world. But when a circulation of the heavenly orbs was completed, which is finished in that space of years, *other genera and species* of animals are propagated, as also of plants and other things, and the first order is destroyed, and so it goes on for ever and ever ‡."

* Omne ex integro animal generabitur, dabiturque terris homo inscius scelerum. Quest. Nat. iii. c. 29.

† This author was Regius Professor of Syriac and Arabic at Paris, where, in 1685, he published a Latin translation of many Arabian MSS. on different departments of philosophy. This work has always been considered of high authority.

‡ Gerbanitæ docebant singulos triginta sex mille annos quadringentos, viginti quinque bina ex singulis animalium speciebus produci, marem scilicet ac feminam, ex quibus animalia propagantur, huncque inferiorem incolunt orbem. Absolutâ autem cœlestium orbium circulatione, quæ illo annorum conficitur spatio, iterum alia producuntur animalium genera et species, quemadmodum et plantarum aliarumque rerum, et primus destruitur ordo, sicque in infinitum producitur.—Histor. Orient. Suppl. per Abrahamum Ecchellensum, Syrum Maronitam, cap. 7 et 8. ad calcem Chronici Oriental. Parisiis, e Typ. regia 1685. fol.

Fortis fell into a singular mistake in rendering this passage, imagining that the number twenty-five referred not to the pairs of every animal created, but to the number of new species created at one time; and hence the doctrine of the Arabian sect appeared to coincide somewhat with his own views; and, to be consistent with his hypothesis, that man and some species of animals and plants are more modern than others.—Fortis, Mem. sur l'Hist. Nat. de l'Italie, vol. i. p. 202.

As we learn much of the tenets of the Egyptian and Oriental schools in the writings of the Greeks, so many speculations of the early Greek authors are made known to us in the works of the Augustan and later ages. Strabo, in particular, enters largely, in the Second Book of his Geography, into the opinions of Eratosthenes and other Greeks on one of the most difficult problems in geology, *viz.*, by what causes marine shells came to be plentifully buried in the earth at such great elevations and distances from the sea. He notices, amongst others the explanation of Xanthus the Lydian, who said that the seas had once been more extensive, and that they had afterwards been partially dried up, as in his own time many lakes, rivers, and wells in Asia had failed during a season of drought. Treating this conjecture with merited disregard, Strabo passes on to the hypothesis of Strato, the natural philosopher, who had observed that the quantity of mud brought down by rivers into the Euxine was so great, that its bed must be gradually raised, while the rivers still continued to pour in an undiminished quantity of water. He therefore conceived, that, originally, when the Euxine was an inland sea, its level had by this means become so much elevated that it burst its barrier near Byzantium, and formed a communication with the Propontis, and this partial drainage had already, he supposed, converted the left side into marshy ground, and that, at last, the whole would be choked up with soil. So, it was argued, the Mediterranean had once opened a passage for itself by the Columns of Hercules into the Atlantic, and perhaps the abundance of sea-shells in Africa, near the Temple of Jupiter Ammon, might also be the deposit of some former inland sea, which had at length forced a passage and escaped. But Strabo rejects this theory as insufficient to account for all the phenomena, and he proposes one of his own, the profoundness of which modern geologists are only beginning to appreciate. "It is not," he says, "because the lands covered by seas were originally at different altitudes, that the waters have risen, or subsided, or receded from some parts and inundated others. But the reason is, that the same land is sometimes raised up and sometimes depressed, and the sea also is simultaneously raised and depressed, so that it either overflows, or returns into its own place again. We must therefore ascribe the cause to the ground, either to that ground which is under the sea, or to that which becomes flooded by it, but rather to that which lies beneath the sea, for this is more moveable, and, on account of its humidity, can be altered with greater celerity*. "*It is proper,*" he observes in continuation, "*to derive our explanations from things which are obvious, and in some measure of daily occurrence, such as deluges, earthquakes, volcanic eruptions, and sudden swellings of the land beneath the sea;* for the last raise up the sea also, and when the same lands subside again, they occasion the sea to be let down. And it is not merely the small, but the large islands also, and not merely the islands but the continents which can be lifted up together with the sea; and both large and small tracts may subside, for habitations and cities, like Bure Bizona, and many others, have been engulfed by earthquakes." In another place, this learned geographer, in alluding to the tradition that Sicily had been separated by a convulsion from Italy, remarks, that at present the land near the sea in those parts was rarely shaken by earthquakes, since there were now open orifices whereby fire and ignited matters and waters escaped; but formerly, when the volcanoes of Etna, the Lipari Islands, Ischia, and others, were closed up, the imprisoned fire and wind might have produced far more vehement movements†. The doctrine, therefore, that volcanoes are safety-valves, and that the subterranean convulsions are probably most violent when first the volcanic energy shifts itself to a new quarter, is not modern.

We learn from a passage in Strabo ‡, that it was a dogma of the Gaulish Druids that the universe was immortal, but destined to survive catastrophes both of fire and water. That this doctrine was communicated to them from the East, with much of their learning, cannot be doubted. Cæsar §, it will be remembered, says, that they made use of Greek letters in arithmetical computations.

* "Quod enim hoc attollitur aut subsidit, et vel inundat quædam loca, vel ab iis recedit, ejus rei causa non est, quod alia aliis sola humiliora sint aut altiora; sed quod idem solum modò attollitur modò deprimitur, simulque etiam modò attollitur modò deprimitur mare: itaque vel exundat vel in suum redit locum."

Posteà, p. 88. "Restat, ut causam adscribamus solo, sive quod mari subest sive quod inundatur; potiùs tamen ei quod mari subest. Hoc enim multò est mobilius et quod ob humiditatem celeriùs mutari possit."—Strabo, lib. ii.

† Strabo, lib. vi. p. 396. ‡ Book iv. § l. vi. ch. 13.

Pliny had no theoretical opinions of his own, concerning changes of the earth's surface; and in this department, as in others, he restricted himself to the task of a compiler, without reasoning on the facts stated by him, or attempting to digest them into regular order. His enumeration of the new islands which had been formed in the Mediterranean, and of other convulsions, shew that the ancients had not been inattentive observers of the changes which had taken place on the earth within the memory of man.

We shall now conclude our remarks on the opinions entertained before the Christian era, concerning the past revolutions of our globe. No particular investigations appear to have been made for the express purpose of interpreting the monuments left by nature of ancient changes, but they were too obvious to be entirely disregarded; and the observation of the present course of nature presented too many proofs of alterations continually in progress on the earth to allow philosophers to believe that nature was in a state of rest, or that the surface had remained, and would continue to remain, unaltered. But they had never compared attentively the results of the destroying and reproductive operations of modern times with those of remote eras, nor had they ever entertained so much as a conjecture concerning the comparative antiquity of the human race, and living species of animals and plants, with those belonging to former conditions of the organic world. They had studied the movements and positions of the heavenly bodies with laborious industry, and made some progress in investigating the animal, vegetable, and mineral kingdoms; but the ancient history of the globe was to them a sealed book, and, although written in characters of the most striking and imposing kind, they were unconscious even of its existence.

CHAPTER III.

Arabian writers of the Tenth century—Persecution of Omar—Cosmogony of the Koran—Early Italian writers—Fracastoro—Controversy as to the real nature of organized fossils—Fossil shells attributed to the Mosaic deluge—Palissy—Steno—Scilla—Quirini—Boyle—Plot—Hooke's Theory of Elevation by earthquakes—His speculations on lost species of animals—Ray—Physico-theological writers—Woodward's Diluvial Theory—Burnet—Whiston—Hutchinson—Leibnitz—Vallisneri—Lazzoro Moro—Generelli—Buffon—His theory condemned by the Sorbonne as unorthodox—Buffon's declaration—Targioni—Arduino—Michell—Catcott—Raspe—Fortis—Testa—Whitehurst—Pallas—Saussure.

AFTER the decline of the Roman empire, the cultivation of physical science was first revived with some success by the Saracens, about the middle of the eighth century of our era. The works of the most eminent classic writers were purchased at great expense from the Christians, and translated into Arabic; and Al Mamûn, son of the famous Harûn-al-Rashid, the contemporary of Charlemagne, received with marks of distinction, at his court at Bagdad, astronomers and men of learning from different countries. This caliph, and some of his successors, encountered much opposition and jealousy from the doctors of the Mahomedan law, who wished the Moslems to confine their studies to the Koran, dreading the effects of the diffusion of a taste for the physical sciences *. Almost all the works of the early Arabian writers are lost. Amongst those of the tenth century, of which fragments are now extant, is a system of mineralogy by Avicenna, a physician, in whose arrangement there is considerable merit. In the same century also, Omar, surnamed "El Aalem," or "the Learned," wrote a work on "the Retreat of the Sea." It appears that on comparing the charts of his own time with those made by the Indian and Persian astronomers two thousand years before, he had satisfied himself that important changes had taken place since the times of history in the form of the coasts of Asia, and that the extension of the sea had been greater at some former periods.

* Mod. Univ. Hist. vol. ii. chap. iv. section iii.

He was confirmed in this opinion by the numerous salt springs and marshes in the interior of Asia; a phenomenon from which Pallas, in more recent times, has drawn the same inference.

Von Hoff has suggested, with great probability, that the changes in the level of the Caspian, (some of which there is reason to believe have happened within the historical era,) and the geological appearances in that district, indicating the desertion by that sea of its ancient bed, had probably led Omar to his theory of a general subsidence. But whatever may have been the proofs relied on, his system was declared contradictory to certain passages in the Koran, and he was called upon publicly to recant his errors; to avoid which persecution he went into voluntary banishment from Samarkand *.

The cosmological opinions expressed in the Koran are few, and merely introduced incidentally; so that it is not easy to understand how they could have interfered so seriously with free discussion on the former changes of the globe. The Prophet declared that the earth was created in two days, and the mountains were then placed on it; and during these, and two additional days, the inhabitants of the earth were formed; and in two more the seven heavens †. There is no more detail of circumstances; and the deluge, which is also mentioned, is discussed with equal brevity. The waters are represented to have poured out of an oven; a strange fable, said to be borrowed from the Persian Magi, who represented them as issuing from the oven of an old woman ‡. All men were drowned, save Noah and his family; and then God said, " O earth, swallow up thy waters; and thou, O heaven, withhold thy rain;" and immediately the waters abated §.

* Von Hoff, Geschichte der Veränderungen der Erdoberfläche, vol. i. p. 406, who cites Delisle, bey Hissmann Welt-und Völkergeschichte. Alte Gesch. 1ter Theil. s. 234.—The Arabian persecutions for heretical dogmas in theology were often very sanguinary. In the same ages wherein learning was most in esteem, the Mahometans were divided into two sects, one of whom maintained that the Koran was increate, and had subsisted in the very essence of God from all eternity; and the other the Motazalites, who, admitting that the Koran was instituted by God, conceived it to have been first made when revealed to the Prophet at Mecca, and accused their opponents of believing in two eternal beings. The opinions of each of these sects were taken up by different caliphs in succession, and the followers of each sometimes submitted to be beheaded, or flogged till at the point of death, rather than renounce their creed.—Mod. Univ. Hist. vol. ii. chap. 4.

† Koran, chap. 41.

‡ Sale's Koran, chap. 11, see note. § Ibid.

We may suppose Omar to have represented the desertion of the land by the sea to have been gradual, and that his hypothesis required a greater lapse of ages than was consistent with Moslem orthodoxy; for it is to be inferred from the Koran, that man and this planet were created at the same time; and although Mahomet did not limit expressly the antiquity of the human race, yet he gave an implied sanction to the Mosaic chronology by the veneration expressed by him for the Hebrew Patriarchs *.

We must now pass over an interval of five centuries, wherein darkness enveloped almost every department of science, and buried in profound oblivion all prior investigations into the earth's history and structure. It was not till the earlier part of the sixteenth century that geological phenomena began to attract the attention of the Christian nations. At that period a very animated controversy sprung up in Italy, concerning the true nature and origin of marine shells, and other organized fossils, found abundantly in the strata of the peninsula †. The excavations made in 1517, for repairing the city of Verona, brought to light a multitude of curious petrifactions, and furnished matter for speculation to different authors, and among the rest to Fracastoro ‡, who declared his opinion, that fossil shells had all belonged to living animals, which had formerly lived and multiplied, where their exuviæ are now found. He exposed the absurdity of having recourse to a certain "plastic force," which it was said had power to fashion stones into organic forms; and, with no less cogent arguments, demonstrated the futility of attributing the situation of the shells in question to the Mosaic deluge, a theory obstinately defended by some. That inundation, he observed, was too transient, it consisted principally of fluviatile waters; and, if it had transported shells to great distances, must have strewed them over the surface, not buried them at vast depths in the interior of mountains. His clear exposition of the

* Kossa, appointed master to the Caliph Al Mamûd, was author of a book, entitled, "The History of the Patriarchs and Prophets, *from the Creation of the World.*"—Mod. Univ. Hist. vol. ii. chap. 4.

† See Brocchi's Discourse on the Progress of the Study of Fossil Conchology in Italy, where some of the following notices on Italian writers will be found more at large.

‡ Museum Calceol.

evidence would have terminated the discussion for ever, if the passions of mankind had not been enlisted in the dispute; and even though doubts should for a time have remained in some minds, they would speedily have been removed by the fresh information obtained almost immediately afterwards, respecting the structure of fossil remains, and of their living analogues. But the clear and philosophical views of Fracastoro were disregarded, and the talent and argumentative powers of the learned were doomed for three centuries to be wasted in the discussion of these two simple and preliminary questions: first, whether fossil remains had ever belonged to living creatures; and, secondly, whether, if this be admitted, all the phenomena could be explained by the Noachian deluge. It had been the consistent belief of the Christian world, down to the period now under consideration, that the origin of this planet was not more remote than a few thousand years; and that since the creation the deluge was the only great catastrophe by which considerable change had been wrought on the earth's surface. On the other hand, the opinion was scarcely less general, that the final dissolution of our system was an event to be looked for at no distant period. The era, it is true, of the expected millennium had passed away; and for five hundred years after the fatal hour, when the annihilation of the planet had been looked for, the monks remained in undisturbed enjoyment of rich grants of land bequeathed to them by pious donors, who, in the preamble of deeds beginning "appropinquante mundi termino"—— "appropinquante magno judicii die," left lasting monuments of the popular delusion *.

But although in the sixteenth century it had become necessary to interpret the prophecies more liberally, and to assign a more distant date to the future conflagration of the world, we find, in the speculations of the early geologists, perpetual allusion to such an approaching catastrophe; while, in all that regarded the antiquity of the earth, no modification whatever of the opinions of the dark ages had been effected. Considerable alarm was at first excited when the attempt was made to

* In the monasteries of Sicily in particular, the title-deeds of many valuable grants of land are headed by such preambles, composed by the testators about the period when the good King Roger was expelling the Saracens from that island.

invalidate by physical proofs an article of faith so generally received, but there was sufficient spirit of toleration and candour amongst the Italian ecclesiastics, to allow the subject to be canvassed with much freedom. They entered warmly themselves into the controversy, often favouring different sides of the question; and however much we may deplore the loss of time and labour devoted to the defence of untenable positions, it must be conceded, that they displayed far less polemic bitterness than certain writers who followed them "beyond the Alps," two centuries and a half later.

The system of scholastic disputations encouraged in the Universities of the middle ages had unfortunately trained men to habits of indefinite argumentation, and they often preferred absurd and extravagant propositions, because greater skill was required to maintain them; the end and object of such intellectual combats being victory and not truth. No theory could be too far-fetched or fantastical not to attract some followers, provided it fell in with popular notions; and as cosmologists were not at all restricted, in building their systems, to the agency of known causes, the opponents of Fracastoro met his arguments by feigning imaginary causes, which differed from each other rather in name than in substance. Andrea Mattioli, for instance, an eminent botanist, the illustrator of Dioscorides, embraced the notion of Agricola, a German miner, that a certain "materia pinguis" or "fatty matter," set into fermentation by heat, gave birth to fossil organic shapes. Yet Mattioli had come to the conclusion from his own observations, that porous bodies, such as bones and shells, might be converted into stone, as being permeable to what he termed the "lapidifying juice." In like manner, Falloppio of Padua conceived that petrified shells had been generated by fermentation in the spots where they were found, or that they had in some cases acquired their form from "the tumultuous movements of terrestrial exhalations." Although not an unskilful professor of anatomy, he taught that certain tusks of elephants dug up in his time at Puglia were mere earthy concretions, and, consistently with these principles, he even went so far as to consider it not improbable, that the vases of Monte Testaceo at Rome were natural impressions stamped in the soil.* In the same spirit, Mercati, who published, in 1574, faithful figures of the fossil shells preserved by Pope Sextus V. in the Museum of the Vatican, expressed an opinion that they were mere stones, which had assumed their peculiar configuration from the influence of the heavenly bodies; and Olivi of Cremona, who described the fossil remains of a rich Museum at Verona, was satisfied with considering them mere " sports of nature."

The title of a work of Cardano's, published in 1552, " De Subtilitate," (corresponding to what would now be called, Transcendental Philosophy,) would lead us to expect in the chapter on minerals, many far-fetched theories characteristic of that age; but, when treating of petrified shells, he decided that they clearly indicated the former sojourn of the sea upon the mountains†.

Some of the fanciful notions of those times were deemed less unreasonable, as being somewhat in harmony with the Aristotelian theory of spontaneous generation, then taught in all the schools. For men who had been instructed in early youth, that a large proportion of living animals and plants were formed from the fortuitous concourse of atoms, or had sprung from the corruption of organic matter, might easily persuade themselves, that organic shapes, often imperfectly preserved in the interior of solid rocks, owed their existence to causes equally obscure and mysterious.

But there were not wanting some, who at the close of this century expressed more sound and sober opinions. Cesalpino, a celebrated botanist, conceived that fossil shells had been left on the land by the retiring sea, and had concreted into stone during the consolidation of the soil‡; and in the following year (1597), Simeone Majoli§ went still further, and, coinciding for the most part with the views of Cesalpino, suggested that the shells and submarine matter of the Veronese, and other districts, might have been cast up, upon the land, by volcanic explosions, like those which gave rise, in 1588, to

* De Fossilib. p. 109 and 176.

† Brocchi, Con. Foss. Subap. Disc. sui Prog. vol. i. p. 5.

‡ De Metallicis.

§ Dies Caniculares.

Monte Nuovo, near Puzzuoli.—This hint was the first imperfect attempt to connect the position of fossil shells with the agency of volcanoes, a system afterwards more fully developed by Hooke, Lazzoro Moro, Hutton, and other writers.

Two years afterwards, Imperati advocated the animal origin of fossilized shells, yet admitted that stones could vegetate by force of "an internal principle;" and, as evidence of this, he referred to the teeth of fish, and spines of echini found petrified*.

Palissy, a French writer on "the Origin of Springs from Rain-water" and of other scientific works, undertook, in 1580, to combat the notions of many of his contemporaries in Italy, that petrified shells had all been deposited by the universal deluge. "He was the first," said Fontenelle, when, in the French Academy, he pronounced his eulogy more than fifty years afterwards, "who dared assert" in Paris, that fossil remains of testacea and fish had once belonged to marine animals.

To enumerate the multitude of Italian writers, who advanced various hypotheses, all equally fantastical, in the early part of the seventeenth century, would be unprofitably tedious, but Fabio Colonna deserves to be distinguished; for, although he gave way to the dogma, that all fossil remains were to be referred to the Noachian deluge, he resisted tne absurd theory of Stelluti, who taught that fossil wood and ammonites were mere clay, altered into such forms by sulphureous waters and subterranean heat; and he pointed out the different states of shells buried in the strata, distinguishing between, first, the mere mould or impression; secondly, the cast or nucleus; and thirdly, the remains of the shell itself. He had also the merit of being the first to point out, that some of the fossils had belonged to marine, and some to terrestrial testacea†. But the most remarkable work of that period was published by Steno, a Dane, once professor of anatomy at Padua, and who afterwards resided many years at the court of the Grand Duke of Tuscany. The treatise bears the quaint title of "De Solido intra Solidum contento naturaliter, (1669,)" by which the author intended to express "On Gems, Crystals, and organic Petrifactions inclosed within solid Rocks." This work

* Storia Naturale.

† Osserv. sugli Animali aquat. e terrest. 1626.

attests the priority of the Italian school in geological research; exemplifying at the same time the powerful obstacles opposed, in that age, to the general reception of enlarged views in the science. Steno had compared the fossil shells with their recent analogues, and traced the various gradations from the state of mere calcination, when their natural gluten only was lost, to the perfect substitution of stony matter. He demonstrated that many fossil teeth found in Tuscany belonged to a species of shark; and he dissected, for the purpose of comparison, one of these fish recently taken from the Mediterranean. That the remains of shells and marine animals found petrified were not of animal origin was still a favourite dogma of many who were unwilling to believe, that the earth could have been inhabited by living beings, long before many of the mountains were formed. By way of compromise, as it were, for dissenting from this opinion, Steno conceded, as Fabio Colonna had done before him, that all marine fossils might have been transported into their present situation at the time of the Noachian deluge. He maintained that fossil vegetables had been once living plants, and he hinted that they might, in some instances, indicate the distinction between fluviatile and marine deposits. He also inferred that the present mountains had not existed ever since the origin of things, suggesting that many strata of submarine origin had been accumulated in the interval between the creation and deluge. Here he displayed his great anxiety to reconcile his theory with the Scriptures; for he at the same time advanced an opinion, which does not seem very consistent with such a doctrine, *viz.* that there was a wide distinction between the shelly, and nearly horizontal beds at the foot of the Apennines, and the older mountains of highly inclined stratification. Both, he observed, were of sedimentary origin; and a considerable interval of time must have separated their formation. Tuscany, according to him, had successively past *through six different states;* and to explain these mighty changes, he called in the agency of inundations, earthquakes, and subterranean fires.

His generalizations were for the most part comprehensive and just; but such was his awe of popular prejudice, that he only ventured to throw them out as mere conjectures, and the timid reserve of his expressions must have raised doubts as to his own confidence in his opinions, and deprived them of some of the authority due to them.

Scilla, a Sicilian painter, published, in 1670, a work on the fossils of Calabria, illustrated by good engravings. This was written in Latin, with great spirit and elegance, and it proves the continued ascendancy of dogmas often refuted; for we find the wit and eloquence of the author chiefly directed against the obstinate incredulity of naturalists, as to the organic nature of fossil shells*. Like many eminent naturalists of his day, Scilla gave way to the popular persuasion that all fossil shells were the effects and proofs of the Mosaic deluge. It may be doubted whether he was perfectly sincere, and some of his contemporaries who took the same course were certainly not so. But so eager were they to root out what they justly considered an absurd prejudice respecting the nature of organized fossils, that they seem to have been ready to make any concessions, in order to establish this preliminary point. Such a compromising policy was short-sighted, since it was to little purpose that the nature of the documents should at length be correctly understood, if men were to be prevented from deducing fair conclusions from them.

The theologians who now entered the field in Italy, Germany, France and England, were innumerable; and henceforward, they who refused to subscribe to the position, that all marine organic remains were proofs of the Mosaic deluge, were exposed to the imputation of disbelieving the whole of the sacred writings. Scarcely any step had been made in approximating to sound theories since the time of Fracastoro, more than a hundred years having been lost, in writing down the dogma that organized fossils were mere sports of nature. An additional period of a century and a half was now destined to be consumed in exploding the hypothesis, that organized fossils had all been buried in the solid strata, by the Noachian flood. Never did a theoretical fallacy, in any branch of science,

* Scilla quotes the remark of Cicero on the story that a stone in Chios had been cleft open, and presented the head of Paniscus in relief—"I believe," said the orator, "that the figure bore some resemblance to Paniscus, but not such that you would have deemed it sculptured by Scopas, for chance never perfectly imitates the truth."

interfere more seriously with accurate observation and the systematic classification of facts. In recent times, we may attribute our rapid progress chiefly to the careful determination of the order of succession in mineral masses, by means of their different organic contents, and their regular superposition. But the old diluvialists were induced by their system to confound all the groups of strata together instead of discriminating,—to refer all appearances to one cause and to one brief period, not to a variety of causes acting throughout a long succession of epochs. They saw the phenomena only as they desired to see them, sometimes misrepresenting facts, and at other times deducing false conclusions from correct data. Under the influence of such prejudices, three centuries were of as little avail, as the same number of years in our own times, when we are no longer required to propel the vessel against the force of an adverse current.

It may be well to forewarn our readers, that in tracing the history of geology from the close of the seventeenth to the end of the eighteenth century, they must expect to be occupied with accounts of the retardation, as well as of the advance of the science. It will be our irksome task to point out the frequent revival of exploded errors, and the relapse from sound to the most absurd opinions. It will be necessary to dwell on futile reasoning and visionary hypothesis, because the most extravagant systems were often invented or controverted by men of acknowledged talent. A sketch of the progress of Geology is the history of a constant and violent struggle between new opinions and ancient doctrines, sanctioned by the implicit faith of many generations, and supposed to rest on scriptural authority. The inquiry, therefore, although highly interesting to one who studies the philosophy of the human mind, is singularly barren of instruction to him who searches for truths in physical science.

Quirini, in 1676 *, contended, in opposition to Scilla, that the diluvian waters could not have conveyed heavy bodies to the summit of mountains, since the agitation of the sea never (as Boyle had demonstrated) extended to great depths†, and

* De Testaceis fossilibus Mus. Septaliani.

† The opinions of Boyle, alluded to by Quirini, were published a few years

still less could the testacea, as some pretended, have lived in these diluvial waters, for "the duration of the flood was brief, and *the heavy rains must have destroyed the saltness of the sea!*" He was the first writer who ventured to maintain that the universality of the Noachian cataclysm ought not to be insisted upon. As to the nature of petrified shells, he conceived that as earthy particles united in the sea to form the shells of mollusca, the same crystallizing process might be effected on the land, and that, in the latter case, the germs of the animals might have been disseminated through the substance of the rocks, and afterwards developed by virtue of humidity. Visionary as was this doctrine, it gained many proselytes even amongst the more sober reasoners of Italy and Germany, for it conceded both that fossil bodies were organic, and that the diluvial theory could not account for them.

In the mean time, the doctrine that fossil shells had never belonged to real animals, maintained its ground in England, where the agitation of the question began at a much later period. Dr. Plot, in his "Natural History of Oxfordshire," (1677,) attributed to "a plastic virtue latent in the earth" the origin of fossil shells and fishes; and Lister, to his accurate account of British shells, in 1678, added the fossil species, under the appellation of *turbinated and bivalve stones.* "Either," said he, "these were terriginous, or, if otherwise, the animals they so exactly represent *have become extinct.*" This writer appears to have been the first who was aware of the continuity over large districts of the principal groups of strata in the British series, and who proposed the construction of regular geological maps.

The "Posthumous Works of Robert Hooke, M.D.," well known as a great mathematician and natural philosopher,

before, in a short article entitled "On the bottom of the Sea." From observations collected from the divers of the pearl fishery, Boyle had ascertained that when the waves were six or seven feet high above the surface of the water, there were no signs of agitation at the depth of fifteen fathoms; and that even during heavy gales of wind, the motion of the water was exceedingly diminished at the depth of twelve or fifteen feet. He had also learnt from some of his informants, that there were currents running in opposite directions at different depths.—Boyle's Works, vol. iii. p. 110. London, 1744. The reader will see, in our chapter on "Marine Currents," that Boyle's doctrine must be received with some modification.

appeared in 1705, containing, "A Discourse of Earthquakes," which, we are informed by his editor, was written in 1668, but revised at subsequent periods *. Hooke frequently refers to the best Italian and English authors who wrote before his time on geological subjects; but there are no passages in his works implying that he participated in the enlarged views of Steno and Lister, or of his contemporary Woodward, in regard to the geographical extent of certain groups of strata. His treatise, however, is the most philosophical production of that age, in regard to the causes of former changes in the organic and inorganic kingdoms of nature.

"However trivial a thing," he says, "a rotten shell may appear to some, yet these monuments of nature are more certain tokens of antiquity than coins or medals, since the best of those may be counterfeited or made by art and design, as may also books, manuscripts, and inscriptions, as all the learned are now sufficiently satisfied has often been actually practised," &c.; "and though it must be granted that it is very difficult to read them (the records of nature) and *to raise a chronology out of them*, and to state the intervals of the time wherein such or such catastrophes and mutations have happened, yet it is not impossible," &c.† Respecting the extinction of species, Hooke was aware that the fossil ammonites, nautili, and many other shells and fossil skeletons found in England, were of different species from any then known; but he doubted whether the species had become extinct, observing that the knowledge of naturalists of all the marine species, especially those inhabiting the deep sea, was very deficient. In some parts of his writings, however, he leans to the opinion that species had been lost; and, in speculating on this subject, he even suggests that there might be some connection between the disappearance of certain kinds of animals and plants, and the changes wrought by earthquakes in former ages: for some species, he observes with great sagacity, are "*peculiar to certain places*, and not to be found elsewhere. If, then, such a place had been swallowed up, it is not improbable but that those animate beings may

* Between the year 1688 and his death, in 1703, he read several memoirs to the Royal Society, and delivered lectures on various subjects, relating to fossil remains and the effects of earthquakes.

† Post. Works, Lecture Feb. 29, 1688.

have been destroyed with it; and this may be true both of aërial and aquatic animals: for those animated bodies, whether vegetables or animals, which were naturally nourished or refreshed by the air, would be destroyed by the water*," &c. Turtles, he adds, and such large ammonites as are found in Portland, seem to have been the productions of the seas of hotter countries, and *it is necessary to suppose that England once lay under the sea within the torrid zone!* To explain this and similar phenomena, he indulges in a variety of speculations concerning changes in the position of the axis of the earth's rotation, a shifting of the earth's centre of gravity, "*analogous to the revolutions of the magnetic pole*," &c. None of these conjectures, however, are proposed dogmatically, but rather in the hope of promoting fresh inquiries and experiments.

In opposition to the prejudices of his age, we find him arguing that nature had not formed fossil bodies, "for no other end than to play the mimic in the mineral kingdom"—that figured stones were "really the several bodies they represent, or the mouldings of them petrified," and "not, as some have imagined, a 'lusus naturæ,' sporting herself in the needless formation of useless beings †."

It was objected to Hooke, that his doctrine of the extinction of species derogated from the wisdom and power of the Omnipotent Creator; but he answered, that, as individuals die, there may be some termination to the duration of a species; and his opinions, he declared, were not repugnant to Holy Writ: for the Scriptures taught that our system was dege-

* Posth. Works, p. 327.

† Posth. Works, Lecture Feb. 15, 1688. Hooke explained, with considerable clearness, the different modes wherein organic substances may become lapidified; and, among other illustrations, he mentions some silicified palm-wood brought from Africa, on which M. de la Hire had read a memoir to the Royal Academy of France, (June, 1692,) wherein he had pointed out not only the tubes running the length of the trunk, but the roots at one extremity. De la Hire, says Hooke, also treated of certain trees found petrified in "the river that passes by Bakan, in the kingdom of *Ava*, and which has for the space of ten leagues the virtue of petrifying wood." It is an interesting fact, that the silicified wood of the Irawadi should have attracted attention more than one hundred years ago. Remarkable discoveries have been recently made there of fossil animals and vegetables by Mr. Crawfurd and Dr. Wallich.—See Geol. Trans. vol. ii. part 3, p. 377, Second Series. De la Hire cites Father Duchatz, in the second volume of "Observations made in the Indies by the Jesuits."

nerating, and tending to its final dissolution; "and as, when that shall happen, all the species will be lost, why not some at one time and some at another *?"

But his principal object was to account for the manner in which shells had been conveyed into the higher parts of "the Alps, Apennines, and Pyrenean hills, and the interior of continents in general." These and other appearances, he said, might have been brought about by earthquakes, "which have turned plains into mountains, and mountains into plains, seas into land, and land into seas, made rivers where there were none before, and swallowed up others that formerly were, &c. &c.; and which, since the creation of the world, have wrought many great changes on the superficial parts of the earth, and have been the instruments of placing shells, bones, plants, fishes, and the like, in those places, where, with much astonishment, we find them †." This doctrine, it is true, had been laid down in terms almost equally explicit by Strabo, to explain the occurrence of fossil shells in the interior of continents, and to that geographer, and other writers of antiquity, Hooke frequently refers; but the revival and developement of the system was an important step in the progress of modern science.

He enumerated all the examples known to him of subterranean disturbance, from "the sad catastrophe of Sodom and Gomorrah" down to the Chilian earthquake of 1646. The elevating of the bottom of the sea, the sinking and submersion of the land, and most of the inequalities of the earth's surface, might, he said, be accounted for by the agency of these subterranean causes. He mentions that the coast near Naples was raised during the eruption of Monte Nuovo; and that, in 1591, land rose in the island of St. Michael, during an eruption; and although it would be more difficult, he says, to prove, he does not doubt but that there had been *as many earthquakes in the parts of the earth under the ocean, as in the parts of the dry land*; in confirmation of which he mentions the immeasurable depth of the sea near some volcanoes. To attest the extent of simultaneous subterranean movements, he refers to an earthquake in the West Indies, in 1690, where

* Posth. Works, Lecture May 29, 1689.

† Posth. Works, p. 312.

the space of earth raised, or "struck upwards" by the shock, exceeded the length of the Alps or the Pyrenees.

As Hooke declared the favourite hypothesis of the day ("that marine fossil bodies were to be referred to Noah's flood") to be wholly untenable, he appears to have felt himself called upon to substitute a diluvial theory of his own, and thus he became involved in countless difficulties and contradictions. "During the great catastrophe," he said, "there might have been a changing of that part which was before dry land into sea by sinking, and of that which was sea into dry land by raising, and marine bodies might have been buried in sediment beneath the ocean, in the interval between the creation and the deluge *." Then followed a disquisition on the separation of the land from the waters, mentioned in Genesis: during which operation some places of the shell of the earth were forced outwards, and others pressed downwards or inwards, &c. His diluvial hypothesis very much resembled that of Steno, and was entirely opposed to the fundamental principles professed by him, that he would explain the former changes of the earth *in a more natural manner* than others had done. When, in despite of this declaration, he required a former "crisis of nature," and taught that earthquakes had become debilitated, and that the Alp, Andes, and other chains, had been lifted up in a few months, his machinery was as extravagant and visionary as that of his most fanciful predecessors; and for this reason, perhaps, his whole theory of earthquakes met with very undeserved neglect.

One of his contemporaries, the celebrated naturalist, Ray, participated in the same desire to explain geological phenomena, by reference to causes less hypothetical than those usually resorted to †. In his Essay on "Chaos and Creation" he proposed a system, agreeing in its outline, and in many of its details, with that of Hooke; but his knowledge of natural history enabled him to elucidate the subject with various original observations. Earthquakes, he suggested, might have

* Posth. Works, p. 410.

† Ray's Physico-theological Discourses were of somewhat later date than Hooke's great work on earthquakes. He speaks of Hooke as one "whom for his learning and deep insight into the mysteries of nature he deservedly honoured.' —*On the Deluge*, chap. 4.

been the second causes employed at the creation, in separating the land from the waters, and in gathering the waters together into one place. He mentions, like Hooke, the earthquake of 1646, which had violently shaken the Andes for some hundreds of leagues, and made many alterations therein. In assigning a cause for the general deluge, he preferred a change in the earth's centre of gravity to the introduction of earthquakes. Some unknown cause, he said, might have forced the subterranean waters outwards, as was, perhaps, indicated by "the breaking up of the fountains of the great deep."

Ray was one of the first of our writers who enlarged upon the effects of running water upon the land, and of the encroachment of the sea upon the shores. So important did he consider the agency of these causes, that he saw in them an indication of the tendency of our system to its final dissolution; and he wondered why the earth did not proceed more rapidly towards a general submersion beneath the sea, when so much matter was carried down by rivers, or undermined in the sea-cliffs. We perceive clearly from his writings, that the gradual decline of our system, and its future consummation by fire, was held to be as necessary an article of faith by the orthodox, as was the recent origin of our planet. His Discourses, like those of Hooke, are highly interesting, as attesting the familiar association in the minds of philosophers, in the age of Newton, of questions in physics and divinity. Ray gave an unequivocal proof of the sincerity of his mind, by sacrificing his preferment in the church, rather than take an oath against the Covenanters, which he could not reconcile with his conscience. His reputation, moreover, in the scientific world placed him high above the temptation of courting popularity, by pandering to the physico-theological taste of his age. It is, therefore, curious to meet with so many citations from the Christian fathers and prophets in his essays on physical science—to find him in one page proceeding by the strict rules of induction, to explain the former changes of the globe, and in the next gravely entertaining the question, whether the sun and stars, and the whole heavens shall be annihilated, together with the earth, at the era of the grand conflagration.

Among the contemporaries of Hooke and Ray, Woodward, a professor of medicine, had acquired the most extensive information respecting the geological structure of the crust of the earth. He had examined many parts of the British strata with minute attention; and his systematic collection of specimens, bequeathed to the University of Cambridge, and still preserved there as arranged by him, shews how far he had advanced in ascertaining the order of superposition. From the great number of facts collected by him we might have expected his theoretical views to be more sound and enlarged than those of his contemporaries; but in his anxiety to accommodate all observed phenomena to the scriptural account of the Creation and Deluge, he arrived at most erroneous results. He conceived "the whole terrestrial globe to have been taken to pieces and dissolved at the flood, and the strata to have settled down from this promiscuous mass as any earthy sediment from a fluid *." In corroboration of these views, he insisted upon the fact, that "marine bodies are lodged in the strata according to the order of their gravity, the heavier shells in stone, the lighter in chalk, and so of the rest †." Ray immediately exposed the unfounded nature of this assertion, remarking truly, that fossil bodies "are often mingled, heavy with light, in the same stratum;" and he even went so far as to say, that Woodward "must have invented the phenomena for the sake of confirming his bold and strange hypothesis ‡"—a strong expression from the pen of a contemporary.

At the same time Burnet published his "Theory of the Earth §." The title is most characteristic of the age,—"The Sacred Theory of the Earth, containing an Account of the Original of the Earth, and of all the general Changes which it hath already undergone, or is to undergo, till the Consummation of all Things." Even Milton had scarcely ventured in his poem to indulge his imagination so freely in painting scenes of the Creation and Deluge, Paradise and Chaos, as this writer, who set forth pretensions to profound philosophy. He explained why the primeval earth enjoyed a perpetual spring before the flood! shewed how the crust of the globe was fissured by "the sun's rays," so that it burst, and thus the dilu-

* Essay towards a Natural History of the Earth, 1695. Preface.

† Ibid. Preface.

‡ Consequences of the Deluge, p. 165.

§ First published in Latin, between the years 1680 and 1690.

vial waters were let loose from a supposed central abyss. Not satisfied with these themes, he derived from the books of the inspired writers, and even from heathen authorities, prophetic views of the future revolutions of the globe, gave a most terrific description of the general conflagration, and proved that a new heaven and a new earth will rise out of a *second chaos*—after which will follow the blessed millennium.

The reader should be informed, that according to the opinion of many respectable writers of that age, there was good scriptural ground for presuming that the garden bestowed upon our first parents was not on the earth itself, but above the clouds, in the middle region between our planet and the moon. Burnet approaches with becoming gravity the discussion of so important a topic. He was willing to concede that the geographical position of Paradise was not in Mesopotamia, yet he maintained that it was upon the earth, and in the southern hemisphere, near the equinoctial line. Butler selected this conceit as a fair mark for his satire, when, amongst the numerous accomplishments of Hudibras, he says—

> He knew the seat of Paradise,
> Could tell in what degree it lies;
> And as he was disposed, could prove it
> Below the moon, or else above it.

Yet the same monarch, who is said never to have slept without Butler's poem under his pillow, was so great an admirer and patron of Burnet's book, that he ordered it to be translated from the Latin into English. The style of the "Sacred Theory" was eloquent, and displayed powers of invention of no ordinary stamp. It was, in fact, a fine historical romance, as Buffon afterwards declared; but it was treated as a work of profound science in the time of its author, and was panegyrized by Addison in a Latin ode, while Steele praised it in the "Spectator," and Warton, in his "Essay on Pope," discovered that Burnet united the faculty of *judgment* with powers of imagination.

Another production of the same school, and equally characteristic of the times, was that of Whiston, entitled, "A New Theory of the Earth, wherein the Creation of the World in six Days, the Universal Deluge, and the General Conflagration, as laid down in the Holy Scriptures, are shewn to be perfectly agreeable to Reason and Philosophy." He was at first a follower of Burnet, but his faith in the infallibility of that writer was shaken by the declared opinion of Newton, that there was every presumption in astronomy against any former change in the inclination of the earth's axis. This was a leading dogma in Burnet's system, though not original, for it was borrowed from an Italian, Alessandro degli Alessandri, who had suggested it in the beginning of the fifteenth century, to account for the former occupation of the present continents by the sea. La Place has since strengthened the arguments of Newton, against the probability of any former revolution of this kind. The remarkable comet of 1680 was fresh in the memory of every one, when Whiston first began his cosmological studies, and the principal novelty of his speculations consisted in attributing the deluge to the near approach to the earth of one of these erratic bodies. Having ascribed an increase of the waters to this source, he adopted Woodward's theory, supposing all stratified deposits to have resulted from the "chaotic sediment of the flood." Whiston was one of the first who ventured to propose that the text of Genesis should be interpreted differently from its ordinary acceptation, so that the doctrine of the earth having existed long previous to the creation of man might no longer be regarded as unorthodox. He had the art to throw an air of plausibility over the most improbable parts of his theory, and seemed to be proceeding in the most sober manner, and by the aid of mathematical demonstration, to the establishment of his various propositions. Locke pronounced a panegyric on his theory, commending him for having explained so many wonderful and before inexplicable things. His book, as well as Burnet's, was attacked and refuted by Keill*. Like all who introduced purely hypothetical causes to account for natural phenomena, he retarded the progress of truth, diverting men from the investigation of the laws of sublunary nature, and inducing them to waste time in speculations on the power of comets to drag the waters of the ocean over the land—on the condensation of the vapours of their tails into water, and other matters equally edifying.

John Hutchinson, who had been employed by Woodward

* An Examination of Dr. Burnet's Theory, &c. 2d edition, 1734.

in making his collection of fossils, published afterwards, in 1724, the first part of his "Moses's Principia," wherein he ridiculed Woodward's hypothesis. He and his numerous followers were accustomed to declaim loudly against human learning, and they maintained that the Hebrew scriptures, when rightly translated, comprised a perfect system of natural philosophy, for which reason they objected to the Newtonian theory of gravitation.

Leibnitz, the great mathematician, published his "Protogæa" in 1680. He imagined this planet to have been originally a burning luminous mass, and that ever since its creation it has been undergoing gradual refrigeration. Nearly all the matter of the earth was at first encompassed by fire. When the outer crust had at length cooled down sufficiently to allow the vapours to be condensed, they fell and formed a universal ocean, investing the globe, and covering the loftiest mountains. Further consolidation produced rents, vacuities, and subterranean caverns, and the ocean, rushing in to fill them, was gradually lowered. The principal feature of this theory, the gradual diminution of the original heat, and of an ancient universal ocean, were adopted by Buffon and De Luc, and entered, under different modifications, into a great number of succeeding systems.

Andrea Celsius, the Swedish astronomer, published, about this time, his remarks on the gradual diminution of the waters in the Baltic, which sea, he imagined, had been sinking from time immemorial at the rate of forty-five inches in a century. His opinions gave rise to a controversy which has lasted even to our own days, and to which we are indebted for correct observations of a variety of facts concerning the gradual filling up of the Baltic by fluviatile and marine sediment. Linnæus * favoured the views of Celsius, because they fell in with his own notions concerning a Paradise, where all the animals were created, and from whence they passed into all other parts of the earth, as these became dry in succession.

In Germany, in the mean time, Scheuchzer laboured to prove, in a work entitled the "Complaint of the Fishes," (1708,) that the earth had been remodelled at the deluge.

* De Telluris habitabilis Incremento, 1743.

Pluche also, in 1732, wrote to the same effect, while Holbach, in 1753, after considering the various attempts to refer all the ancient formations to the Noachian flood, exposed the insufficiency of the cause.

We return with pleasure to the geologists of Italy, who preceded, as we before saw, the naturalists of other countries in their investigations into the ancient history of the earth, and who still maintained a decided pre-eminence. They refuted and ridiculed the physico-theological systems of Burnet, Whiston, and Woodward*, while Vallisneri †, in his comments on the Woodwardian theory, remarked how much the interests of religion as well as those of sound philosophy had suffered, by perpetually mixing up the sacred writings with questions in physical science. The works of this author were rich in original observations. He attempted the first general sketch of the marine deposits of Italy, their geographical extent and most characteristic organic remains. In his treatise "On the Origin of Springs," he explained their dependence on the order, and often on the dislocations of the strata, and reasoned philosophically against the opinions of those who regarded the disordered state of the earth's crust as exhibiting signs of the wrath of God for the sins of man. He found himself under the necessity of contending in his preliminary chapter against St. Jerome, and four other principal interpreters of scripture, besides several professors of divinity, "that springs did not flow by subterranean syphons and cavities from the sea upwards, losing their saltness in the passage," for this theory had been made to rest on the infallible testimony of Holy Writ.

Although reluctant to generalize on the rich materials accumulated in his travels, Vallisneri had been so much struck with the remarkable continuity of the more recent marine strata, from one end of Italy to the other, that he came to the conclusion that the ocean formerly extended over the whole earth, and abode there for a long time. This opinion, how-

* Ramazzini even asserted, that the ideas of Burnet were mainly borrowed from a dialogue of one Patrizio; but Brocchi, after reading that dialogue, assures us, that there was scarcely any other correspondence between these systems, except that both were equally whimsical.

† Dei Corpi marini, Lettere critiche, &c. 1721.

ever untenable, was a great step beyond Woodward's diluvian hypothesis, against which Vallisneri, and after him all the Tuscan geologists, uniformly contended, while it was warmly supported by the members of the Institute of Bologna*.

Among others of that day, Spada, a priest of Grezzana, in 1737, wrote to prove that the petrified marine bodies near Verona were not diluvian†. Mattani drew similar inference, from the shells of Volterra, and other places; while Costantini, on the other hand, whose observations on the valley of the Brenta and other districts were not without value, undertook to vindicate the truth of the deluge, as also to prove that Italy had been peopled by the descendants of Japhet‡.

Lazzoro Moro, in his work (published in 1740), "On the Marine Bodies which are found in the Mountains§," attempted to apply the theory of earthquakes, as expounded by Strabo, Pliny, and other ancient authors, with whom he was familiar, to the geological phenomena described by Vallisneri‖. His attention was awakened to the elevating power of subterranean forces, by a remarkable phenomenon which happened in his own time, and which had also been noticed by Vallisneri in his letters. A new island rose in 1707, from a deep part of the sea near Santorino in the Mediterranean, during continued shocks of an earthquake, and increasing rapidly in size, grew in less than a month to be half a mile in circumference, and about twenty-five feet above high-water mark. It was soon afterwards covered by volcanic ejections, but when first examined it was found to be a white rock, bearing on its surface living oysters and crustacea. In order to ridicule the various theories then in vogue, Moro ingeniously supposes the arrival on this new isle of a party of naturalists ignorant of its recent origin. One immediately points to the marine shells, as proofs of the universal deluge; another argues, that they demonstrate the former residence of the sea upon the mountains; a third dismisses them as mere *sports of nature*;

* Brocchi, p. 28. † Ibid. p. 33. ‡ Ibid. p. 37.

§ Sui Crostacei ed altri Corpi marini che si trovano sui Monti.

‖ Moro does not cite the works of Hooke and Ray, and although so many of his views were in accordance with theirs, he was probably ignorant of their writings, for they had not been translated. As he always refers to the Latin edition of Burnet, and a French translation of Woodward, we may presume that he did not read English.

while a fourth affirms, that they were born and nourished within the rock in ancient caverns, into which salt water had been raised in the shape of vapour, by the action of subterranean heat.

Moro pointed with great judgment to the *faults* and dislocations of the strata described by Vallisneri, in the Alps and other chains, in confirmation of his doctrine, that the continents had been heaved up by subterranean movements. He objected, on solid grounds, to the hypotheses of Burnet and of Woodward; yet he ventured so far to disregard the protest of Vallisneri, as to undertake the adaptation of every part of his own system to the Mosaic account of the creation. On the third day, he said the globe was every where covered to the same depth by fresh water, and when it pleased the Supreme Being that the dry land should appear, volcanic explosions broke up the smooth and regular surface of the earth composed of primary rocks. These rose in mountain masses above the waves, and allowed melted metals and salts to ascend through fissures. *The sea gradually acquired its saltness from volcanic exhalations*, and, while it became more circumscribed in area, increased in depth. Sand and ashes ejected by volcanoes were regularly disposed along the bottom of the ocean and formed the secondary strata, which in their turn were lifted up by earthquakes. We shall not attempt to follow him in tracing the progress of the creation of vegetables, and animals on the other days of creation; but, upon the whole, we may remark that few of the old cosmological theories had been conceived with so little violation of known analogies.

The style of Moro was extremely prolix, and, like Hutton, who, at a later period, advanced many of the same views, he stood in need of an illustrator. The Scotch geologist was not more fortunate in the advocacy of Playfair, than was Moro in numbering amongst his admirers Cirillo Generelli, who, nine years afterwards, delivered at a sitting of Academicians at Cremona a spirited exposition of his theory. This learned Carmelitan friar does not pretend to have been an original observer, but he had studied sufficiently to be enabled to confirm the opinions of Moro by arguments from other writers; and his selection of the doctrines then best established is so judicious, that we shall present a brief abstract of them to our readers, as illustrating the state of geology in Europe, and in

Italy in particular, before the middle of the last century. The bowels of the earth, says he, have carefully preserved the memorials of past events, and this truth the marine productions so frequent in the hills attest. From the reflections of Lazzoro Moro we may assure ourselves, that these are the effects of earthquakes in past times, which have changed vast spaces of sea into terra firma, and inhabited lands into seas. In this, more than in any other department of physics, are observations and experiments indispensable, and we must diligently consider facts. The land is known, wherever we make excavations, to be composed of different strata or soils placed one above the other, some of sand, some of rock, some of chalk, others of marl, coal, pumice, gypsum, lime, and the rest. These ingredients are sometimes pure, and sometimes confusedly intermixed. Within are often imprisoned different marine fishes, like dried mummies, and more frequently shells, crustacea, corals, plants, &c., not only in Italy, but in France, Germany, England, Africa, Asia, and America. Sometimes in the lowest, sometimes in the loftiest beds of the earth, some upon the mountains, some in deep mines, others near the sea, and others hundreds of miles distant from it. But there are in some districts rocks, wherein no marine bodies are found. The remains of animals consist chiefly of their more solid parts, and the most rocky strata must have been soft when such exuviæ were inclosed in them. Vegetable productions are found in different states of maturity, indicating that they were imbedded in different seasons. Elephants, elks, and other terrestrial quadrupeds, have been found in England and elsewhere, in superficial strata, never covered by the sea. Alternations are rare, yet not without example, of marine strata, and those which contain marshy and terrestrial productions. Marine animals are arranged in the subterraneous beds with admirable order, in distinct groups, oysters here, dentalia, or corals there, &c., as now, according to Marsilli*, on the shores of the Adriatic. We must abandon the doctrine once so popular, that organized fossils have not been derived from living beings, and we cannot account for their present position by the ancient theory of Strato, nor by that of Leibnitz, nor by the

* Saggio fisico intorno alla Storia del Mare, part i. p. 24.

universal deluge, as explained by Woodward and others, "nor is it reasonable to call the Deity capriciously upon the stage, and to make him work miracles, for the sake of confirming our preconceived hypotheses."—"I hold in utter abomination, most learned Academicians! those systems which are built with their foundations in the air, and cannot be propped up without a miracle; and I undertake, with the assistance of Moro, to explain to you, how these marine animals were transported into the mountains by natural causes*." A brief abstract then follows of Moro's theory, by which, says Generelli, we may explain all the phenomena, as Vallisneri so ardently desired, "*without violence, without fictions, without hypotheses, without miracles*†." The Carmelitan then proceeds to struggle against an obvious objection to Moro's system, considered as a method of explaining the revolutions of the earth, *naturally*. If earthquakes have been the agents of such mighty changes, how does it happen that their effects since the times of history have been so inconsiderable? This same difficulty had, as we have seen, presented itself to Hooke, half a century before, and forced him to resort to a former "crisis of nature;" but Generelli defended his position by shewing how numerous were the accounts of eruptions and earthquakes, of new islands, and of elevations and subsidences of land, and yet how much greater a number of like events must have been unattested and unrecorded during the last six thousand years. He also appealed to Vallisneri as an authority to prove that the mineral masses containing shells bore, upon the whole, but a small proportion to those rocks which were destitute of organic remains; and the latter, says the learned monk, might have been created as they now exist, *in the beginning*. He then describes the continual waste of mountains and continents, by the action of rivers and torrents, and concludes with these eloquent and original observations: "Is it possible that this waste should have continued for six thousand, and *perhaps* a greater number of years, and that the mountains should remain so great, unless their ruins have been repaired? Is it credible that the Author of

* Abbomino al sommo qualsivoglia sistema, che sia di pianta fabbricato in aria; massime quando è tale, che non possa sostenersi senza un miracolo, &c. De' Crostacei e di altre produz. del Mare, &c. 1749.

† Senza violenze, senza finzioni, senza supposti, senza miracoli.—*Ib.*

nature should have founded the world upon such laws, as that the dry land should for ever be growing smaller, and at last become wholly submerged beneath the waters? Is it credible that, amid so many created things, the mountains alone should daily diminish in number and bulk, without there being any repair of their losses? This would be contrary to that order of Providence which is seen to reign in all other things in the universe. Wherefore I deem it just to conclude, that the same cause which, in the beginning of time, raised mountains from the abyss, has, down to the present day, continued to produce others, in order to restore from time to time the losses of all such as sink down in different places, or are rent asunder, or in other ways suffer disintegration. If this be admitted, we can easily understand why there should now be found upon many mountains so great a number of crustacea and other marine animals."

The reader will remark, that although this admirable essay embraces so large a portion of the principal objects of geological research, it makes no allusion to the extinction of certain classes of animals; and it is evident that no opinions on this head had, at that time, gained a firm footing in Italy. That Lister and other English naturalists should long before have declared in favour of the loss of species, while Scilla and most of his countrymen hesitated, was natural, since the Italian museums were filled with fossil shells, belonging to species of which a great portion did actually exist in the Mediterranean, whereas the English collectors could obtain no recent species from their own strata.

The weakest point in Moro's system consisted in deriving *all* the stratified rocks from volcanic ejections, an absurdity which his opponents took care to expose, especially Vito Amici*. Moro seems to have been misled by his anxious desire to represent the formation of secondary rocks as having occupied an extremely short period, while at the same time he wished to employ known agents in nature. To imagine torrents, rivers, currents, partial floods, and all the operations of moving water, to have gone on exerting an energy many thousand times greater than at present, would have appeared preposterous and incredible, and would have

* Sui Testacei della Sicilia.

required a hundred violent hypotheses; but we are so unacquainted with the true sources of subterranean disturbances, that their former violence may in theory be multiplied indefinitely, without its being possible to prove the same manifest contradiction or absurdity in the conjecture. For this reason, perhaps, Moro preferred to derive the materials of the strata from volcanic ejections, rather than from transportation by running water.

Marsilli, in the work above alluded to by Generelli, had been prompted to institute inquiries into the bed of the Adriatic, by discovering in the territory of Parma, (what Spada had observed near Verona, and Schiavo in Sicily,) that fossil shells were not scattered through the rocks at random, but disposed in regular order, according to families. But with a view of throwing further light upon these questions, Donati, in 1750, undertook a more extensive investigation of the Adriatic, and discovered, by numerous soundings, that deposits of sand, marl, and tufaceous incrustations, most strictly analogous to those of the Subapennine hills, were in the act of accumulating there. He ascertained that there were no shells in some of the submarine tracts, while in other places they lived together in families, particularly the genera Arca, Pecten, Venus, Murex, and some others. A contemporary naturalist, Baldassari, had shewn the same grouping of organic remains in the tertiary marls of the Sienese territory.

Buffon first made known his theoretical views concerning the former changes of the earth in his Natural History, published in 1749. His opinions were directly opposed to the systems of Hooke, Ray, and Moro, for he attributed no influence whatever to subterranean movements and volcanoes, but returned to the universal ocean of Leibnitz. By this aqueous envelope the highest mountains were once covered. Marine currents then acted violently, and formed horizontal strata, by washing away land in some parts, and depositing it in others; they also excavated deep submarine valleys. He was greatly at a loss for some machinery to depress the level of the ocean, and cause the land to be left dry. He therefore speculated on the possibility of subterranean caverns having opened, into which the water entered, so that he involuntarily approximated to Hooke's theory of subsidences by earthquakes.

Buffon had never profited, like Moro, by the observations of Vallisneri, or he never could have imagined that the strata were generally horizontal, and that those which contain organic remains had never been disturbed since the era of their formation. He was conscious of the great power annually exerted by rivers and marine currents in transporting earthy materials to lower levels, and he even contemplated the period when they would destroy all the present continents. Although in geology he was not an original observer, his genius enabled him to render his hypothesis attractive; and by the eloquence of his style, and the boldness of his speculations, he awakened curiosity and provoked a spirit of inquiry amongst his countrymen.

Soon after the publication of his "Natural History," in which was included his "Theory of the Earth," he received an official letter (dated January, 1751), from the Sorbonne or Faculty of Theology in Paris, informing him that fourteen propositions in his works "were reprehensible and contrary to the creed of the church." The first of these obnoxious passages, and the only one relating to geology, was as follows. "The waters of the sea have produced the mountains and valleys of the land—the waters of the heavens, reducing all to a level, will at last deliver the whole land over to the sea, and the sea, successively prevailing over the land, will leave dry new continents like those which we inhabit." Buffon was invited by the College in very courteous terms, to send in an explanation, or rather a recantation, of his unorthodox opinions. To this he submitted, and a general assembly of the Faculty having approved of his "Declaration," he was required to publish it in his next work. The document begins with these words—"I declare that I had no intention to contradict the text of Scripture; that I believe most firmly all therein related about the creation, both as to order of time and matter of fact; and *I abandon everything in my book respecting the formation of the earth*, and generally all which may be contrary to the narration of Moses*."

The grand principle which Buffon was called upon to renounce was simply this, "that the present mountains and valleys of the earth are due to secondary causes, and that the same causes will in time destroy all the continents, hills

* Hist. Nat. tom. v. Ed. de l'Imp. Royale, Paris, 1769.

and valleys, and reproduce others like them." Now, whatever may be the defects of many of his views, it is no longer controverted, that the present continents are of secondary origin. The doctrine is as firmly established as the earth's rotation on its axis; and that the land now elevated above the level of the sea will not endure for ever, is an opinion which gains ground daily, in proportion as we enlarge our experience of the changes now in progress.

Hollmann was the author of a Memoir in the Transactions of the Royal Society of Gottingen in 1753, wherein he proposed an hypothesis closely corresponding to the opinions of Buffon; and devoted the rest of his work to refuting certain diluvial theories of his day.

Targioni, in his voluminous "Travels in Tuscany, 1751 and 1754," laboured to fill up the sketch of the geology of that region, left by Steno sixty years before. Notwithstanding a want of arrangement and condensation in his memoirs, they contained a rich store of faithful observations. He has not indulged in many general views, but in regard to the origin of valleys he was opposed to the theory of Buffon, who attributed them principally to submarine currents. The Tuscan naturalist laboured to shew that both the larger and smaller valleys of the Apennines were excavated by rivers, and floods, caused by the bursting of the barriers of lakes, after the retreat of the ocean. He also maintained that the elephants, and other quadrupeds so frequent in the lacustrine and alluvial deposits of Italy, had inhabited that peninsula; and had not been transported thither, as some had conceived, by Hannibal, or the Romans, nor by what they were pleased to term "a catastrophe of nature."

Arduino *, in his memoirs on the mountains of Padua, Vicenza, and Verona, first recognized the distinction between primary, secondary, and tertiary rocks, and shewed that in those districts there had been a succession of submarine volcanic eruptions. In the very same year the treatise of Lehman †, a German mineralogist, and director of the Prussian mines, appeared, who also divided mountains into three classes: the

* Giornale del Griselini, 1759.

† Essai d'une Hist. Nat. de Couches de la Terre, 1759.

first, which were formed with the world and prior to the creation of animals, and which contained no fragments of other rocks; the second class, of mountains which resulted from the partial destruction of the primary rocks by a general revolution; and the third class, which resulted from local revolutions, and, in part, from the Noachian deluge.

In the following year (1760) the Rev. John Michell, Woodwardian Professor of Mineralogy at Cambridge, published in the Philosophical Transactions, an Essay on the Cause and Phenomena of Earthquakes. His attention had been drawn to this subject by the great earthquake of Lisbon in 1755. He advanced many original and philosophical views respecting the propagation of subterranean movements, and the caverns and fissures wherein steam might be generated. In order to point out the application of his theory to the structure of the globe, he was led to describe the arrangement and disturbance of the strata, their usual horizontality in low countries, and their contortions and fractured state in the neighbourhood of mountain chains. He also explained, with surprising accuracy, the relations of the central ridges of older rocks to the "long narrow slips of similar earths, stones, and minerals," which are parallel to these ridges. In his generalizations, derived in great part from his own observations on the geological structure of Yorkshire, he anticipated many of the views more fully developed by later naturalists *.

Michell's papers were entirely free from all physico-theological disquisitions, but some of his contemporaries were still earnestly engaged in defending or impugning the Woodwardian hypothesis. We find many of these writings referred to by Catcott, an Hutchinsonian, who published a "Treatise on the Deluge" in 1761. He laboured particularly to refute an explanation offered by his contemporary, Bishop Clayton, of the Mosaic writings. That prelate had declared that the Deluge "could not be literally true, save in respect to that part where Noah lived before the flood." Catcott insisted on the univer-

* Some of Michell's observations anticipate in so remarkable a manner the theories established forty years afterwards, that his writings would probably have formed an era in the science, if his researches had been uninterrupted. He held, however, his professorship only eight years, when he succeeded to a benefice, and from that time he appears to have entirely discontinued his scientific pursuits.

sality of the deluge, and referred to traditions of inundations mentioned by ancient writers, or by travellers in the East-Indies, China, South America, and other countries. This part of his book is valuable, although it is not easy to see what bearing the traditions have, if admitted to be authentic, on the Bishop's argument, since no evidence is adduced to prove that the catastrophes were contemporaneous events, while some of them are expressly represented by ancient authors to have occurred in succession.

The doctrines of Arduino, above adverted to, were afterwards confirmed by Fortis and Desmarest, in their travels in the same country, and they, as well as Baldassari, laboured to complete the history of the Subapennine strata. In the work of Odoardi *, there was also a clear argument in favour of the distinct ages of the older Apennine strata, and the Subapennine formations of more recent origin. He pointed out that the strata of these two groups were *unconformable*, and must have been the deposits of different seas at distant periods of time.

A history of the new islands by Raspe, an Hanoverian, appeared in 1763, in Latin. In this work, all the authentic accounts of earthquakes which had produced permanent changes on the solid parts of the earth were collected together and examined with judicious criticism. The best systems which had been proposed concerning the ancient history of the globe, both by ancient and modern writers, are reviewed. The merits and defects of the systems of Hooke, Ray, Moro, Buffon, and others, are fairly estimated. Great admiration is expressed for the hypothesis of Hooke, and his explanation of the origin of the strata is shewn to have been more correct than Moro's, while their theory of the effects of earthquakes was the same. Raspe had not seen Michell's memoir, and his views concerning the geological structure of the earth were perhaps less enlarged, yet he was able to add many additional arguments in favour of Hooke's theory, and to render it, as he said, a nearer approach to what Hooke would have written had he lived in later times. As to the periods wherein all the earthquakes happened, to which we owe the elevation of various parts of our continents and islands, Raspe says he pretends not to assign their duration,

* Sui Corpi Marini del Feltrino, 1761.

still less to defend Hooke's suggestion, that the convulsions almost all took place during the Noachian deluge. He adverts to the apparent indications of the former tropical heat of the climate of Europe, and the changes in the species of animals and plants, as among the most obscure and difficult problems in geology. In regard to the islands raised from the sea, within the times of history or tradition, he declares that some of them were composed of strata containing organic remains, and that they were not, as Buffon had asserted, made of mere volcanic matter. His work concludes with an eloquent exhortation to naturalists, to examine the isles which rose in 1707, in the Grecian Archipelago, and in 1720 in the Azores, and not to neglect such splendid opportunities of studying nature "in the act of parturition." That Hooke's writings should have been neglected for more than half a century, was matter of astonishment to Raspe; but, it is still more wonderful that his own luminous exposition of that theory should, for more than another half century, have excited so little interest.

Gustavus Brander published, in 1766, his "Fossilia Hantoniensia," containing excellent figures of fossil shells from the more modern marine strata of our island. "Various opinions," he says in the preface, "had been entertained concerning the time when and how these bodies became deposited. Some there are who conceive that it might have been effected in a wonderful length of time by a gradual changing and shifting of the sea, &c. But the most common cause assigned is that of "the deluge." This conjecture, he says, even if the universality of the flood be not called in question, is purely hypothetical. In his opinion fossil animals and testacea were, for the most part, of unknown species, and of such as were known, the living analogues now belonged to southern latitudes.

Soldani * applied successfully his knowledge of zoology to illustrate the history of stratified masses. He explained that microscopic testacea and zoophytes inhabited the depths of the Mediterranean, and that the fossil species were, in like manner, found in those deposits wherein the fineness of their particles, and the absence of pebbles, implied that they were accumulated in a deep sea far from any shore. This author first remarked the alternation of marine and fresh-water strata in the Paris

* Saggio orittografico, &c. 1780, and other Works.

basin. A lively controversy arose between Fortis and another Italian naturalist, Testa, concerning the fish of Monte Bolca, in 1793. Their letters*, written with great spirit and elegance, shew that they were aware that a large proportion of the Sub-apennine shells were identical with living species, and some of them with species now living in the torrid zone. Fortis conjectured that when the volcanos of the Vicentin were burning, the waters of the Adriatic had a higher temperature; and in this manner, he said that the shells of warmer regions may once have peopled their own seas. But Testa was disposed to think, that these species of testacea were still common to their own and to equinoctial seas, for many, he said, once supposed to be confined to hotter regions, had been afterwards discovered in the Mediterranean †.

While these Italian naturalists, together with Cortesi and Spallanzani, were busily engaged in pointing out the analogy between the deposits of modern and ancient seas, and the habits and arrangement of their organic inhabitants, and while some progress was making in the same country, in investigating the ancient and modern volcanic rocks, the most original observers among the English and German writers, Wallerius and Whitehurst ‡, were wasting their strength in contending, according to the old Woodwardian hypothesis, that all the strata were formed by the Noachian deluge. But Whitehurst's description of the rocks of Derbyshire was most faithful, and he atoned for false theoretical views, by providing data for their refutation.

The mathematician, Boscovich, of Ragusa in Dalmatia, in his letters, published at Venice in 1772, declared his persuasion, that the effects of earthquakes, although insensible in the course of a few years, do nevertheless raise, from time to time,

* Lett. sui Pesci Fossili di Bolca. *Milan*, 1793.

† This argument of Testa has been strengthened of late years by the discovery, that dealers in shells had long been in the habit of selling Mediterranean species as shells of more southern and distant latitudes, for the sake of enhancing their price. It appears, moreover, from several hundred experiments made by that distinguished hydrographer Captain Smyth, on the water within eight fathoms of the surface, that the temperature of the Mediterranean is on an average $3\frac{1}{2}$° of Fahrenheit higher than the western part of the Atlantic ocean; an important fact which in some degree may help to explain why many species are common to tropical latitudes, and to the Mediterranean.

‡ Inquiry into the Original State and Formation of the Earth. 1778.

and let down different parts of the crust of our globe, and sometimes fold and twist them. Like Hooke, Ray, and Moro, he conceived the subterranean movements to have acted with greater energy at former epochs.

Towards the close of the eighteenth century, the idea of distinguishing the mineral masses on our globe into separate groups, and studying their relations, began to be generally diffused. Pallas and Saussure were among the most celebrated whose labours contributed to this end. After an attentive examination of the two great mountain chains of Siberia, Pallas announced the result that the granitic rocks were in the middle, the schistose at their sides, and the limestones again on the outside of these; and this he conceived would prove a general law in the formation of all chains composed chiefly of primary rocks *.

In his "Travels in Russia," in 1793 and 1794, he made many geological observations on the recent strata near the Wolga and the Caspian, and adduced proofs of the greater extent of the latter sea at no distant era in the earth's history. His memoir on the fossil bones of Siberia attracted attention to some of the most remarkable phenomena in geology. He stated that he had found a rhinoceros entire in the frozen soil, with its skin and flesh: an elephant, found afterwards in a mass of ice on the shore of the north sea, removed all doubt as to the accuracy of so wonderful a discovery †.

The subjects relating to natural history which engaged the attention of Pallas were too multifarious to admit of his devoting a large share of his labours exclusively to geology. Saussure, on the other hand, employed the chief portion of his time in studying the structure of the Alps and Jura, and he provided valuable data for those who followed him. We cannot enter into the details of these observations, and he did not pretend to have arrived at any general system. The few theoretical observations which escaped from him are, like those of Pallas, mere modifications of the old cosmological doctrines.

* Observ. on the Formation of Mountains, Act. Petrop. ann. 1778, part i.

† Nov. comm. Petr. XVII. Cuvier, Eloge de Pallas.

CHAPTER IV.

Werner's application of Geology to the art of Mining—Excursive character of his Lectures—Enthusiasm of his pupils—His authority—His theoretical errors—Desmarest's Map and Description of Auvergne—Controversy between the Vulcanists and Neptunists—Intemperance of the rival Sects.—Hutton's Theory of the Earth—His discovery of Granite Veins—Originality of his Views—Why opposed.—Playfair's Illustrations—Influence of Voltaire's Writings on Geology—Imputations cast on the Huttonians by Williams, Kirwan, and De Luc—Smith's Map of England—Geological Society of London—Progress of the Science in France—Growing Importance of the Study of Organic Remains.

The art of mining has long been taught in France, Germany, and Hungary, in scientific institutions established for that purpose, where mineralogy has always been a principal branch of instruction *.

Werner was named, in 1775, professor of that science in the "School of Mines" at Freyberg in Saxony. He directed his attention not merely to the composition and external characters of minerals, but also to what he termed "geognosy," or the natural position of minerals in particular rocks, together with the grouping of those rocks, their geographical distribution, and various relations. The phenomena observed in the structure of the globe had hitherto served for little else than to furnish interesting topics for philosophical discussion; but when Werner pointed out their application to the practical purposes of mining, they were instantly regarded by a large class of men as an essential part of their professional education, and from that time the science was cultivated in Europe more ardently and systematically. Werner's mind was at once ima-

* Our miners have been left to themselves, almost without the assistance of scientific works in the English language, and without any "school of mines," to blunder their own way into a certain degree of practical skill. The inconvenience of this want of system in a country where so much capital is expended, and often wasted, in mining adventures, has been well exposed by an eminent practical miner.—See "Prospectus of a School of Mines in Cornwall, by J. Taylor, 1825."

ginative and richly stored with miscellaneous knowledge. He associated everything with his favourite science, and in his excursive lectures he pointed out all the economical uses of minerals, and their application to medicine; the influence of the mineral composition of rocks upon the soil, and of the soil upon the resources, wealth, and civilization of man. The vast sandy plains of Tartary and Africa he would say retained their inhabitants in the shape of wandering shepherds; the granitic mountains and the low calcareous and alluvial plains gave rise to different manners, degrees of wealth and intelligence. The history even of languages, and the migrations of tribes had, according to him, been determined by the direction of particular strata. The qualities of certain stones used in building would lead him to descant on the architecture of different ages and nations, and the physical geography of a country frequently invited him to treat of military tactics. The charm of his manners and his eloquence kindled enthusiasm in the minds of all his pupils, many of whom only intended at first to acquire a slight knowledge of mineralogy; but, when they had once heard him, they devoted themselves to it as the business of their lives. In a few years a small school of mines, before unheard of in Europe, was raised to the rank of a great university, and men already distinguished in science studied the German language, and came from the most distant countries to hear the great oracle of geology *.

Werner had a great antipathy to the mechanical labour of writing, and he could never be persuaded to pen more than a few brief memoirs, and those containing no development of his general views. Although the natural modesty of his disposition was excessive, approaching even to timidity, he indulged in the most bold and sweeping generalizations, and he inspired all his scholars with a most implicit faith in his doctrines. Their admiration of his genius, and the feelings of gratitude and friendship which they all felt for him, were not undeserved; but the supreme authority usurped by him over the opinions of his contemporaries, was eventually prejudicial to the progress of the science, so much so, as greatly to counterbalance the advantages which it derived from his exertions. If it be true that delivery be the first, second, and third requisite in a

* Cuvier, Eloge de Werner.

popular orator, it is no less certain that to travel is of threefold importance to those who desire to originate just and comprehensive views concerning the structure of our globe, and Werner had never travelled to distant countries. He had merely explored a small portion of Germany, and conceived, and persuaded others to believe, that the whole surface of our planet, and all the mountain chains in the world, were made after the model of his own province. It was a ruling object of ambition in the minds of his pupils to confirm the generalizations of their great master, and to discover in the most distant parts of the globe his "universal formations," which he supposed had been each in succession simultaneously precipitated over the whole earth from a common menstruum, or "chaotic fluid." Unfortunately, the limited district examined by the Saxon professor was no type of the world, nor even of Europe; and, what was still more deplorable, when the ingenuity of his scholars had tortured the phenomena of distant countries, and even of another hemisphere, into conformity with his theoretical standard, it was discovered that "the master" had misinterpreted many of the appearances in the immediate neighbourhood of Freyberg.

Thus, for example, within a day's journey of his school, the porphyry, called by him primitive, has been found not only to send forth veins or dikes through strata of the coal formation, but to overlie them in mass. The granite of the Hartz mountains, on the other hand, which he supposed to be the nucleus of the chain, is now well known to traverse and breach the other beds, penetrating even into the plain (as near Goslar); and nearer Freyberg, in the Erzgebirge, the mica slate does not mantle round the granite, as the professor supposed, but abuts abruptly against it. But it is still more remarkable, that in the Hartz mountains all his flötz rocks, which he represented as horizontal, are highly inclined, and often nearly vertical, as the chalk at Goslar, and the green sand near Blankenberg.

The principal merit of Werner's system of instruction consisted in steadily directing the attention of his scholars to the constant relations of certain mineral groups, and their regular order of superposition. But he had been anticipated, as we have shewn in the last chapter, in the discovery of this general law, by several geologists in Italy and elsewhere; and his leading divisions of the secondary strata were at the same time made the basis of an arrangement of the British strata by our countryman, William Smith, to whose work we shall return by-and-by. In regard to basalt and other igneous rocks, Werner's theory was original, but it was also extremely erroneous. The basalts of Saxony and Hesse, to which his observations were chiefly confined, consisted of tabular masses capping the hills, and not connected with the levels of existing valleys, like many in Auvergne and the Vivarais. These basalts, and all other rocks of the same family in other countries, were, according to him, chemical precipitates from water. He denied that they were the products of submarine volcanos, and even taught that, in the primeval ages of the world, there were no volcanos. His theory was opposed, in a two-fold sense, to the doctrine of uniformity in the course of nature; for not only did he introduce, without scruple, many imaginary causes supposed to have once effected great revolutions in the earth, and then to have become extinct, but new ones also were feigned to have come into play in modern times; and, above all, that most violent instrument of change, the agency of subterranean fire. So early as 1768, before Werner had commenced his mineralogical studies, Raspe had truly characterized the basalts of Hesse as of igneous origin. Arduino, as we have already seen, had pointed out numerous varieties of trap-rock in the Vicentin, as analogous to volcanic products, and as distinctly referrible to ancient submarine eruptions. Desmarest, as we stated, had, in company with Fortis, examined the Vicentin in 1766, and confirmed Arduino's views. In 1772, Banks, Solander, and Troil, compared the columnar basalt of Hecla with that of the Hebrides. Collini, in 1774, recognised the true nature of the igneous rocks on the Rhine, between Andernach and Bonn. In 1775, Guettard visited the Vivarais, and established the relation of basaltic currents to lavas. Lastly, in 1779, Faujas published his description of the volcanos of the Vivarais and Velay, and shewed how the streams of basalt had poured out from craters which still remain in a perfect state *.

When sound opinions had for twenty years prevailed in Europe concerning the true nature of the ancient trap-rocks,

* Cuvier, Éloge de Desmarest.

Werner by his dictum caused a retrograde movement, and not only overturned the true theory, but substituted for it one of the most unphilosophical ever advanced in any science. The continued ascendancy of his dogmas on this subject was the more astonishing, because a variety of new and striking facts were daily accumulated in favour of the correct opinions first established. Desmarest, after a careful examination of Auvergne, pointed out first the most recent volcanos which had their craters still entire, and their streams of lava conforming to the level of the present river-courses. He then shewed that there were others of an intermediate epoch, whose craters were nearly effaced, and whose lavas were less intimately connected with the present valleys; and, lastly, that there were volcanic rocks still more ancient, without any discernible craters or scoriæ, and bearing the closest analogy to rocks in other parts of Europe, the igneous origin of which was denied by the school of Freyberg*.

Desmarest's map of Auvergne was a work of uncommon merit. He first made a trigonometrical survey of the district, and delineated its physical geography with minute accuracy and admirable graphic power. He contrived, at the same time, to express, without the aid of colours, a vast quantity of geological detail, the different ages, and sometimes even the structure of the volcanic rocks, distinguishing them from the fresh-water and the granitic. They alone who have carefully studied Auvergne, and traced the different lava streams from their craters to their termination,—the various isolated basaltic cappings,—the relation of some lavas to the present valleys,—the absence of such relations in others,—can appreciate the extraordinary fidelity of this elaborate work. No other district of equal dimensions in Europe exhibits, perhaps, so beautiful and varied a series of phenomena; and, fortunately, Desmarest possessed at once the mathematical knowledge required for the construction of a map, skill in mineralogy, and a power of original generalization.

Dolomieu, another of Werner's contemporaries, had found prismatic basalts among the ancient lavas of Etna, and in 1784

* Journ. de Phys., vol. xiii. p. 115; and Mem. de l'Inst., Sciences, Mathémat. et Phys., vol. vi. p. 219.

had observed the alternations of submarine and calcareous strata in the Val di Noto in Sicily *. In 1790, he also described similar phenomena in the Vicentin and in the Tyrol †. Montlosier also published, in 1788, an elegant and spirited essay on the volcanos of Auvergne, combining accurate local observations with comprehensive views. In opposition to this mass of evidence, the scholars of Werner were prepared to support his opinions to their utmost extent, maintaining in the fulness of their faith that even obsidian was an aqueous precipitate. As they were blinded by their veneration for the great teacher, they were impatient of opposition, and soon imbibed the spirit of a faction; and their opponents, the Vulcanists, were not long in becoming contaminated with the same intemperate zeal. Ridicule and irony were weapons more frequently employed than argument by the rival sects, till at last the controversy was carried on with a degree of bitterness, almost unprecedented in questions of physical science. Desmarest alone, who had long before provided ample materials for refuting such a theory, kept aloof from the strife, and whenever a zealous Neptunist wished to draw the old man into an argument, he was satisfied with replying, "Go and see.‡"

It would be contrary to all analogy, in matters of graver import, that a war should rage with such fury on the continent, and that the inhabitants of our island should not mingle in the affray. Although in England the personal influence of Werner was wanting to stimulate men to the defence of the weaker side of the question, they contrived to find good reason for espousing the Wernerian errors with great enthusiasm. In order to explain the peculiar motives which led many to enter, even with party feeling, into this contest, we must present the reader with a sketch of the views unfolded by Hutton, a contemporary of the Saxon geologist. That naturalist had been educated as a physician, but, declining the practice of medicine, he resolved, when young, to remain content with the small independence inherited from his father, and thenceforth to give his undivided attention to scientific pursuits. He resided at Edinburgh, where he enjoyed the society of many

* Journ. de Phys., tom. xxv. p. 191. † Ib. tom. xxxvii. part ii. p. 200.
‡ Cuvier, Éloge de Desmarest.

men of high attainments, who loved him for the simplicity of his manners and the sincerity of his character. His application was unwearied, and he made frequent tours through different parts of England and Scotland, acquiring considerable skill as a mineralogist, and constantly arriving at grand and comprehensive views in geology. He communicated the results of his observations unreservedly, and with the fearless spirit of one who was conscious that love of truth was the sole stimulus of all his exertions. When at length he had matured his views, he published, in 1788, his "Theory of the Earth *," and the same, afterwards more fully developed in a separate work, in 1795. This treatise was the first in which geology was declared to be in no way concerned about "questions as to the origin of things;" the first in which an attempt was made to dispense entirely with all hypothetical causes, and to explain the former changes of the earth's crust, by reference exclusively to natural agents. Hutton laboured to give fixed principles to geology, as Newton had succeeded in doing to astronomy; but in the former science too little progress had been made towards furnishing the necessary data to enable any philosopher, however great his genius, to realize so noble a project.

"The ruins of an older world," said Hutton, "are visible in the present structure of our planet, and the strata which now compose our continents have been once beneath the sea, and were formed out of the waste of pre-existing continents. The same forces are still destroying, by chemical decomposition or mechanical violence, even the hardest rocks, and transporting the materials to the sea, where they are spread out, and form strata analogous to those of more ancient date. Although loosely deposited along the bottom of the ocean, they become afterwards altered and consolidated by volcanic heat, and then heaved up, fractured and contorted." Although Hutton had never explored any region of active volcanos, he had convinced himself that basalt and many other trap-rocks were of igneous origin, and that many of them had been injected in a melted state through fissures in the older strata. The compactness of these rocks, and their different

* Ed. Phil. Trans., 1788.

aspect from that of ordinary lava, he attributed to their having cooled down under the pressure of the sea, and in order to remove the objections started against this theory, his friend Sir James Hall instituted a most curious and instructive series of chemical experiments, illustrating the crystalline arrangement and texture assumed by melted matter cooled down under high pressure. The absence of stratification in granite, and its analogy in mineral character to rocks which he deemed of igneous origin, led Hutton to conclude that granite must also have been formed from matter in fusion, and this inference he felt could not be fully confirmed, unless he discovered at the contact of granite and other strata a repetition of the phenomena exhibited so constantly by the trap-rocks. Resolved to try his theory by this test, he went to the Grampians and surveyed the line of junction of the granite and superincumbent stratified masses, and found in Glen Tilt in 1785 the most clear and unequivocal proofs in support of his views. Veins of red granite are there seen branching out from the principal mass, and traversing the black micaceous schist and primary limestone. The intersected stratified rocks are so distinct in colour and appearance as to render the example in that locality most striking, and the alteration of the limestone in contact was very analogous to that produced by trap veins on calcareous strata. This verification of his system filled him with delight, and called forth such marks of joy and exultation, that the guides who accompanied him, says his biographer, were convinced that he must have discovered a vein of silver or gold *. He was aware that the same theory would not explain the origin of the primary schists, but these he called primary, rejecting the term primitive, and was disposed to consider them as sedimentary rocks altered by heat, and that they originated in some other form from the waste of previously existing rocks.

By this important discovery of granite veins to which he had been led by fair induction from an independent class of facts, Hutton prepared the way for the greatest innovation on the systems of his predecessors. Vallisneri had pointed out the general fact, that there were certain fundamental rocks which contained no organic remains, and which he supposed to have been formed

* Playfair's Works, vol. iv., p. 75.

before the creation of living beings. Moro, Generelli, and other Italian writers embraced the same doctrine, and Lehman regarded the mountains called by him primitive, as parts of the original nucleus of the globe. The same tenet was an article of faith in the school of Freyberg; and if any one ventured to doubt the possibility of our being enabled to carry back our researches to the creation of the present order of things, the granitic rocks were triumphantly appealed to. On them seemed written in legible characters, the memorable inscription

> Dinanzi a me non fur cose create
> Se non eterne,

and no small sensation was excited when Hutton seemed, with unhallowed hand, desirous to erase characters already regarded by many as sacred. "In the economy of the world," said the Scotch geologist, "I can find no traces of a beginning, no prospect of an end;" and the declaration was the more startling when coupled with the doctrine, that all past changes on the globe had been brought about by the slow agency of existing causes. The imagination was first fatigued and overpowered by endeavouring to conceive the immensity of time required for the annihilation of whole continents by so insensible a process. Yet when the thoughts had wandered through these interminable periods, no resting place was assigned in the remotest distance. The oldest rocks were represented to be of a derivative nature, the last of an antecedent series, and that perhaps one of many pre-existing worlds. Such views of the immensity of past time, like those unfolded by the Newtonian philosophy in regard to space, were too vast to awaken ideas of sublimity unmixed with a painful sense of our incapacity to conceive a plan of such infinite extent. Worlds are seen beyond worlds immeasurably distant from each other, and beyond them all innumerable other systems are faintly traced on the confines of the visible universe.

The characteristic feature of the Huttonian theory was, as before hinted, the exclusion of all causes not supposed to belong to the present order of nature. Its greatest defect consisted in the undue influence attributed to subterranean heat, which was supposed necessary for the consolidation of all submarine deposits. Hutton made no step beyond Hooke, Moro, and

Raspe, in pointing out in what manner the laws now governing earthquakes, might bring about geological changes, if sufficient time be allowed. On the contrary, he seems to have fallen far short of some of their views. He imagined that the continents were first gradually destroyed, and when their ruins had furnished materials for new continents, they were upheaved by violent and paroxysmal convulsions. He therefore required alternate periods of disturbance and repose, and such he believed had been, and would for ever be, the course of nature. Generelli, in his exposition of Moro's system, had made a far nearer approximation towards reconciling geological appearances with the state of nature as known to us, for while he agreed with Hutton, that the decay and reproduction of rocks were always in progress, proceeding with the utmost uniformity, the learned Carmelitan represented the repairs of mountains by elevation from below, to be effected by an equally constant and synchronous operation. Neither of these theories considered singly, satisfies all the conditions of the great problem, which a geologist, who rejects cosmological causes, is called upon to solve; but they probably contain together the germs of a perfect system. There can be no doubt, that periods of disturbance and repose have followed each other in succession in every region of the globe, but it may be equally true, that the energy of the subterranean movements has been always uniform as regards the *whole earth*. The force of earthquakes may for a cycle of years have been invariably confined, as it is now, to large but determinate spaces, and may then have gradually shifted its position, so that another region, which had for ages been at rest, became in its turn the grand theatre of action.

Although Hutton's knowledge of mineralogy and chemistry was considerable, he possessed but little information concerning organic remains. They merely served him as they did Werner to characterize certain strata, and to prove their marine origin. The theory of former revolutions in organic life was not yet fully recognized, and without this class of proofs in support of the antiquity of the globe, the indefinite periods demanded by the Huttonian hypothesis appeared visionary to many, and some, who deemed the doctrine inconsistent with revealed truths, indulged very uncharitable suspicions of the motives of its author. They accused him of a deliberate design of reviving the heathen dogma of an "eternal succession," and of denying that this world ever had a beginning. Playfair, in the biography of his friend, has the following comment on this part of their theory:—"In the planetary motions, where geometry has carried the eye so far, both into the future and the past, we discover no mark either of the commencement or termination of the present order. It is unreasonable, indeed, to suppose that such marks should anywhere exist. The Author of nature has not given laws to the universe, which, like the institutions of men, carry in themselves the elements of their own destruction. He has not permitted in His works any symptom of infancy or of old age, or any sign by which we may estimate either their future or their past duration. *He may put an end, as he no doubt gave a beginning,* to the present system at some determinate period of time; but we may rest assured that this great catastrophe will not be brought about by the laws now existing, and that it is not indicated by any thing which we perceive *."

The party feeling excited against the Huttonian doctrines, and the open disregard of candour and temper in the controversy, will hardly be credited by our readers, unless we recall to their recollection that the mind of the English public was at that time in a state of feverish excitement. A class of writers in France had been labouring industriously for many years, to diminish the influence of the clergy, by sapping the foundation of the Christian faith, and their success, and the consequences of the Revolution, had alarmed the most resolute minds, while the imagination of the more timid was continually haunted by dread of innovation, as by the phantom of some fearful dream.

Voltaire had used the modern discoveries in physics as one of the numerous weapons of attack and ridicule directed by him against the Scriptures. He found that the most popular systems of geology were accommodated to the sacred writings, and that much ingenuity had been employed to make every fact coincide exactly with the Mosaic account of the creation and deluge. It was, therefore, with no friendly feelings, that he contemplated the cultivators of geology in general, regarding the science as one which had been successfully enlisted by

* Playfair's Works, vol. iv. p. 55.

theologians as an ally in their cause*. He knew that the majority of those who were aware of the abundance of fossil shells in the interior of continents, were still persuaded that they were proofs of the universal deluge; and as the readiest way of shaking this article of faith, he endeavoured to inculcate scepticism, as to the real nature of such shells, and to recall from contempt the exploded dogma of the sixteenth century, that they were sports of nature. He also pretended that vegetable impressions were not those of real plants†. Yet he was perfectly convinced that the shells had really belonged to living testacea, as may be seen in his essay, "On the formation of Mountains‡." He would sometimes, in defiance of all consistency, shift his ground when addressing the vulgar; and admitting the true nature of the shells collected in the Alps, and other places, pretend that they were eastern species, which had fallen from the hats of pilgrims coming from Syria. The numerous essays written by him on geological subjects were all calculated to strengthen prejudices, partly because he was ignorant of the real state of the science, and partly from his bad faith §. On the other hand, they who knew that his attacks were directed by a desire to invalidate scripture, and who were unacquainted with the true merits of the question, might well deem the old diluvian hypothesis incontrovertible, if Voltaire could adduce no better argument against it, than to deny the true nature of organic remains.

* In allusion to the theories of Burnet, Woodward, and other physico-theological writers, he declared that they were as fond of changes of scene on the face of the globe, as were the populace at a play. "Every one of them destroys and renovates the earth after his own fashion, as Descartes framed it: for philosophers put themselves without ceremony in the place of God, and think to create a universe with a word."—Dissertation envoyée à l'Académie de Bologne, sur les changemens arrivés dans notre Globe. Unfortunately this and similar ridicule directed against the cosmologists was too well deserved.

† See the chapter on "Des pierres figurées."

‡ In that essay he lays it down "that all naturalists are now agreed that deposits of shells in the midst of the continents, are monuments of the continued occupation of these districts by the ocean." In another place also, when speaking of the fossil shells of Touraine, he admits their true origin.

§ As an instance of his desire to throw doubt indiscriminately on all geological data, we may recall the passage, where he says that "the bones of a rein-deer and hippopotamus discovered near Etampes, did not prove, as some would have it, that Lapland and the Nile were once on a tour from Paris to Orleans, but merely that a lover of curiosities once preserved them in his cabinet."

It is only by careful attention to impediments originating in extrinsic causes, that we can explain the slow and reluctant adoption of the simplest truths in geology. First, we find many able naturalists adducing the fossil remains of marine animals, as proofs of an event related in Scripture. The evidence is deemed conclusive by the multitude for a century or more; for it favours opinions which they entertained before, and they are gratified by supposing them confirmed by fresh and unexpected proofs. Many, who see through the fallacy, have no wish to undeceive those who are influenced by it, approving the effect of the delusion, and conniving at it as a pious fraud; until finally, an opposite party, who are hostile to the sacred writings, labour to explode the erroneous opinion, by substituting for it another dogma which they know to be equally unsound.

The heretical vulcanists were now openly assailed in England, by imputations of the most illiberal kind. We cannot estimate the malevolence of such a persecution, by the pain which similar insinuations might now inflict; for although charges of infidelity and atheism must always be odious, they were injurious in the extreme at that moment of political excitement: and it was better perhaps for a man's good reception in society, that his moral character should have been traduced, than that he should become a mark for these poisoned weapons. We shall pass over the works of numerous divines, who may be excused for sensitiveness on points which then excited so much uneasiness in the public mind; and we shall say nothing of the amiable poet Cowper *, who could hardly be expected to have inquired into the merits of doctrines in physics. But we find in the foremost ranks of the intolerant, several laymen who had high claims to scientific reputation. Amongst these, appears Williams, a mineral surveyor of Edinburgh, who published a "Natural History of the Mineral Kingdom" in 1789, a work of great merit for that day, and of practical utility, as containing the best account of the coal strata. In his preface he misrepresents Hutton's theory altogether, and charges him with considering all rocks to be lavas of different colours and structure; and also with "warping

* The Task, book iii. "The Garden."

every thing to support the eternity of the world *." He descants on the pernicious influence of such sceptical notions, as leading to downright infidelity and atheism, "and as being nothing less than to depose the Almighty Creator of the universe from his office †."

Kirwan, president of the Royal Academy of Dublin, a chemist and mineralogist of some merit, but who possessed much greater authority in the scientific world than he was entitled by his talents to enjoy, in the introduction to his "Geological Essays, 1799," said "that *sound* geology *graduated* into religion, and was required to dispel certain systems of atheism or infidelity, of which they had had recent experience ‡." He was an uncompromising defender of the aqueous theory of all rocks, and was scarcely surpassed by Burnet and Whiston, in his desire to adduce the Mosaic writings in confirmation of his opinions.

De Luc, in the preliminary discourse to his Treatise on Geology §, says, "the weapons have been changed by which revealed religion is attacked; it is now assailed by geology, and this science has become essential to theologians." He imputes the failure of former geological systems to their having been anti-mosaical, and directed against a "sublime tradition." These and similar imputations, reiterated in the works of De Luc, seem to have been taken for granted by some modern writers: it is therefore necessary to state, in justice to the numerous geologists of different nations, whose works we have considered, that none of them were guilty of endeavouring, by arguments drawn from physics, to invalidate scriptural tenets. On the contrary, the majority of them, who were fortunate enough "to discover the true causes of things," did not deserve another part of the poet's panegyric, "*Atque metus omnes subjecit pedibus*." The caution, and even timid reserve, of many eminent Italian authors of the earlier period is very apparent; and there can hardly be a doubt that they subscribed to certain dogmas, and particularly to the first diluvian theory, out of deference to popular prejudices, rather than from conviction. If they were guilty of dissimulation, we must not blame their want of moral courage, but reserve our condemnation for the intolerance of the times, and

* p. 577. † p. 59. ‡ Introd. p. 2. § London, 1809.

that inquisitorial power which forced Galileo to abjure, and the two Jesuits to disclaim the theory of Newton *.

Hutton answered Kirwan's attacks with great warmth, and with the indignation excited by unmerited reproach. He had always displayed, says Playfair, "the utmost disposition to admire the beneficent design manifested in the structure of the world, and he contemplated with delight those parts of his theory which made the greatest additions to our knowledge of final causes." We may say with equal truth, that in no scientific works in our language can more eloquent passages be found, concerning the fitness, harmony, and grandeur of all parts of the creation, than in those of Playfair. They are evidently the unaffected expressions of a mind, which contemplated the study of nature, as best calculated to elevate our conceptions of the attributes of the First Cause. At any other time the force and elegance of Playfair's style must have insured popularity to the Huttonian doctrines; but, by a singular coincidence, neptunianism and orthodoxy were now associated in the same creed; and the tide of prejudice ran so strong, that the majority were carried far away into the chaotic fluid, and other cosmological inventions of Werner. These fictions the Saxon Professor had borrowed with little modification, and without any improvement, from his predecessors. They had not the smallest foundation, either in Scripture, or in common sense, but were perhaps approved of by many as being so ideal and unsubstantial, that they could never come into violent collision with any preconceived opinions.

The great object of De Luc's writings was to disprove the

* I observe that, in a most able and interesting article "the Life of Galileo," recently published in the "Library of Useful Knowledge," it is asserted that both Galileo's work, and the book of Copernicus "Nisi corrigatur," were still to be seen on the forbidden list of the Index at Rome in 1828. But I was assured in the same year, by Professor Scarpellini, at Rome, that Pius VII., a pontiff distinguished for his love of science, procured in 1818 a repeal of the edicts against Galileo and the Copernican system. He assembled the Congregation, and the late cardinal Toriozzi, assessor of the Sacred Office, proposed "that they should wipe off this scandal from the church." The repeal was carried, with the dissentient voice of one Dominican only. Long before this time the Newtonian theory had been taught in the Sapienza, and all catholic universities in Europe (with the exception, I am told, of Salamanca); but it was always required of professors, in deference to the decrees of the church, to use the term *hypothesis*, instead of theory. They now speak of the Copernican *theory*.

high antiquity attributed by Hutton to our present continents, and particularly to seek out some cause for the excavation of valleys more speedy and violent than the action of ordinary rivers. Hutton had said, that the erosion of rivers, and such floods as occur in the usual course of nature, might progressively, if time be allowed, hollow out great valleys, but he had also observed, "that on our continents there is no spot on which a river may not formerly have run *." De Luc generally reasoned against him as if he had said, that the existing rivers flowing *at their present levels* had caused all these inequalities of the earth's surface; and Playfair, in his zeal to prove how much De Luc underrated the force of running water, did not sufficiently expose his misstatement of the Huttonian proposition. But we must defer the full consideration of this controverted question for the present.

While the tenets of the rival schools of Freyberg and Edinburgh were warmly espoused by devoted partisans, the labours of an individual, unassisted by the advantages of wealth or station in society, were almost unheeded. Mr. William Smith, an English surveyor, published his "Tabular View of the British Strata" in 1790, wherein he proposed a classification of the secondary formations in the west of England. Although he had not communicated with Werner, it appeared by this work that he had arrived at the same views respecting the laws of superposition of stratified rocks; that he was aware that the order of succession of different groups was never inverted; and that they might be identified at very distant points by their peculiar organized fossils.

From the time of the appearance of the "Tabular View," he laboured to construct a geological map of the whole of England, and, with the greatest disinterestedness of mind, communicated the results of his investigations to all who desired information, giving such publicity to his original views, as to enable his contemporaries almost to compete with him in the race. The execution of his map was completed in 1815, and remains a lasting monument of original talent and extraordinary perseverance, for he had explored the whole country on foot without the guidance of previous observers, or the aid of fellow-labour-

* Theory of the Earth, vol. ii. p. 296; and Playfair's "Illustrations," note 16, p. 352.

ers, and had succeeded in throwing into natural divisions the whole complicated series of British rocks*. D'Aubuisson, a distinguished pupil of Werner, paid a just tribute of praise to this remarkable performance, observing that "what many celebrated mineralogists had only accomplished for a small part of Germany in the course of half a century, had been effected by a single individual for the whole of England."

We have now arrived at the era of living authors, and shall bring to a conclusion our sketch of the progress of opinion in geology. The contention of the rival factions of the Vulcanists and Neptunists had been carried to such a height, that these names had become terms of reproach, and the two parties had been less occupied in searching for truth, than for such arguments as might strengthen their own cause, or serve to annoy their antagonists. A new school at last arose who professed the strictest neutrality, and the utmost indifference to the systems of Werner and Hutton, and who were resolved diligently to devote their labours to observation. The reaction, provoked by the intemperance of the conflicting parties, now produced a tendency to extreme caution. Speculative views were discountenanced, and through fear of exposing themselves to the suspicion of a bias towards the dogmas of a party, some geologists became anxious to entertain no opinion whatever on the causes of phenomena, and were inclined to scepticism even where the conclusions deducible from observed facts scarcely admitted of reasonable doubt. But although the reluctance to theorize was carried somewhat to excess, no measure could be more salutary at such a moment than a suspension of all attempts to form what were termed "theories of the earth." A great body of new data were required, and the Geological Society of London, founded in 1807, conduced greatly to the attainment of this desirable end. To multiply and record observations, and patiently to await the result at some future period, was the object proposed by them, and it was their favourite maxim that the time was not

* Werner invented a new language to express his divisions of rocks, and some of his technical terms, such as grauwacke, gneiss, and others, passed current in every country in Europe. Smith adopted for the most part English provincial terms, often of barbarous sound, such as gault, cornbrash, clunch clay, &c., and affixed them to subdivisions of the British series. Many of these still retain their place in our scientific classifications, and attest his priority of arrangement.

yet come for a general system of geology, but that all must be content for many years to be exclusively engaged in furnishing materials for future generalizations. By acting up to these principles with consistency, they in a few years disarmed all prejudice, and rescued the science from the imputation of being a dangerous, or at best but a visionary pursuit.

Inquiries were at the same time prosecuted with great success by the French naturalists, who devoted their attention especially to the study of organic remains. They shewed that the specific characters of fossil shells and vertebrated animals might be determined with the utmost precision, and by their exertions a degree of accuracy was introduced into this department of science, of which it had never before been deemed susceptible. It was found that, by the careful discrimination of the fossil contents of strata, the contemporary origin of different groups could often be established, even where all identity of mineralogical character was wanting, and where no light could be derived from the order of superposition. The minute investigation, moreover, of the relics of the animate creation of former ages, had a powerful effect in dispelling the illusion which had long prevailed concerning the absence of analogy between the ancient and modern state of our planet. A close comparison of the recent and fossil species, and the inferences drawn in regard to their habits, accustomed the geologist to contemplate the earth as having been at successive periods the dwelling place of animals and plants of different races, some of which were discovered to have been terrestrial and others aquatic—some fitted to live in seas, others in the waters of lakes and rivers. By the consideration of these topics, the mind was slowly and insensibly withdrawn from imaginary pictures of catastrophes and chaotic confusion, such as haunted the imagination of the early cosmogonists. Numerous proofs were discovered of the tranquil deposition of sedimentary matter and the slow development of organic life. If many still continued to maintain, that " the thread of induction was broken," yet in reasoning by the strict rules of induction from recent to fossil species, they virtually disclaimed the dogma which in theory they professed. The adoption of the same generic, and, in some cases, even the same specific names for the exuviæ of fossil animals, and their living analogues, was an important step towards familiarizing the mind with the idea of the identity and unity of the system in distant eras. It was an acknowledgment, as it were, that a considerable part of the ancient memorials of nature were written in a living language. The growing importance then of the natural history of organic remains, and its general application to geology, may be pointed out as the characteristic feature of the progress of the science during the present century. This branch of knowledge has already become an instrument of great power in the discovery of truths in geology, and is continuing daily to unfold new data for grand and enlarged views respecting the former changes of the earth.

When we compare the result of observations in the last thirty years with those of the three preceding centuries, we cannot but look forward with the most sanguine expectations to the degree of excellence to which geology may be carried, even by the labours of the present generation. Never, perhaps, did any science, with the exception of astronomy, unfold, in an equally brief period, so many novel and unexpected truths, and overturn so many preconceived opinions. The senses had for ages declared the earth to be at rest, until the astronomer taught that it was carried through space with inconceivable rapidity. In like manner was the surface of this planet regarded as having remained unaltered since its creation, until the geologist proved that it had been the theatre of reiterated change, and was still the subject of slow but never ending fluctuations. The discovery of other systems in the boundless regions of space was the triumph of astronomy—to trace the same system through various transformations—to behold it at successive eras adorned with different hills and valleys, lakes and seas, and peopled with new inhabitants, was the delightful meed of geological research. By the geometer were measured the regions of space, and the relative distances of the heavenly bodies—by the geologist myriads of ages were reckoned, not by arithmetical computation, but by a train of physical events—a succession of phenomena in the animate and inanimate worlds—signs which convey to our minds more definite ideas than figures can do, of the immensity of time.

Whether our investigation of the earth's history and structure will eventually be productive of as great practical benefits to mankind, as a knowledge of the distant heavens, must remain for the decision of posterity. It was not till astronomy had been enriched by the observations of many centuries, and had made its way against popular prejudices to the establishment of a sound theory, that its application to the useful arts was most conspicuous. The cultivation of geology began at a later period; and in every step which it has hitherto made towards sound theoretical principles, it has had to contend against more violent prepossessions. The practical advantages already derived from it have not been inconsiderable: but our generalizations are yet imperfect, and they who follow may be expected to reap the most valuable fruits of our labour. Meanwhile the charm of first discovery is our own, and as we explore this magnificent field of inquiry, the sentiment of a great historian of our times may continually be present to our minds, that "he who calls what has vanished back again into being, enjoys a bliss like that of creating *."

* Niebuhr's Hist. of Rome, vol. i. p. 5. Hare and Thirlwall's translation.

CHAPTER V.

Review of the causes which have retarded the progress of Geology—Effects of prepossessions in regard to the duration of past time—Of prejudices arising from our peculiar position as inhabitants of the land—Of those occasioned by our not seeing subterranean changes now in progress—All these causes combine to make the former course of Nature appear different from the present—Several objections to the assumption, that existing causes have produced the former changes of the earth's surface, removed by modern discoveries.

We have seen that, during the progress of geology, there have been great fluctuations of opinion respecting the nature of the causes to which all former changes of the earth's surface are referrible. The first observers conceived that the monuments which the geologist endeavours to decipher, relate to a period when the physical constitution of the earth differed entirely from the present, and that, even after the creation of living beings, there have been causes in action distinct in kind or degree from those now forming part of the economy of nature. These views have been gradually modified, and some of them entirely abandoned in proportion as observations have been multiplied, and the signs of former mutations more skilfully interpreted. Many appearances, which for a long time were regarded as indicating mysterious and extraordinary agency, are finally recognized as the necessary result of the laws now governing the material world; and the discovery of this unlooked for conformity has induced some geologists to infer that there has never been any interruption to the same uniform order of physical events. The same assemblage of general causes, they conceive, may have been sufficient to produce, by their various combinations, the endless diversity of effects, of which the shell of the earth has preserved the memorials, and, consistently with these principles, the recurrence of analogous changes is expected by them in time to come.

Whether we coincide or not in this doctrine, we must admit that the gradual progress of opinion concerning the succession of phenomena in remote eras, resembles in a singular manner that which accompanies the growing intelligence of every people, in regard to the economy of nature in modern times. In an early stage of advancement, when a great number of natural appearances are unintelligible, an eclipse, an earthquake, a flood, or the approach of a comet, with many other occurrences afterwards found to belong to the regular course of events, are regarded as prodigies. The same delusion prevails as to moral phenomena, and many of these are ascribed to the intervention of demons, ghosts, witches, and other immaterial and supernatural agents. By degrees, many of the enigmas of the moral and physical world are explained, and, instead of being due to extrinsic and irregular causes, they are found to depend on fixed and invariable laws. The philosopher at last becomes convinced of the undeviating uniformity of secondary causes, and, guided by his faith in this principle, he determines the probability of accounts transmitted to him of former occurrences, and often rejects the fabulous tales of former ages, on the ground of their being irreconcilable with the experience of more enlightened ages.

As a belief in want of conformity in the physical constitution of the earth, in ancient and modern times, was for a long time universally prevalent, and that too amongst men who were convinced that the order of nature is *now* uniform, and has continued so for several thousand years; every circumstance which could have influenced their minds and given an undue bias to their opinions deserves particular attention. Now the reader may easily satisfy himself, that, however undeviating the course of nature may have been from the earliest epochs, it was impossible for the first cultivators of geology to come to such a conclusion, so long as they were under a delusion as to the age of the world, and the date of the first creation of animate beings. However fantastical some theories of the sixteenth century may now appear to us,—however unworthy of men of great talent and sound judgment, we may rest assured that, if the same misconceptions now prevailed in regard to the memorials of human transactions, it would give rise to a similar train of absurdities. Let us imagine, for example, that Champollion, and the French and Tuscan literati now engaged in exploring the antiquities of Egypt, had visited that country with a firm belief that the banks of the Nile were never peopled by the human race before the beginning of the ninetteenth century, and that their faith in this dogma was as difficult to shake as the opinion of our ancestors, that the earth was never the abode of living beings until the creation of the present continents, and of the species now existing,—it is easy to perceive what extravagant systems they would frame, while under the influence of this delusion, to account for the monuments discovered in Egypt. The sight of the pyramids, obelisks, colossal statues, and ruined temples, would fill them with such astonishment, that for a time they would be as men spell-bound—wholly incapacitated to reason with sobriety. They might incline at first to refer the construction of such stupendous works to some superhuman powers of a primeval world. A system might be invented resembling that so gravely advanced by Manetho, who relates that a dynasty of gods originally ruled in Egypt, of whom Vulcan, the first monarch, reigned nine thousand years. After them came Hercules and other demi-gods, who were at last succeeded by human kings. When some fanciful speculations of this kind had amused the imagination for a time, some vast repository of mummies would be discovered and would immediately undeceive those antiquaries who enjoyed an opportunity of personally examining them, but the prejudices of others at a distance, who were not eye-witnesses of the whole phenomena, would not be so easily overcome. The coneurrent report of many travellers would indeed render it necessary for them to accommodate ancient theories to some of the new facts, and much wit and ingenuity would be required to modify and defend their old positions. Each new invention would violate a greater number of known analogies; for if a theory be required to embrace some false principle, it becomes more visionary in proportion as facts are multiplied, as would be the case if geometers were now required to form an astronomical system on the assumption of the immobility of the earth.

Amongst other fanciful conjectures concerning the history of Egypt, we may suppose some of the following to be started. 'As the banks of the Nile have been so recently colonized, the 'curious subtances called mummies could never in reality have 'belonged to men. They may have been generated by some '*plastic virtue* residing in the interior of the earth, or they may 'be abortions of nature produced by her incipient efforts in the 'work of creation. For if deformed beings are sometimes born 'even now, when the scheme of the universe is fully devoloped, 'many more may have been "sent before their time, scarce 'half made up," when the planet itself was in the embryo state. 'But if these notions appear to derogate from the perfection 'of the Divine attributes, and if these mummies be in all their 'parts true representations of the human form, may we not 'refer them to the future rather than the past? May we not 'be looking into the womb of nature, and not her grave? may 'not these images be like the shades of the unborn in Virgil's 'Elysium—the archetypes of men not yet called into existence?'

These speculations, if advocated by eloquent writers, would not fail to attract many zealous votaries, for they would relieve men from the painful necessity of renouncing preconceived opinions. Incredible as such scepticism may appear, it would be rivalled by many systems of the sixteenth and seventeenth centuries, and among others by that of the learned Falloppio, who regarded the tusks of fossil elephants as earthy concretions, and the vases of Monte Testaceo, near Rome, as works of nature, and not of art. But when one generation had passed away, and another not compromised to the support of antiquated dogmas had succeeded, they would review the evidence afforded by mummies more impartially, and would no longer controvert the preliminary question, that human beings had lived in Egypt before the nineteenth century: so that when a hundred years perhaps had been lost, the industry and talents of the philosopher would be at last directed to the elucidation of points of real historical importance.

But we have adverted to one only of many prejudices with which the earlier geologists had to contend. Even when they conceded that the earth had been peopled with animate beings at an earlier period than was at first supposed, they had no conception that the quantity of time bore so great a proportion to the historical era as is now generally conceded. How fatal every error as to the quantity of time must prove to the introduction of rational views concerning the state of things in former ages, may be conceived by supposing that the annals of the civil and military transactions of a great nation were perused under the impression that they occurred in a period of one hundred instead of two thousand years. Such a portion of history would immediately assume the air of a romance; the events would seem devoid of credibility, and inconsistent with the present course of human affairs. A crowd of incidents would follow each other in thick succession. Armies and fleets would appear to be assembled only to be destroyed, and cities built merely to fall in ruins. There would be the most violent transitions from foreign or intestine war to periods of profound peace, and the works effected during the years of disorder or tranquillity would be alike superhuman in magnitude.

He who should study the monuments of the natural world under the influence of a similar infatuation, must draw a no less exaggerated picture of the energy and violence of causes, and must experience the same insurmountable difficulty in reconciling the former and present state of nature. If we could behold in one view all the volcanic cones thrown up in Iceland, Italy, Sicily, and other parts of Europe, during the last five thousand years, and could see the lavas which have flowed during the same period; the dislocations, subsidences and elevations caused by earthquakes; the lands added to various deltas, or devoured by the sea, together with the effects of devastation by floods, and imagine that all these events had happened in one year, we must form most exalted ideas of the activity of the agents, and the suddenness of the revolutions. Were an equal amount of change to pass before our eyes in the next year, could we avoid the conclusion that some great crisis of nature was at hand? If geologists, therefore, have misinterpreted the signs of a succession of events, so as to conclude that centuries were implied where the characters imported thousands of years, and thousands of years where the language of nature signified millions, they could not, if they reasoned logically from such false premises, come to any other conclusion, than that the system of the natural world had undergone a complete revolution.

We should be warranted in ascribing the erection of the great pyramid to superhuman power, if we were convinced that it was raised in one day; and if we imagine, in the same manner, a mountain chain to have been elevated, during an equally small fraction of the time which was really occupied in upheaving it, we might then be justified in inferring, that the subterranean movements were once far more energetic than in our own times. We know that one earthquake may raise the coast of Chili for a hundred miles to the average height of about five feet. A repetition of two thousand shocks of equal violence might produce a mountain chain one hundred miles long, and ten thousand feet high. Now, should one only of these convulsions happen in a century, it would be consistent with the order of events experienced by the Chilians from the earliest times; but if the whole of them were to occur in the next hundred years, the entire district must be depopulated, scarcely any animals or plants could survive, and the surface would be one confused heap of ruin and desolation.

One consequence of undervaluing greatly the quantity of past time is the apparent coincidence which it occasions of events necessarily disconnected, or which are so unusual, that it would be inconsistent with all calculation of chances to suppose them to happen at one and the same time. When the unlooked for association of such rare phenomena is witnessed in the present course of nature, it scarcely ever fails to excite a suspicion of the preternatural in those minds which are not firmly convinced of the uniform agency of secondary causes;—as if the death of some individual in whose fate they are interested, happens to be accompanied by the appearance of a luminous meteor, or a comet, or the shock of an earthquake. It would be only necessary to multiply such coincidences indefinitely, and the mind of every philosopher would be disturbed. Now it would be difficult to exaggerate the number of physical events, many of them most rare and unconnected in their nature, which were imagined by the Woodwardian hypothesis to have happened in the course of a few months; and numerous other examples might be found of popular geological theories, which require us to imagine that a long succession of events happened in a brief and almost momentary period.

The sources of prejudice hitherto considered may be deemed as in a great degree peculiar to the infancy of the science; but others are common to the first cultivators of geology and to ourselves, and are all singularly calculated to produce the same deception, and to strengthen our belief that the course of nature in the earlier ages differed widely from that now established. Although we cannot fully explain all these circumstances, without assuming some things as proved, which it will be the object of another part of this work to demonstrate, we must briefly allude to them in this place.

The first and greatest difficulty, then, consists in our habitual unconsciousness that our position as observers is essentially unfavourable, when we endeavour to estimate the magnitude of the changes now in progress. In consequence of our inattention to this subject, we are liable to the greatest mistakes in contrasting the present with former states of the globe. We inhabit about a fourth part of the surface; and that portion is almost exclusively the theatre of decay and not of reproduction. We know, indeed, that new deposits are annually formed in seas and lakes, and that every year some new igneous rocks are produced in the bowels of the earth, but we cannot watch the progress of their formation; and, as they are only present to our minds by the aid of reflection, it requires an effort both of the reason and the imagination to appreciate duly their importance. It is, therefore, not surprising that we imperfectly estimate the result of operations invisible to us; and that, when analogous results of some former epoch are presented to our inspection, we cannot recognise the analogy. He who has observed the quarrying of stone from a rock, and has seen it shipped for some distant port, and then endeavours to conceive what kind of edifice will be raised by the materials, is in the same predicament as a geologist, who, while he is confined to the land, sees the decomposition of rocks, and the transportation of matter by rivers to the sea, and then endeavours to picture to himself the new strata which Nature is building beneath the waters. Nor is his position less unfavourable when, beholding a volcanic eruption, he tries to conceive what changes the column of lava has produced, in its passage upwards, on the intersected strata; or what form the melted matter may assume at great depths on cooling down; or what may be the extent of the subterranean rivers and reservoirs of liquid matter far beneath the surface. It should, therefore, be remembered, that the task imposed on those who study the earth's history requires no ordinary share of discretion, for we are precluded from collating the corresponding parts of a system existing at two different periods. If we were inhabitants of another element—if the great ocean were our domain, instead of the narrow limits of the land, our difficulties would be considerably lessened;

while, on the other hand, there can be little doubt, although the reader may, perhaps, smile at the bare suggestion of such an idea, that an amphibious being, who should possess our faculties, would still more easily arrive at sound theoretical opinions in geology, since he might behold, on the one hand, the decomposition of rocks in the atmosphere, and the transportation of matter by running water; and, on the other, examine the deposition of sediment in the sea, and the imbedding of animal remains in new strata. He might ascertain, by direct observation, the action of a mountain torrent, as well as of a marine current; might compare the products of volcanos on the land with those poured out beneath the waters; and might mark, on the one hand, the growth of the forest, and on the other that of the coral reef. Yet, even with these advantages, he would be liable to fall into the greatest errors when endeavouring to reason on rocks of subterranean origin. He would seek in vain, within the sphere of his observation, for any direct analogy to the process of their formation, and would therefore be in danger of attributing them, wherever they are upraised to view, to some "primeval state of nature." But if we may be allowed so far to indulge the imagination, as to suppose a being, entirely confined to the nether world—some "dusky melancholy sprite," like Umbriel, who could "flit on sooty pinions to the central earth," but who was never permitted to "sully the fair face of light," and emerge into the regions of water and of air; and if this being should busy himself in investigating the structure of the globe, he might frame theories the exact converse of those usually adopted by human philosophers. He might infer that the stratified rocks, containing shells and other organic remains, were the oldest of created things, belonging to some original and nascent state of the planet. "Of these masses," he might say, "whether they consist of loose incoherent sand, soft clay, or solid rock, none have been formed in modern times. Every year some part of them are broken and shattered by earthquakes, or melted up by volcanic fire; and, when they cool down slowly from a state of fusion, they assume a crystalline form, perfectly distinct from those inexplicable rocks which are so regularly bedded, and contain stones full of curious impressions and fantastic markings. This process cannot have been carried on for an indefinite time, for in that case all the stratified rocks would long ere this have been fused and crystallized. It is therefore probable that the whole planet once consisted of these curiously-bedded formations, at a time when the volcanic fire had not yet been brought into activity. Since that period there seems to have been a gradual development of heat, and this augmentation we may expect to continue till the whole globe shall be in a state of fluidity and incandescence."

Such might be the system of the Gnome at the very same time that the followers of Leibnitz, reasoning on what they saw on the outer surface, would be teaching the doctrine of gradual refrigeration, and averring that the earth had begun its career as a fiery comet, and would hereafter become a frozen icy mass. The tenets of the schools of the nether and of the upper world would be directly opposed to each other, for both would partake of the prejudices inevitably resulting from the continual contemplation of one class of phenomena to the exclusion of another. Man observes the annual decomposition of crystalline and igneous rocks, and may sometimes see their conversion into stratified deposits; but he cannot witness the reconversion of the sedimentary into the crystalline by subterranean fire. He is in the habit of regarding all the sedimentary rocks as more recent than the unstratified, for the same reason that we may suppose him to fall into the opposite error if he saw the origin of the igneous class only.

It is only by becoming sensible of our natural disadvantages that we shall be roused to exertion, and prompted to seek out opportunities of discovering the operations now in progress, such as do not present themselves readily to view. We are called upon, in our researches into the state of the earth, as in our endeavours to comprehend the mechanism of the heavens, to invent means for overcoming the limited range of our vision. We are perpetually required to bring, as far as possible, within the sphere of observation, things to which the eye, unassisted by art, could never obtain access. It was not an impossible contingency that astronomers might have been placed, at some period, in a situation much resembling that in which the geologist seems to stand at present. If the Italians, for example, in the early part of the twelfth century, had discovered at Amalphi, instead of the pandects of Justinian, some ancient manuscripts filled with astronomical observations relating to a period of three thousand years, and made by some ancient geometers who possessed optical instruments as perfect as any in modern Europe, they would probably, on consulting these memorials, have come to a conclusion that there had been a great revolution in the solar and sidereal systems. "Many primary and secondary planets," they might say, "are enumerated in these tables, which exist no longer. Their positions are assigned with such precision, that we may assure ourselves that there is nothing in their place at present but the blue ether. Where one star is visible to us, these documents represent several thousands. Some of those which are now single, consisted then of two separate bodies, often distinguished by different colours, and revolving periodically round a common centre of gravity. There is no analogy to them in the universe at present, for they were neither fixed stars nor planets, but stood in the mutual relation of sun and planet to each other. We must conclude, therefore, that there has occurred, at no distant period, a tremendous catastrophe, whereby thousands of worlds have been annihilated at once, and some heavenly bodies absorbed into the substance of others." When such doctrines had prevailed for ages, the discovery of one of the worlds, supposed to have been lost, by aid of the first rude telescope, would not dissipate the delusion, for the whole burden of proof would now be thrown on those who insisted on the stability of the system from the beginning of time, and these philosophers would be required to demonstrate the existence of *all* the worlds said to have been annihilated. Such popular prejudices would be most unfavourable to the advancement of astronomy; for, instead of persevering in the attempt to improve their instruments, and laboriously to make and record observations, the greater number would despair of verifying the continued existence of the heavenly bodies not visible to the naked eye. Instead of confessing the extent of their ignorance, and striving to remove it by bringing to light new facts, they would be engaged in the indolent employment of framing imaginary theories concerning catastrophes and mighty revolutions in the system of the universe.

For more than two centuries the shelly strata of the Subapennine hills afforded matter of speculation to the early geologists of Italy, and few of them had any suspicion that similar deposits were then forming in the neighbouring sea. They were as unconscious of the continued action of causes still producing similar effects, as the astronomers, in the case supposed by us, of the existence of certain heavenly bodies still giving and reflecting light, and performing their movements as in the olden time. Some imagined that the strata, so rich in organic remains, instead of being due to secondary agents, had been so created in the beginning of things by the fiat of the Almighty; and others ascribed the imbedded fossil bodies to some plastic power which resided in the earth in the early ages of the world. At length Donati explored the bed of the Adriatic, and found the closest resemblance between the new deposits there forming, and those which constituted hills above a thousand feet high in various parts of the peninsula. He ascertained that certain genera of living testacea were grouped together at the bottom of the sea in precisely the same manner as were their fossil analogues in the strata of the hills, and that some species were common to the recent and fossil world. Beds of shells, moreover, in the Adriatic, were becoming incrusted with calcareous rock; and others were recently enclosed in deposits of sand and clay, precisely as fossil shells were found in the hills. This splendid discovery of the identity of modern and ancient submarine operations was not made without the aid of artificial instruments, which, like the telescope, brought phenomena into view not otherwise within the sphere of human observation.

In like manner, in the Vicentin, a great series of volcanic and marine sedimentary rocks were examined in the early part of the last century; but no geologist suspected, before the time of Arduino, that these were partly composed of ancient submarine lavas. If, when these enquiries were first made, geologists had been told that the mode of formation of such rocks might be fully elucidated by the study of processes then going on in certain parts of the Mediterranean, they would have been as incredulous as geometers would have been before the time of Newton, if any one had informed them that, by making experiments on the motion of bodies on the earth, they might discover the laws which regulated the movements of distant planets.

The establishment, from time to time, of numerous points of identification, drew at length from geologists a reluctant admission, that there was more correspondence between the physical constitution of the globe, and more uniformity in the laws regulating the changes of its surface, from the most remote eras to the present, than they at first imagined. If, in this state of the science, they still despaired of reconciling every class of geological phenomena to the operations of ordinary causes, even by straining analogy to the utmost limits of credibility, we might have expected, that the balance of probability at least would now have been presumed to incline towards the identity of the causes. But, after repeated experience of the failure of attempts to speculate on different classes of geological phenomena, as belonging to a distinct order of things, each new sect persevered systematically in the principles adopted by their predecessors. They invariably began, as each new problem presented itself, whether relating to the animate or inanimate world, to assume in their theories, that the economy of nature was formerly governed by rules quite independent of those now established. Whether they endeavoured to account for the origin of certain igneous rocks, or to explain the forces which elevated hills or excavated valleys, or the causes which led to the extinction of certain races of animals, they first presupposed an original and dissimilar order of nature; and when at length they approximated, or entirely came round to an opposite opinion, it was always with the feeling, that they conceded what they were justified *à priori* in deeming improbable. In a word, the same men who, as natural philosophers, would have been greatly surprised to find any deviation from the usual course of Nature *in their own time*, were equally surprised, as geologists, not to find such deviations at every period of the past.

The Huttonians were conscious that no check could be given to the utmost licence of conjecture in speculating on the causes of geological phenomena, unless we can assume invariable constancy in the order of Nature. But when they asserted this uniformity without any limitation as to time, they were considered, by the majority of their contemporaries, to have been carried too far, especially as they applied the same principle to the laws of the organic, as well as of the inanimate world *.

* Playfair, after admitting the extinction of some species, says, "The inhabitants of the globe, then, like all other parts of it, are subject to change. It is not only the individual that perishes, but whole *species*, and even perhaps *genera*, are extinguished."——"A change in the animal kingdom seems to be a *part of the order of nature*, and is visible in instances to which human power cannot have extended."—Illustrations of the Huttonian Theory, § 413.

We shall first advert briefly to many difficulties which formerly appeared insurmountable, but which, in the last forty years, have been partially or entirely removed by the progress of science; and shall afterwards consider the objections that still remain to the doctrine of absolute uniformity.

In the first place, it was necessary for the supporters of this doctrine to take for granted incalculable periods of time, in order to explain the formation of sedimentary strata by causes now in diurnal action. The time which they required theoretically, is now granted, as it were, or has become absolutely requisite, to account for another class of phenomena brought to light by more recent investigations. It must always have been evident to unbiassed minds, that successive strata, containing, in regular order of superposition, distinct beds of shells and corals, arranged in families as they grow at the bottom of the sea, could only have been formed by slow and insensible degrees in a great lapse of ages; yet, until organic remains were minutely examined and specifically determined, it was rarely possible to prove that the series of deposits met with in one country was not formed simultaneously with that found in another. But we are now able to determine, in numerous instances, the relative dates of sedimentary rocks in distant regions, and to show, by their organic remains, that they were not of contemporary origin, but formed in succession. We often find, that where an interruption in the consecutive formation in one district is indicated by a sudden transition from one assemblage of fossil species to another, the chasm is filled up, in some other district, by other important groups of strata. The more attentively we study the European continent, the greater we find the extension of the whole series of geological formations. No sooner does the calendar appear to be completed, and the signs of a succession of physical events arranged in chronological order, than we are called upon to intercalate, as it were, some new periods of vast duration. A geologist, whose observations have been confined to England, is accustomed to consider the superior and newer groups of marine strata in our island as modern, and such they are, comparatively speaking; but when he has travelled through the Italian peninsula and in Sicily, and has seen strata of more recent origin forming mountains several thousand feet high, and has marked a long series both of volcanic and submarine operations, all newer than any of the regular strata which enter largely into the physical structure of Great Britain, he returns with more exalted conceptions of the antiquity of some of those modern deposits, than he before entertained of the oldest of the British series. We cannot reflect on the concessions thus extorted from us, in regard to the duration of past time, without foreseeing that the period may arrive when part of the Huttonian theory will be combated on the ground of its departing too far from the assumption of uniformity in the order of nature. On a closer investigation of extinct volcanos, we find proofs that they broke out at successive eras, and that the eruptions of one group were often concluded long before others had commenced their activity. Some were burning when one class of organic beings were in existence, others came into action when different races of animals and plants existed,—it follows, therefore, that the convulsions caused by subterranean movements, which are merely another portion of the volcanic phenomena, occurred also in succession, and their effects must be divided into separate sums, and assigned to separate periods of time; and this is not all:—when we examine the volcanic products, whether they be lavas which flowed out under water or upon dry land, we find that intervals of time, often of great length, intervened between their formation, and that the effects of one eruption were not greater in amount than that which now results during ordinary volcanic convulsions. The accompanying or preceding earthquakes, therefore, may be considered to have been also successive, and to have been in like manner interrupted by intervals of time, and not to have exceeded in violence those now experienced in the ordinary course of nature. Already, therefore, may we regard the doctrine of the sudden elevation of whole continents by paroxysmal eruptions as invalidated; and there was the greatest inconsistency in the adoption of such a tenet by the Huttonians, who were anxious to reconcile former changes to the present economy of the world. It was contrary to analogy to suppose, that Nature had been at any former epoch parsimonious of time and prodigal of violence—to imagine that one district was not at rest while another was convulsed—that the disturbing forces were not kept under subjection, so as never to carry simultaneous havoc and desolation over the whole earth, or even over one great region. If it could have been shown, that a certain combination of circumstances would at some future period produce a crisis in the subterranean action, we should certainly have had no right to oppose our experience for the last three thousand years as an argument against the probability of such occurrences in past ages; but it is not pretended that such a combination can be foreseen. In speculating on catastrophes by water, we may certainly anticipate great floods in future, and we may therefore presume that they have happened again and again in past times. The existence of enormous seas of fresh-water, such as the North American lakes, the largest of which is elevated more than six hundred feet above the level of the ocean, and is in parts twelve hundred feet deep, is alone sufficient to assure us, that the time will come, however distant, when a deluge will lay waste a considerable part of the American continent. No hypothetical agency is required to cause the sudden escape of the confined waters. Such changes of level, and opening of fissures, as have accompanied earthquakes since the commencement of the present century, or such excavation of ravines as the receding cataract of Niagara is now effecting, might breach the barriers. Notwithstanding, therefore, that we have not witnessed within the last three thousand years the devastation by deluge of a large continent, yet, as we may predict the future occurrence of such catastrophes, we are authorized to regard them as part of the present order of Nature, and they may be introduced into geological speculations respecting the past, provided we do not imagine them to have been more frequent or general than we expect them to be in time to come.

The great contrast in the aspect of the older and newer rocks, in their texture, structure, and in the derangement of the strata, appeared formerly one of the strongest grounds for presuming that the causes to which they owed their origin were perfectly dissimilar from those now in operation. But this incongruity may now be regarded as the natural result of subsequent modifications, since the difference of relative age is demonstrated to have been so immense, that, however slow and insensible the change, it must have become important in the course of so many ages. In addition to volcanic heat, to which the Vulcanists formerly attributed too much influence, we must allow for the effect of mechanical pressure, of chemical affinity, of percolation by mineral waters, of permeation by elastic fluids, and the action, perhaps, of many other forces less understood, such as electricity and magnetism. In regard to the signs of upraising and sinking, of fracture and contortion in rocks, it is evident that newer strata cannot be shaken by earthquakes, unless the subjacent rocks are also affected; so that the contrast in the relative degree of disturbance in the more ancient and the newer strata, is one of many proofs that the convulsions have happened in different eras, and the fact confirms the uniformity of the action of subterranean forces, instead of their greater violence in the primeval ages.

The popular doctrine of universal formations, or the unlimited geographical extent of strata, distinguished by similar mineral characters, appeared for a long time to present insurmountable objections to the supposition, that the earth's crust had been formed by causes now acting. If it had merely been assumed, that rocks originating from fusion by subterranean fire presented in all parts of the globe a perfect correspondence in their mineral composition, the assumption would not have been extravagant; for, as the elementary substances that enter largely into the composition of rocks are few in number, they may be expected to arrange themselves invariably in the same forms, whenever the elementary particles are freely exposed to the action of chemical affinities. But when it was imagined that sedimentary mixtures, including animal and vegetable remains, and evidently formed in the beds of ancient seas, were of homogeneous nature throughout a whole hemisphere, or even farther, the dogma precluded at once all hope of recognizing the slightest analogy between the ancient and modern causes of decay and reproduction. For we know that existing rivers carry down from different mountain-chains sediment of distinct colours and composition; where the chains are near the sea, coarse sand and gravel is swept in; where they are distant, the finest mud. We know, also, that the matter introduced by springs into lakes and seas is very diversified in mineral composition; in short, contemporaneous strata now in the progress of formation are greatly varied in their composition, and could never afford formations of homogeneous mineral ingredients co-extensive with the greater part of the earth's surface. This theory, however, is as inapplicable to the effects of those operations to which the formation of the earth's crust is due, as to the effects of existing causes. The first investigators of sedimentary rocks had never reflected on the great areas occupied by modern deltas of large rivers; still less on the much greater areas over which marine currents, preying alike on river-deltas, and continuous lines of sea-coast, might be diffusing homogeneous mixtures. They were ignorant of the vast spaces over which calcareous and other mineral springs abound upon the land and in the sea, especially in and near volcanic regions, and of the quantity of matter discharged by them. When, therefore, they ascertained the extent of the geographical distribution of certain groups of ancient strata—when they traced them continuously from one extremity of Europe to the other, and found them flanking, throughout their entire range, great mountain-chains, they were astonished at so unexpected a discovery; and, considering themselves at liberty to disregard all modern analogy, they indulged in the sweeping generalization, that the law of continuity prevailed throughout strata of contemporaneous origin over the whole planet. The difficulty of dissipating this delusion was extreme, because some rocks, formed under similar circumstances at different epochs, present the same external characters, and often the same internal composition; and all these were assumed to be contemporaneous until the contrary could be shown, which, in the absence of evidence derived from direct superposition, and in the scarcity of organic remains, was often impossible.

Innumerable other false generalizations have been derived from the same source; such, for instance, as the former universality of the ocean, now disproved by the discovery of the remains of terrestrial vegetation, contemporary with every successive race of marine animals; but we shall dwell no longer on exploded errors, but proceed at once to contend against weightier objections, which will require more attentive consideration.

CHAPTER VI.

Proofs that the climate of the Northern hemisphere was formerly hotter—Direct proofs from the organic remains of the Sicilian and Italian strata—Proofs from analogy derived from extinct Quadrupeds—Imbedding of Animals in Icebergs—Siberian Mammoths—Evidence in regard to temperature, from the fossil remains of tertiary and secondary rocks—From the Plants of the Coal formation.

THAT the climate of the northern hemisphere has undergone an important change, and that its mean annual temperature must once have resembled that now experienced within the tropics, was the opinion of some of the first naturalists who investigated the contents of ancient strata. Their conjecture became more probable when the shells and corals of the secondary rocks were more carefully examined, for these organic remains were found to be intimately connected by generic affinity with species now living in warmer latitudes. At a later period, many reptiles, such as turtles, tortoises, and large saurian animals, were discovered in the European strata in great abundance; and they supplied new and powerful arguments, from analogy, in support of the doctrine, that the heat of the climate had been great when our secondary formations were deposited. Lastly, when the botanist turned his attention to the specific determination of fossil plants, the evidence acquired the fullest confirmation, for the flora of a country is peculiarly influenced by temperature; and the ancient vegetation of the earth might, more readily than the forms of animals, have afforded conflicting proofs, had the popular theory been without foundation. When the examination of animal and vegetable remains was extended to rocks in the most northern parts of Europe and North America, and even to the Arctic regions, indications of the same revolution in climate were discovered.

It cannot be said, that in this, as in many other departments of geology, we have investigated the phenomena of former eras, and neglected those of the present state of things. On the contrary, since the first agitation of this interesting question, the accessions to our knowledge of living animals and plants have been immense, and have far surpassed all the data previously obtained for generalizing, concerning the relation of certain types of organization to particular climates. The tropical and temperate zones of South America and of Australia have been explored; and, on close comparison, it has been found, that scarcely any of the species of the animate creation in these extensive continents are identical with those inhabiting the old world. Yet the zoologist and botanist, well acquainted with the geographical distribution of organic beings in other parts of the globe, would have been able, if distinct groups of species had been presented to them from these regions, to recognise those which had been collected from latitudes within, and those which were brought from without the tropics.

Before we attempt to explain the probable causes of great vicissitudes of temperature on the earth's surface, we shall take a rapid view of some of the principal data which appear to warrant, to the utmost extent, the popular opinions now entertained on the subject. To insist on the soundness of the inference, is the more necessary, because some zoologists have of late undertaken to vindicate the uniformity of the laws of nature, not by accounting for former fluctuations in climate, but by denying the value of the evidence on this subject *.

It is not merely by reasoning from analogy that we are led to infer a diminution of temperature in the climate of Europe; there are direct proofs in confirmation of the same doctrine, in the only countries hitherto investigated by expert geologists where we could expect to meet with direct proofs. It is not in England or Northern France, but around the borders of the Mediterranean, from the South of Spain to Calabria, and in the islands of the Mediterranean, that we must look for conclusive evidence on this question; for it is not in strata, where the

* See two articles by the Rev. Dr. Fleming, in the Edinburgh New Phil. Journ. No. 12, p. 277, April, 1829; and No. 15, p. 65, Jan. 1830.

organic remains belong to extinct species, but where living species abound in a fossil state, that a theory of climate can be subjected to the experimentum crucis. In Sicily, Ischia, and Calábria, where the fossil testacea of the more recent strata belong almost entirely to species now known to inhabit the Mediterranean, the conchologist remarks, that individuals in the inland deposits exceed in their average size their living analogues *. Yet no doubt can be entertained, on the ground of such difference in their dimensions, of their specific identity, because the living individuals attain sometimes, though rarely, the average size of the fossils; and so perfect is the preservation of the latter, that they retain, in some instances, their colour, which affords an additional element of comparison.

As we proceed northwards in the Italian peninsula, and pass from the region of active, to that of extinct volcanos—from districts now violently convulsed from time to time, to those which are comparatively undisturbed by earthquakes, we find the assemblage of fossil shells, in the modern (Subapennine) strata, to depart somewhat more widely from the type of the neighbouring seas. The proportion of species, identifiable with those now living in the Mediterranean, is still considerable; but it no longer predominates, as in the South of Italy, over the unknown species. Although occurring in localities which are removed several degrees farther from the equator (as at Sienna, Parma, Asti, &c.), the shells yield clear indications of a hotter climate. Many of them are common to the Subapennine hills, to the Mediterranean, and to the Indian Ocean. Those in the fossil state, and their living analogues from the tropics, correspond in size; whereas the individuals of the same species from the Mediterranean are dwarfish and

* I collected several hundred species of shells in Sicily, some from an elevation of several thousand feet, and forty species or more in Ischia, partly from an elevation of above one thousand feet, and these were carefully compared with recent shells procured by Professor O. G. Costa, from the Neapolitan seas. Not only were the fossil species for the most part identical with those now living, but the relative abundance in which different species occur in the strata and in the sea corresponds in a remarkable manner. Yet the larger average size of the fossil individuals of many species was very striking. A comparison of the fossil shells of the more modern strata of Calabria and Otranto, in the collection of Professor Costa, afforded similar results.

degenerate, and stunted in their growth, for want of conditions which the Indian Ocean still supplies *.

This evidence amounts to demonstration, and is not neutralized by any facts of a conflicting character; such, for instance, as the association, in the same group, of individuals referrible to species now confined to arctic regions. On the contrary, whenever any of the fossils shells are identified with living species foreign to the Mediterranean, it is not in the Northern Ocean, but between the tropics, that they must be sought †. On the other hand, the associated unknown species belong, for the most part, to genera which are either exclusively limited to equinoctial regions, or are now most largely developed there. Of the former, the genus Pleurotoma ‡ is a remarkable example; of the latter, the genus Cypræa §.

When we proceed to the central and northern parts of Europe, far from the modern theatres of volcanic action, and where there is no evidence of considerable inequalities of the earth's surface having been produced since the present species

* Professor Guidotti, of Parma, whose collection of Subapennine shells is unrivalled, and who has obtained from the North of Italy above twelve hundred species, showed me numerous suites of specimens in a fossil state, as well as from the Mediterranean and Indian seas, illustrating these views. Among other examples, the Bulla lignaria, a very common shell, is invariably found fossil of the same magnitude as it now reaches in the Indian sea, and much smaller in a living state in the Mediterranean. The common Orthoceras of the Mediterranean, O. raphanista, attains larger average dimensions in a fossil, than in a recent state. Professor Bonelli, of Turin, who has above eight hundred species of shells from the Subapennines in the public museum, pointed out to me many examples, in confirmation of the same point.

† Thus, for example, Rostellaria curvirostris, found fossil by Signor Bonelli near Turin, is only known at present as an Indian shell. Murex cornutus, fossil at Asti, is now only known recent in warmer latitudes, the only localities given by Linnæus and Lamarck being the African and Great Indian Oceans. Conus antediluvianus cannot be distinguished from a shell now brought from Owhyhee. Among other familiar instances mentioned to me by Italian naturalists, in confirmation of the same point, Buccinum clathratum, Lam. was cited; but Professor Costa assured me that this shell, although extremely rare, still occurs in the Mediterranean.

‡ Of the genus Pleurotoma, no living representative has yet been found in the Mediterranean; yet no less than twenty-five species are now to be seen in the museum at Turin, all procured by Professor Bonelli from the Subapennine strata of northern Italy. In a fossil state, they are associated with many shells, specifically identical with testacea, now living in the Mediterranean.

§ The genus Cypræa is represented by many large fossil species in the Subapennine hills, with which are associated one small, and two minute species of the same genus, which alone are now found in the Mediterranean.

were in existence, our opportunities are necessarily more limited of procuring evidence from the contents of marine strata. It is only in lacustrine deposits, or in ancient river-beds, or in the sand and gravel of land-floods, or the stalagmite of ancient caverns once inhabited by wild beasts, that we can obtain access to proofs of the changes which animal life underwent during those periods when the marine strata already adverted to were deposited farther to the south. As far, however, as proofs from analogy can be depended upon, nothing can be more striking than the harmony of the testimony derived from the last-mentioned sources. We often find, in such situations, the remains of extinct species of quadrupeds, such as the elephant, rhinoceros, hippopotamus, hyæna, and tiger, which belong to genera now confined to warmer regions. Some of the accompanying fossil species, which are identifiable with those now living, belong to animals which inhabit the same latitudes at present.* It seems, therefore, fair to infer, that the same change of climate which has caused certain Indian species of testacea to become rare, or to degenerate in size, or to disappear from the Mediterranean, and certain genera of the Subapennine hills, now exclusively tropical, to retain no longer any representatives in the adjoining seas, has also contributed to the annihilation of certain genera of land-mammifera, which inhabited the continents at about the same epoch. The mammoth (Elephas primigenius), and other extinct animals of the same era, may not have required the same temperature as their living congeners within the tropics; but we may infer, that the climate was milder

* Bones of the mammoth have been recently found at North Cliff, in the county of York, in a lacustrine formation, in which all the land and fresh-water shells, thirteen in number, have been accurately identified with species and varieties now existing in that country. Bones of the Bison, an animal now inhabiting a cold or temperate climate, have also been found in the same place. That these quadrupeds, and the indigenous species of testacea associated with them, were all contemporary inhabitants of Yorkshire (a fact of the greatest importance in geology), has been established by unequivocal proofs, by the Rev. W. V. Vernon, who caused a pit to be sunk to the depth of more than two hundred feet, through undisturbed strata, in which the remains of the mammoth were found imbedded together with the shells, in a deposit which had evidently resulted from tranquil waters.—Phil. Mag. Sept. 1829, and Jan. 1830. These facts, as Mr. Vernon observes, indicate that there has been little alteration in the temperature of these latitudes since the mammoth lived there.

than that now experienced in some of the regions once inhabited by them, because, in Northern Russia, where their bones are found in immense numbers, it would be difficult, if not impossible, for such animals to obtain subsistence at present, during an arctic winter *. It has been said, that as the modern northern animals migrate, the Siberian elephant may also have shifted his place during the inclemency of the season†, but this conjecture seems forced, even in regard to the elephant, and still more so, when applied to the Siberian rhinoceros, found in the frozen gravel of that country; as animals of this genus are heavy and slow in their motions, and can hardly be supposed to have accomplished great periodical migrations to southern latitudes. That the mammoth, however, continued for a long time to exist in Siberia after the winters had become extremely cold, is demonstrable, since their bones are found in icebergs, and in the frozen gravel, in such abundance as could only have been supplied by many successive generations. So many skeletons could not have belonged to herds which lived at one time in the district, even if those northern countries had once been clothed with vegetation as luxuriant as that of an Indian jungle. But, if we suppose the change to have been extremely slow, and to have consisted, not so much in a diminution of the mean annual temperature, as in an alteration from what has been termed an "insular" to an "excessive" climate, from one in which the temperature of winter and summer were nearly equalized to one wherein the seasons were violently contrasted, we may, perhaps, explain the phenomenon. Siberia and other arctic regions, after having possessed for ages a more uniform temperature, may, after certain changes in the form of the arctic land, have become occasionally exposed

* I fully agree with Dr. Fleming, that the kind of food which the existing species of elephant prefers will not enable us to determine, or even to offer a feasible conjecture, concerning that of the extinct species. No one, as he observes, acquainted with the gramineous character of the food of our fallow-deer, stag, or roe, would have assigned a lichen to the rein-deer. But, admitting that the trees and herbage on which the fossil elephants and rhinoceroses may have fed were not of a tropical character, but such perhaps as now grow in the temperate zone, it is still highly improbable that the vegetation which nourished these great quadrupeds was as scanty as that of our arctic regions, or that it was covered during the greater part of every year by snow.

† Dr. Fleming, Edin. New Phil. Journ. No. xii. p. 285. April, 1829.

to extremely severe winters. When these first occurred at distant intervals, the drift snow would fill the valleys, and herds of herbivorous quadrupeds would be surprised and buried in a frozen mass, as often happens to cattle and human beings, overwhelmed, in the Alpine valleys of Switzerland, by avalanches. When valleys have become filled with ice, as those of Spitzbergen, the contraction of the mass causes innumerable deep rents, such as are seen in the mer de glace on Mont Blanc. These deep crevices usually become filled with loose snow, but sometimes a thin covering is drifted across the mouth of the chasm, capable of sustaining a certain weight. Such treacherous bridges are liable to give way when heavy animals are crossing, which are then precipitated at once into the body of a glacier, which slowly descends to the sea, and becomes a floating iceberg *. As bears, foxes, and deer now abound in Spitzbergen, we may confidently assume that the imbedding of animal remains in the glaciers of that island must be an event of almost annual occurrence †. The conversion of drift snow into permanent glaciers and icebergs, when it happens to become covered over with alluvial matter, transported by torrents and floods, is by no means a rare phenomenon in the arctic regions ‡. During a series of milder seasons inter-

* See Dr. Latta's account of his escape, when the covering of a crevice in a glacier of Spitzbergen gave way with him as he passed. Ed. New Phil. Journ. No. v. p. 95. June, 1827.

† Dr. Richardson tells me, that in North America, about lat. 65°, he found the carcase of a deer, which had fallen into a fissure in a rock. It had become buried in snow, and the flesh, after the animal had been buried three months, had only become slightly putrescent. In the innumerable fissures, traversing a slippery glacier, these accidents must be far more frequent, so often as herbivorous animals pass over them in their migrations, or when they hastily cross them when pursued by beasts of prey.

‡ Along the coast, in particular, E. and W. of the Mackenzie river, when the sea is frozen over, the drift snow from the land forms a talus abutting against a perpendicular cliff. On the melting of the snow, torrents rush down from the land, charged with gravel and soil, and, falling over the edge of the cliff, cover the snow, which is often of considerable depth, with alluvium. Water, if any infiltration takes place, is frozen before it penetrates to the bottom of the mass, which is at last consolidated into a compact iceberg, protected from the heat of the sun, by a covering of alluvium, on which vegetation often flourishes. I am indebted to Dr. Richardson for this information, who has seen permanent glaciers, forming in this manner, in districts of North America now inhabited by many large herbivorous animals. The same process must evidently take place under river cliffs, as well as along the sea-shore.

vening between the severe winters, the mammoths may have recovered their numbers, and the rhinoceroses may have multiplied again, so that the repetition of such catastrophes may have been indefinite. The increasing cold, and greater frequency of inclement winters, would at last thin their numbers, and their final extirpation would be consummated by the rapid augmentation of other herbivorous quadrupeds, more fitted for the new climate.

That the greater part of the elephants lived in Siberia after it had become subject to intense cold, is confirmed, among other reasons, by the state of the ivory, which has been so largely exported in commerce. Its perfect preservation indicates, that from the period when the individuals died, their remains were either buried in a frozen soil, or at least were not exposed to decay in a warm atmosphere. The same conclusion may be deduced from the clothing of the mammoth, of which the entire carcase was discovered by Mr. Adams on the shores of the frozen ocean, near the mouth of the river Lena, inclosed in a mass of ice. The skin of that individual was covered with long hair and with thick wool, about an inch in length. Bishop Heber informs us, that along the lower range of the Himalaya mountains, in the north-eastern borders of the Delhi territory, between lat. 29° and 30°, he saw an Indian elephant covered with shaggy hair. In that region, where, within a short space, a nearly tropical, and a cold climate meet, dogs and horses become covered, in the course of a winter or two, with shaggy hair, and many other species become, in as short a time, clothed with the same fine short shawl-wool, which distinguishes the indigenous species of the country. Lions, tigers, hyænas, are there found with elks, chamois, and other species of genera usually abundant in colder latitudes*

If we pass from the consideration of these more modern deposits, whether of marine or continental origin, in which existing species are intermixed with the extinct, to strata of somewhat higher antiquity, (older tertiary strata, Calcaire Grossier, London clay, fresh-water formations of Paris and Isle of Wight, &c.,) we can only reason from analogy, since the species,

* Narrative of a Journey through the Upper Provinces of India, vol. ii. pp. 166—219.

whether of mammalia, reptiles, or testacea, are scarcely in any instance identifiable with any now in being *. In these strata, whether they were formed in seas or lakes, we find the remains of many animals, analogous to those of hot climates, such as the crocodile, turtle, and tortoise, and many large shells of the genus nautilus, and plants indicating such a temperature as is now found along the borders of the Mediterranean. A great interval of time appears to have elapsed between the deposition of the last mentioned (tertiary) strata, and the *secondary* formations, which constitute the principal portion of the more elevated land in Europe. In these secondary rocks a very distinct assemblage of organized fossils are entombed, all of unknown species, and many of them referrible to genera, and families now most abundant between the tropics. Among the most remarkable, are many gigantic reptiles, some of them herbivorous, others carnivorous, and far exceeding in size any now known even in the torrid zone. The genera are for the most part extinct, but some of them, as the crocodile and monitor, have still representatives in the warmest parts of the earth. Coral reefs also were evidently numerous in the seas of the same period, and composed of species belonging to genera now characteristic of a tropical climate. The number of immense chambered shells also leads us to infer an elevated temperature; and the associated fossil plants, although imperfectly known, tend to the same conclusion, the Cycadeæ constituting the most numerous family. But the study of the fossil flora of the coal deposits of still higher antiquity, has yielded the most extraordinary evidence of an extremely hot climate, for it consisted almost exclusively of large vascular cryptogamic plants. We learn, from the labours of M. Ad. Brongniart, that there existed, at that epoch, Equiseta upwards of ten feet high, and from five to six inches in diameter; tree

* In the London clay, I believe, no recent species are yet discovered. But of twelve hundred species of shells, collected from the different fresh-water and marine formations of the Paris basin, M. Deshayes informs me, that there are some, but not perhaps exceeding one in a hundred, which he regards as perfectly identical with living species. Among these are Melanopsis buccinoides, from Epernais, now living in the Grecian archipelago, and Melania inquinata, now found between the tropics in the Phillippine islands. Venus divaricata is not uncommon in the calcaire grossier at Grignon.

ferns of from forty to fifty feet in height, and arborescent Lycopodiaceæ, of from sixty to seventy feet high*. Of the above classes of vegetables, the species are all small at present in cold climates; while in tropical regions, there occur, together with small species, many of a much greater size, but their development at present, even in the hottest parts of the globe, is inferior to that indicated by the petrified forms of the coal formation. An elevated and uniform temperature, and great humidity in the air, are the causes most favourable for the numerical predominance, and the great size of these plants within the torrid zone at present.† If the gigantic size and form of these fossil plants are remarkable, still more so is the extent of their geographical distribution; for impressions of arborescent ferns, such as characterize our English carboniferous strata, have been brought from Melville island, in latitude 75°‡. The corals and chambered shells, which occur in beds interstratified with the coal (as in mountain limestone), afford also indications of a warm climate,—the gigantic orthocerata of this era being, to recent multilocular shells, what the fossil ferns, equiseta, and other plants of the coal strata, are in comparison with plants now growing within the tropics. These shells also, like the vegetable impressions, have been brought from rocks in very high latitudes in North America.

* Consid. Générales sur la Nature de la Végétation, &c. Ann. des Sci. Nat. Nov. 1828.

† Humboldt, in speaking of the vegetation of the present era, considers the laws which govern the distribution of vegetable forms to be sufficiently constant to enable a botanist, who is informed of the number of one class of plants, to conjecture, with tolerable accuracy, the relative number of all others. It is premature, perhaps, to apply this law of proportion to the fossil botany of strata, between the coal formation and the chalk, as M. Adolphe Brongniart has attempted, as the number of species hitherto procured is so inconsiderable, that the quotient would be materially altered by the addition of one or two species. It may also be objected, that the fossil flora consists of such plants as may accidentally have been floated into seas, lakes, or estuaries, and may often, perhaps always, give a false representation of the numerical relations of families, then living on the land. Yet, after allowing for all liability to error on these grounds, the argument founded on the comparative numbers of the fossil plants of the carboniferous strata is very strong.

Martius informs us, that on seeing the tesselated surface of the stems of arborescent ferns in Brazil, he was reminded of their prototypes, in the impressions which he had seen in the coal-mines of Germany.

‡ Mr. Konig's description of the rocks brought home by Captain Parry, Journ of Science, vol. xv. p. 20.

In vain should we attempt to explain away the phenomena of the carboniferous and other secondary formations, by supposing that the plants were drifted from equatorial seas. During the accumulation, and consolidation of so many sedimentary deposits, and the various movements and dislocations to which they were subjected at different periods, rivers and currents must often have changed their direction, and wood might as often be floated from the arctic towards tropical seas, as in an opposite direction. It is undeniable, that the materials for future beds of lignite and coal are now amassed in high latitudes far from the districts where the forests grew, and on shores where scarcely a stunted shrub can now exist. The Mackenzie, and other rivers of North America, carry pines with their roots attached for many hundred miles towards the north, into the arctic sea, where they are imbedded in deltas, and some of them drifted still farther, by currents towards the pole. But such agency, although it might account for some partial anomalies in the admixture of vegetable remains of different climes, can by no means weaken the arguments deduced from the general character of fossil vegetable remains. We cannot suppose the leaves of tree ferns to be transported by water for thousands of miles, without being injured; nor, if this were possible, would the same hypothesis explain the presence of uninjured corals and multilocular shells of contemporary origin, for these must have lived in the same latitudes where they are now inclosed in rocks. The plants, moreover, whose remains have given rise to the coal beds, must be supposed to have grown upon the same land, the destruction of which provided materials for the sandstones and conglomerates of that group of strata. The coarseness of the particles of many of these rocks attests that they were not borne from very remote localities, but were most probably derived from islands in a vast sea, which was continuous, at that time, over a great part of the northern hemisphere, as is demonstrated by the great extent of the mountain and transition limestone formations. The same observation is applicable to many secondary strata of a later epoch. There must have been dry land in these latitudes, to provide materials by its disintegration for sandstones,—to afford a beach whereon the oviparous reptiles deposited their eggs,—to furnish an habitation for the opossum of

Stonesfield, and the insects of Solenhofen. The vegetation of the same lands, therefore, must in general have imparted to fossil floras their prevailing character.

From the considerations above enumerated, we must infer, that the remains both of the animal and vegetable kingdom preserved in strata of different ages, indicate that there has been a great diminution of temperature throughout the northern hemisphere, in the latitudes now occupied by Europe, Asia, and America. The change has extended to the arctic circle, as well as to the temperate zone. The heat and humidity of the air, and the uniformity of climate, appear to have been most remarkable when the oldest strata hitherto discovered were formed. The approximation to a climate similar to that now enjoyed in these latitudes, does not commence till the era of the formations termed tertiary, and while the different tertiary rocks were deposited in succession, the temperature seems to have been still farther lowered, and to have continued to diminish gradually, even after the appearance of a great portion of existing species upon the earth.

CHAPTER VII.

On the causes of vicissitudes in climate—Remarks on the present diffusion of heat over the globe—On the dependence of the mean temperature on the relative position of land and sea—Isothermal lines—Currents from equatorial regions—Drifting of Icebergs—Different temperature of Northern and Southern hemispheres—Combination of causes which might produce the extreme cold of which the earth's surface is susceptible—On the conditions necessary for the production of the extreme of heat, and its probable effects on organic life.

As the proofs enumerated in the last chapter indicate that the earth's surface has experienced great changes of climate since the deposition of the older sedimentary strata, we have next to inquire, how such vicissitudes can be reconciled with the existing order of nature. The cosmogonist has availed himself of this, as of every obscure problem in geology, to confirm his views concerning a period when the laws of the animate and inanimate world were wholly distinct from those now established; and he has in this, as in all other cases, succeeded so far, as to divert attention from that class of facts, which, if fully understood, might probably lead to an explanation of the phenomenon. At first, it was imagined that the earth's axis had been for ages perpendicular to the plane of the ecliptic, so that there was a perpetual equinox, and unity of seasons throughout the year:—that the planet enjoyed this '*paradisiacal*' state until the era of the great flood; but in that catastrophe, whether by the shock of a comet, or some other convulsion, it lost its equal poize, and hence the obliquity of its axis, and with that the varied seasons of the temperate zone, and the long nights and days of the polar circles. When the advancement of astronomical science had exploded this theory, it was assumed, that the earth at its creation was in a state of fluidity, and red hot, and that ever since that era it had been cooling down, contracting its dimensions, and acquiring a solid crust,—an hypothesis equally arbitrary, but more calculated for lasting popularity, because, by referring the mind directly to the beginning of things, it requires no support from observations, nor from any ulterior hypothesis. They who are satisfied with this solution are relieved from all necessity of inquiry into the present laws which regulate the diffusion of heat over the surface, for however well these may be ascertained, they cannot possibly afford a full and exact elucidation of the internal changes of an embryo world. As well might an ornithologist study the plumage and external form of a full-fledged bird, in the hope of divining the colour of its egg, or the mysterious metamorphoses of the yolk during incubation.

But if, instead of vague conjectures as to what might have been the state of the planet at the era of its creation, we fix our thoughts steadily on the connection at present between climate and the distribution of land and sea; and if we then consider what influence former fluctuations in the physical geography of the earth must have had on superficial temperature, we may perhaps approximate to a true theory. If doubt still remain, it should be ascribed to our ignorance of the laws of Nature, not to revolutions in her economy;—it should stimulate us to farther research, not tempt us to indulge our fancies in framing imaginary systems for the government of infant worlds.

In considering the laws which regulate the diffusion of heat over the globe, says Humboldt, we must beware not to regard the climate of Europe as a type of the temperature which all countries placed under the same latitudes enjoy. The physical sciences, observes this philosopher, always bear the impress of the places where they began to be cultivated; and, as in geology, an attempt was at first made to refer all the volcanic phenomena to those of the volcanos in Italy, so in meteorology, a small part of the old world, the centre of the primitive civilization of Europe, was for a long time considered a type to which the climate of all corresponding latitudes might be referred. But this region, constituting only one-seventh of the whole globe, proved eventually to be the exception to the general rule; and for the same reason we may warn the geologist to be on his guard, and not hastily to assume that the temperature of the earth in the present era is a type of that which most usually obtains, since he contemplates far mightier alterations in the position of land and sea, than those which now cause the climate of Europe to differ from that of other countries in the same parallels.

It is now well ascertained that zones of equal warmth, both in the atmosphere and in the waters of the ocean, are neither parallel to the equator nor to each other *. It is also discovered that the same mean annual temperature may exist in two places which enjoy very different climates, for the seasons may be nearly equalized or violently contrasted. Thus the lines of equal winter temperature do not coincide with the lines of equal annual heat, or the isothermal lines. The deviations of all these lines from the same parallel of latitude, are determined by a multitude of circumstances, among the principal of which are the position, direction, and elevation of the continents and islands, the position and depth of the sea, and the direction of currents and of winds.

It is necessary to go northwards in Europe in order to find the same mean quantity of annual heat as in North America. On comparing these two continents, it is found that places situated in the same latitudes, have sometimes a mean difference of temperature amounting to 11° or even sometimes 17° of Fahrenheit; and places on the two continents which have the same mean temperature, have sometimes a difference in latitude of from 7° to 13° †. The principal cause of greater intensity of cold in corresponding latitudes of North America and Europe, is the connexion of the former country with the polar circle, by a large tract of land, some of which is from three to five thousand feet in height, and, on the other hand, the separation of Europe from the arctic circle by an ocean. The ocean has a tendency to preserve every where a mean temperature, which it communicates to the contiguous land, so that it tempers the climate, moderating alike an excess of heat or cold. The elevated land, on the other hand, rising to the

* We are indebted to Baron Alex. Humboldt for collecting together in a beautiful essay, the scattered data on which some approximation to a true theory of the distribution of heat over the globe may be founded. Many of these data are derived from the author's own observations, and many from the works of M. Prevost on the radiation of heat, and other writers. See Humboldt on Isothermal Lines, Mémoires d'Arcueil, tom. iii. translated in the Edin. Phil. Journ. vol. iii. July, 1820.

† Humboldt's tables, Essay on Isothermal Lines, &c.

colder regions of the atmosphere, becomes a great reservoir of ice and snow, attracts, condenses, and congeals vapour, and communicates its cold to the adjoining country. For this reason, Greenland, forming part of a continent which stretches northward to the 82nd degree of latitude, experiences under the 60th parallel a more rigorous climate than Lapland under the 72nd parallel.

But if land be situated between the 40th parallel and the equator, it produces exactly the opposite effect, unless it be of extreme height, for it then warms the tracts of land or sea that intervene between it and the polar circle. For the surface being in this case exposed to the vertical, or nearly vertical rays of the sun, absorbs a large quantity of heat, which it diffuses by radiation into the atmosphere. For this reason, the western parts of the old continent derive warmth from Africa, "which, like an immense furnace," says Malte-Brun *, "distributes its heat to Arabia, to Turkey in Asia, and to Europe." On the contrary, Asia in its north-eastern extremity, experiences in the same latitude extreme cold, for it has land on the north between the 60th and 70th parallel, while to the south it is separated from the equator by the North Pacific.

In consequence of the more equal temperature of the waters of the ocean, the climate of islands and coasts differs essentially from that of the interior of continents, the former being characterized by mild winters and less temperate summers; for the sea breezes moderate the cold of winter, as well as the summer heat. When, therefore, we trace round the globe those belts in which the mean annual temperature is the same, we often find great differences in climate; for there are *insular* climates where the seasons are nearly equalized, and *excessive* climates as they have been termed, where the temperature of winter and summer is strongly contrasted. The whole of Europe, compared with the eastern parts of America and Asia, has an insular climate. The northern part of China, and the Atlantic region of the United States, exhibit "excessive climates." We find at New York, says Humboldt, the summer of Rome and the winter of Copenhagen; at Quebec, the summer of Paris and the winter of Petersburgh. At Pekin, in China,

* Phys. Geog. Book xvii.

where the mean temperature of the year is that of the coasts of Brittany, the scorching heats of summer are greater than at Cairo, and the winters as rigorous as at Upsal *.

If lines be drawn round the globe through all those places which have the same winter temperature, they are found to deviate from the terrestrial parallels much farther than the lines of equal mean annual heat. For the lines of equal winter in Europe are often curved so as to reach parallels of latitude 9° or 10° distant from each other, whereas the isothermal lines only differ from 4° to 5°.

Among other influential causes, both of remarkable diversity in the mean annual heat, and of unequal division of heat in the different seasons, are the direction of currents and the accumulation and drifting of ice in high latitudes. That most powerful current, the Gulf stream, after doubling the Cape of Good Hope, flows to the northward along the western coast of Africa, then crosses the Atlantic, and accumulates in the Gulf of Mexico. It then issues through the Straits of Bahama, running northwards at the rate of four miles an hour, and retains in the parallel of 38°, nearly one thousand miles from the above strait, a temperature 10° Fahr. warmer than the air. The general climate of Europe is materially affected by the volume of warmer water thus borne northwards, for it maintains an open sea free from ice in the meridian of East Greenland and Spitzbergen, and thus moderates the cold of all the lands lying to the south. Until the waters of the great current reach the circumpolar sea, their specific gravity is less than that of the lower strata of water; but when they arrive near Spitzbergen, they meet with the water of melted ice which is still lighter, for it is a well known law of this fluid, that it passes the point of greatest density when cooled down below 40°, and between that and the freezing point expands again. The warmer current, therefore, being now the heavier, sinks below the surface, so that in the lower regions it is found to be from 16° to 20° Fahrenheit, above the mean temperature of the climate. The movements of the sea, however, cause this under current sometimes to appear at the surface, and greatly to moderate the cold †.

* On Isothermal Lines.

† Scoresby's Arctic Regions, vol. i. p. 210.

The great glaciers generated in the valleys of Spitzbergen, in the 79° of north latitude, are almost all cut off at the beach, being melted by the feeble remnant of heat retained by the Gulf stream. In Baffin's Bay, on the contrary, on the east coast of old Greenland, where the temperature of the sea is not mitigated by the same cause, and where there is no warmer under-current, the glaciers stretch out from the shore, and furnish repeated crops of mountainous masses of ice which float off into the ocean * The number and dimensions of these bergs is prodigious. Capt. Ross saw several of them together in Baffin's Bay aground in water fifteen hundred feet deep! Many of them are driven down into Hudson's Bay, and, accumulating there, diffuse excessive cold over the neighbouring continent, so that Captain Franklin reports, that at the mouth of Hayes river, which lies in the same latitude as the north of Prussia or the south of Scotland, ice is found every where in digging wells at the depth of four feet!

When we compare the climate of the northern and southern hemispheres, we obtain still more instruction in regard to the influence of the distribution of land and sea on climate. The dry land in the southern hemisphere, is to that of the northern in the ratio only of one to three, excluding from our consideration that part which lies between the pole and the 74° of south latitude, which has hitherto proved inaccessible. The predominance of ice in the antarctic over the arctic zone is very great; for that which encircles the southern pole, extends to lower latitudes by ten degrees than that around the north pole †. It is probable that this remarkable difference is partly attributable, as Cook conjectured, to the existence of a considerable tract of high land between the 70th parallel of south latitude and the pole. There is, however,

* Scoresby's Arctic Region, vol. i. p. 208.—Dr. Latta's observations on the Glaciers of Spitzbergen, &c. Edin. New Phil. Journ. vol. iii. p. 97.

† Captain Weddell, in 1823, reached 3° farther than Captain Cook, and arrived at 74° 15′ longitude, 34° 17′ west. After having passed through a sea strewed with numerous ice-islands, he arrived, in that high latitude, at an open ocean; but even if he had sailed 6° farther south, he would not have penetrated to higher latitudes than Captain Parry in the arctic circle, who reached lat. 81° 10′ north. The important discovery, therefore, of Captain Weddell, does not destroy the presumption, that the general prevalence of ice, in low latitudes in the southern hemisphere, arises from the existence of greater tracts of land in the antarctic, than in the arctic ocean.

another reason suggested by Humboldt, to which great weight is due,—the small quantity of land in the tropical and temperate zones south of the line. If Africa and New Holland extended farther to the south, a diminution of ice would take place in consequence of the radiation of heat from these continents during summer, which would warm the contiguous sea and rarefy the air. The heated aërial currents would then ascend and flow more rapidly towards the south pole, and moderate the winter. In confirmation of these views, it is stated that the cap of ice, which extends as far as the 68° and 71° of south latitude, advances more towards the equator whenever it meets a free sea; that is, wherever the extremities of the present continents are not opposite to it; and this circumstance seems explicable only on the principle above alluded to, of the radiation of heat from the lands so situated.

Before the amount of difference between the temperature of the two hemispheres was ascertained, it was referred by astronomers to the acceleration of the earth's motion in its perihelium; in consequence of which the spring and summer of the southern hemisphere are shorter, by nearly eight days, than those seasons north of the equator. A sensible effect is probably produced by this source of disturbance, but it is quite inadequate to explain the whole phenomenon. It is, however, of importance to the geologist to bear in mind, that in consequence of the procession of the equinoxes the two hemispheres receive alternately, each for a period of upwards of 10,000 years, a greater share of solar light and heat. This cause may sometimes tend to counterbalance inequalities resulting from other circumstances of a far more influential nature; but, on the other hand, it must sometimes tend to increase the extreme of deviation which certain combinations of causes produce at distant epochs. But, whatever may now be the inferiority of heat in the temperate and arctic zones south of the line, it is quite evident that the cold would be far more intense if there happened, instead of open sea, to be tracts of elevated land between the 55th and 70th parallel; for, in Sandwich land, in 54° and 58° of south latitude, the perpetual snow and ice reach to the sea beach; and what is still more astonishing, in the island of Georgia, which is in the 53° south latitude, or the same parallel as the central counties of England, the perpetual snow descends to the level of the ocean. When we consider this fact, and then recollect that the highest mountains in Scotland do not attain the limit of perpetual snow on this side of the equator, we learn that latitude is one only of many powerful causes, which determine the climate of particular regions of the globe. The permanence of the snow, in this instance, is partly due to the floating ice, which chills the atmosphere and condenses the vapour, so that in summer the sun cannot pierce through the foggy air. The distance to which icebergs float from the polar regions on the opposite sides of the line, is, as might have been anticipated, very different. Their extreme limit in the northern hemisphere appears to be the Azores (north latitude 42°), to which isles they are sometimes drifted from Baffin's Bay *. But in the other hemisphere they have been seen, within the last two years, at different points off the Cape of Good Hope, between latitude 36° and 39° †. One of these was two miles in circumference, and 150 feet high ‡. Others rose from 250 to 300 feet above the level of the sea, and were, therefore, of great volume below, since it is ascertained, by experiments on the buoyancy of ice floating in sea-water, that for every solid foot seen above, there must at least be eight feet below water §. If ice islands from the north polar regions floated as far, they might reach Cape St. Vincent, and, then being drawn by the current that always sets in from the Atlantic through the Straits of Gibraltar, be drifted into the Mediterranean, where clouds and mists would immediately deform the serene sky of spring and summer.

The great extent of sea gives a particular character to climates south of the equator, the winters being mild, and the summers cold. Thus, in Van Dieman's land, corresponding nearly in latitude to Rome, the winters are more mild than at Naples, and the summers not warmer than those at Paris, which is 7° farther from the equator ‖. The effect on vegetation is very remarkable:—tree ferns, for instance, which require abundance of moisture, and an equalization of the seasons,

* Examples will be given in Major Rennell's forthcoming work on Currents.

† On Icebergs in low Latitudes in the Southern Hemisphere, by Captain Hosburg, Hydrographer to the East India Company; read to the Royal Society, February, 1830.

‡ Edin. New Phil. Journ. No. xv. p. 193; January, 1830.

§ Scoresby's Arctic Regions, vol. i., p. 234.

‖ Humboldt, ib.

are found in Van Dieman's land in latitude 42°, and in New Zealand in south latitude 45°. The orchideous parasites also advance towards the 38° and 42° of south latitude *.

Having offered these brief remarks on the diffusion of heat over the globe in the present state of the surface, we shall now proceed to speculate on the vicissitudes of climate, which must attend those endless variations in the geographical features of our planet, which are contemplated in geology. In order to confine ourselves within the strict limits of analogy, we shall assume, 1st, That the proportion of dry land to sea continues always the same. 2dly, That the volume of the land rising above the level of the sea, is a constant quantity; and not only that its mean, but that its extreme height, are only liable to trifling variations. 3dly, That both the mean and extreme depth of the sea are equal at every epoch; and, 4thly, It will be consistent, with due caution, to assume, that the grouping together of the land in great continents is a necessary part of the economy of nature; for it is possible, that the laws which govern the subterranean forces, and which act simultaneously along certain lines, cannot but produce, at every epoch, continuous mountain-chains; so that the subdivision of the whole land into innumerable islands may be precluded. If it be objected, that the maximum of elevation of land and depth of sea are probably not constant, nor the gathering together of all the lands in certain parts, nor even perhaps the relative extent of land and water; we reply, that the arguments which we shall adduce will be greatly strengthened, if, in these peculiarities of the surface, there be considerable deviations from the present type. If, for example, all other circumstances being the same, the land is at one time more divided into islands than at another, a greater uniformity of climate might be produced, the mean temperature remaining unaltered; or if, at another era, there were mountains higher than the Himalaya, these, when placed in high

* These forms of vegetation might perhaps be developed in still higher latitudes, if the ice in the antarctic circle did not extend farther from the pole than in the arctic. Humboldt observes, that it is in the *mountainous, temperate, humid*, and *shady* parts of the equatorial regions, that the family of ferns produces the greatest number of species. As we know, therefore, that elevation often compensates the effect of latitude in plants, we may easily understand that a class of vegetables which grow at a certain height in the torrid zone, would flourish on the plains far from the equator, provided the temperature throughout the year was equally uniform.

latitudes would cause a greater excess of cold. So if we suppose, that at certain periods no chain of hills in the world rose beyond the height of 10,000 feet, a greater heat might then have prevailed than is compatible with the existence of mountains thrice that elevation.

However constant we believe the relative proportion of sea and land to continue, we know that there is annually some small variation in their respective geographical positions, and that in every century the land is in some parts raised, and in others depressed by earthquakes, and so likewise is the bed of the sea. By these and other ceaseless changes, the configuration of the earth's surface has been remodelled again and again since it was the habitation of organic beings, and the bed of the ocean has been lifted up to the height of some of the loftiest mountains. The imagination is apt to take alarm, when called upon to admit the formation of such irregularities of the crust of the earth, after it had become the habitation of living creatures; but if time be allowed, the operation need not subvert the ordinary repose of nature, and the result is insignificant, if we consider how slightly the highest mountain chains cause our globe to differ from a perfect sphere. Chimborazo, although it rises to more than 21,000 feet above the surface of the sea, would only be represented on an artificial globe, of about six feet in diameter, by a grain of sand less than one-twentieth of an inch in thickness*. The superficial inequalities of the earth, then, may be deemed minute in quantity, and their distribution at any particular epoch must be regarded in geology as temporary peculiarities, like the height and outline of the cone of Vesuvius in the interval between two eruptions. But, although the unevenness of the surface is so unimportant, in reference to the magnitude of the globe, it is on the position and direction of these small inequalities that the state of the atmosphere and both the local and general climate are mainly dependent.

Before we consider the effect which a material change in the distribution of land and sea must occasion, it may be well to remark, how greatly organic life may be affected by those minor mutations, which need not in the least degree alter the

* Malte-Brun's System of Geography, book i. p. 6.

general temperature. Thus, for example, if we suppose, by a series of convulsions, a certain part of Greenland to become sea, and, in compensation, a tract of land to rise and connect Spitzbergen with Lapland,—an accession not greater in amount than one which the geologist can prove to have occurred in certain districts bordering the Mediterranean, within a comparatively modern period,—this altered form of the land might occasion an interchange between the climate of certain parts of North America and of Europe, which lie in corresponding latitudes. Many European species would probably perish in consequence, because the mean temperature would be greatly lowered; and others would fail in America because it would there be raised. On the other hand, in places where the mean annual heat remained unaltered, some species which flourish in Europe, where the seasons are more uniform, would be unable to resist the great heat of the North American summer, or the intense cold of the winter; while others, now fitted by their habits for the great contrast of the American seasons, would not be fitted for the *insular* climate of Europe*. Many plants, for instance, will endure a severe frost, but cannot ripen their seeds without a certain intensity of summer heat and a certain quantity of light; others cannot endure the same intensity of heat or cold. It is now established, that many species of animals, which are at present the contemporaries of man, have survived great changes in the physical geography of the globe. If such species be termed modern, in comparison to races which preceded them, their remains, nevertheless, enter into submarine deposits many hundred miles in length, and which have since been raised from the deep to no inconsiderable altitude. When, therefore, it is shewn that changes of the temperature of the atmosphere may be the consequence of such physical revolutions of the surface, we ought no longer to wonder that we find the distribution of existing species to be *local*, in regard to *longitude* as well as latitude. If all species were now, by an exertion of creative power, to be diffused uniformly throughout those zones where there is an equal degree of heat, and in all respects a similar climate, they would begin from this moment to

* According to Humboldt, the vine can be cultivated with advantage 10° farther north in Europe, than in North America.

depart more and more from their original distribution. Aquatic and terrestrial species would be displaced, as Hooke long ago observed, so often as land and water exchanged places; and there would also, by the formation of new mountains and other changes, be transpositions of climate, contributing, in the manner before alluded to, to the local extermination of species.

If we now proceed to consider the circumstances required for a *general* change of temperature, it will appear, from the facts and principles already laid down, that whenever a greater extent of high land is collected in the polar regions, the cold will augment; and the same result will be produced when there shall be more sea between or near the tropics; while, on the contrary, so often as the above conditions are reversed, the heat will be greater. If this be admitted, it will follow as a corollary, that unless the superficial inequalities of the earth be fixed and permanent, there must be never-ending fluctuations in the mean temperature of every zone, and that the climate of one era can no more be a type of every other, than is one of our four seasons of all the rest. It has been well said, that the earth is covered by an ocean, and in the midst of this ocean there are two great islands, and many smaller ones; for the whole of the continents and islands occupy an area scarcely exceeding one-fourth of the whole superficies of the spheroid. Now, on a fair calculation, we may expect that at any given epoch, there will not be more than about one-fourth dry land in a particular region; such, for example, as the arctic and antarctic circles. If, therefore, at present there should happen in the only one of these regions which we can explore, to be much more than this average proportion of land, and some of it above five thousand feet in height, this alone affords ground for concluding, that in the present state of things, the mean heat of the climate is below that which the earth's surface, in its more ordinary state, would enjoy. This presumption is heightened, when we remember that the mean depth of the Atlantic ocean is calculated to be about three miles, and that of the Pacific four miles *; so that we might look not only for more than two-thirds sea in the frigid zones, but for

* See Young's Nat. Phil. Lect. 47. Laplace seems often to have changed his opinion, reasoning from the depth required to account for the phenomena of the tides; but his final conclusion respecting the sea was "que sa profondeur moyenne

water of great depth, which could not readily be reduced to the freezing point. The same opinion is farther confirmed, when we compare the quantity of land lying between the poles and the 30th parallels of north and south latitude, and the quantity placed between those parallels and the equator; for it is clear, that at present we must have not only more than the usual degree of cold in the polar regions, but also less than the average quantity of heat generated in the intertropical zone.

In order to simplify our view of the various changes in climate, which different combinations of geographical circumstances may produce, we shall first consider the conditions necessary for bringing about the extreme of cold, or what may be termed the winter of the "great year," or geological cycle, and afterwards, the conditions requisite for producing the maximum of heat, or the summer of the same year.

To begin with the northern hemisphere. Let us suppose those hills of the Italian peninsula and of Sicily, which are of comparatively modern origin, and contain many fossil shells identical with living species, to subside again into the sea, from which they have been raised, and that an extent of land of equal area and height (varying from one to three thousand feet) should rise up in the Arctic ocean, between Siberia and the north pole. In speaking of such changes, we need not allude to the manner in which we conceive it possible that they may be brought about, nor of the time required for their accomplishment,—reserving for a future occasion, not only the proofs that revolutions of equal magnitude have taken place, but that analogous mutations are still in gradual progress. The alteration now supposed in the physical geography of the northern regions would cause additional snow and ice to accumulate where now there is usually an open sea; and the temperature of the greater part of Europe would be somewhat lowered, so as to resemble more nearly that of corresponding latitudes of North America; or, in other words, it might be necessary to travel about 10° farther south, in order to meet with the same climate which we now enjoy. There would be no compensation derived from the disappearance of land in the Mediterranean countries; for, on the contrary, the mean heat of

est du même ordre que la hauteur moyenne des continens et des îles au-dessus de son niveau, hauteur qui ne surpasse pas mille mètres (3280 ft.)." Mec. Céleste. Bk. 11. et Syst. du Monde, p. 254.

the soil so situated, is probably far above that which would belong to the sea, by which we imagine it to be replaced. But let the configuration of the surface be still further varied, and let some large district within or near the tropics, such as Mexico for example, with its mountains rising to the height of twelve thousand feet and upwards, be converted into sea, while lands of equal elevation and extent are transferred to the arctic circle. From this change there would, in the first place, result a sensible diminution of temperature near the tropic, for the soil of Mexico would no longer be heated by the sun; so that the atmosphere would be less warm, as also the Atlantic, and the Gulf stream. On the other hand, the whole of Europe, Northern Asia, and North America, would feel the influence of the enormous quantity of ice and snow, now generated at vast heights on the new arctic continent. If, as we have already seen, there are some points in the southern hemisphere where snow is perpetual to the level of the sea, in latitudes as low as central England, such might now assuredly be the case throughout a great part of Europe. If at present the extreme limits of drifted icebergs are the Azores, they might easily reach the equator after the changes above supposed. To pursue the subject still farther, let the Himalaya mountains, with the whole of Hindostan, sink down, and their place be occupied by the Indian ocean, and then let an equal extent of territory and mountains, of the same vast height, stretch from North Greenland to the Orkney islands. It seems difficult to exaggerate the amount to which the climate of the northern hemisphere would now be cooled down. But, notwithstanding the great refrigeration which would thus be produced, it is probable that the difference of mean temperature between the arctic and equatorial latitudes would not be increased in a very high ratio, for no great disturbance can be brought about in the climate of a particular region, without immediately affecting all other latitudes, however remote. The heat and cold which surround the globe are in a state of constant and universal flux and reflux. The heated and rarefied air is always rising and flowing from the equator towards the poles in the higher regions of the atmosphere, and, in the lower, the colder air is flowing back to restore the equilibrium. That this circulation is constantly going on in the aërial currents is not disputed *, and that a corresponding interchange takes place in the seas, is demonstrated, according to Humboldt, by the cold which is found to exist at great depths between the tropics; and, among other proofs, may be mentioned the great volume of water which the Gulf stream is constantly bearing northwards, while another current flows *from* the north along the coast of Greenland and Labrador, and helps to restore the equilibrium †.

Currents of heavier and colder water pass from the poles towards the equator, which cool the inferior parts of the ocean; so that the heat of the torrid zone, and the cold of the polar circle, balance each other. The refrigeration, therefore, of the polar regions, resulting from the supposed alteration in the distribution of land and sea, would be immediately communicated to the tropics, and from them would extend to the antarctic circle, where the atmosphere and the ocean would be cooled, so that ice and snow would augment. Although the mean temperature of higher latitudes in the southern hemisphere is, as we have stated, for the most part lower than that of the same parallels in the northern, yet for a considerable space on each side of the line, the mean annual heat of the waters is found to be the same in corresponding parallels. When, therefore, by the new position of the land, the generating of icebergs had become of frequent occurrence in the temperate zone, and when they were frequently drifted as far as the equator, the same degree of cold would immediately be communicated as far as the tropic of Capricorn, and from thence to the lands or ocean to the south. The freedom, then, of the circulation of heat and cold from pole to pole being duly considered, it will be evident that the mean quantity of heat

* The trade wind continually blows with great force from the Island of St. Vincent to that of Barbadoes; notwithstanding which, during the eruption of the volcano in the Island of St. Vincent, in 1812, ashes fell in profusion from a great height in the atmosphere upon Barbadoes. This apparent transportation of matter against the wind, confirmed the opinion of the existence of a counter-current in the higher regions, which had previously rested on theoretical conclusions. Daniell's Meteorological Essays, &c., p. 103.

† In speaking of the circulation of air and water in this chapter, no allusion is made to the trade winds, or to irregularities in the direction of currents, caused by the rotatory motion of the earth. These causes prevent the movements from being direct from north to south, or from south to north, but they do not affect the theory of a constant circulation.

which at two different periods visits the same point, may differ far more widely than the mean quantity which any two points receive in the same parallels of latitude, at one and the same period. For the range of temperature in a given zone, or in other words, the curves of the isothermal lines, must always be circumscribed within narrow limits, the climate of each place in that zone being controlled by the combined influence of the geographical peculiarities of all other parts of the earth. But, when we compare the state of things as existing at two distinct epochs, a particular zone may at one time be under the influence of one class of disturbing causes, as for example those of a refrigerating nature, and at another time may be affected by a combination of opposite circumstances. The lands to the north of Greenland cause the present climate of North America to be colder than that of Europe in the same latitudes, but they also affect, to a certain extent, the temperature of the atmosphere in Europe; and the entire removal from the northern hemisphere of that great source of refrigeration would not assimilate the mean temperature of America to that now experienced in Europe, but would render the continents on both sides of the Atlantic much warmer.

To return to the state of the earth, after the changes before supposed by us, we must not omit to dwell on the important effects to which a wide expanse of perpetual snow would give rise. It is probable that nearly the whole sea, from the poles to the parallels of 45°, would be frozen over, for it is well known that the immediate proximity of land, is not essential to the formation and increase of field ice, provided there be in some part of the same zone a sufficient quantity of glaciers generated on or near the land, to cool down the sea *. Field ice is almost always covered with snow, through which the sun's rays are unable to penetrate †, and thus not only land as extensive as our existing continents, but immense tracts of sea in the frigid and temperate zones, would now present a

* See Scoresby's Arctic Regions, vol. i. p 320.

† Captain Scoresby, in his account of the arctic regions, observes, that when the sun's rays "fall upon the snow-clad surface of the ice or land, they are in a great measure reflected, without producing any material elevation of temperature; but when they impinge on the black exterior of a ship, the pitch on one side occasionally becomes fluid, while ice is rapidly generated at the other." vol. i. p. 378.

solid surface covered with snow, and reflecting the sun's rays for the greater part of the year. Within the tropics, moreover, where we suppose the ocean to predominate, the sky would no longer be serene and clear, as in the present era; but the melting of floating ice would cause quick condensations of vapour, and fogs and clouds would deprive the vertical rays of the sun of half their power. The whole planet, therefore, would receive annually a smaller proportion of solar influence, and the external crust would part, by radiation, with some of the heat which had been accumulated in it, during a different state of the surface. This heat would be dissipated into the spaces surrounding our atmosphere, which, according to the calculations of M. Fourier, have a temperature much inferior to that of freezing water.

At this period, the climate of equinoctial lands might resemble that of the present temperate zone, or perhaps be far more wintery. They who should then inhabit the small isles and coral reefs, which are now seen in the Indian ocean and South Pacific, would wonder that zoophytes of such large dimensions had once been so prolific in those seas; or if, perchance, they found the wood and fruit of the cocoa-nut tree or the palm silicified by the waters of some mineral spring, or incrusted with calcareous matter, they would muse on the revolutions that had annihilated such genera, and replaced them by the oak, the chestnut, and the pine. With equal admiration would they compare the skeletons of their small lizards with the bones of fossil alligators and crocodiles more than twenty feet in length, which, at a former epoch, had multiplied between the tropics; and when they saw a pine included in an iceberg, drifted from latitudes which we now call temperate, they would be astonished at the proof thus afforded, that forests had once grown where nothing could be seen in their own times but a wilderness of snow.

As we have not yet supposed any mutations to have taken place in the relative position of land and sea in the southern hemisphere, we might still increase greatly the intensity of cold, by transferring the land still remaining in the equatorial and contiguous regions, to higher southern latitudes; but it is unnecessary to pursue the subject farther, as we are too ignorant of the laws governing the direction of subterranean forces, to determine whether such a crisis be within the limits of possibility. At the same time we may observe, that the distribution of land at present is so remarkably irregular, and appears so capricious, if we may so express ourselves, that the two extremes of terrestrial heat and cold are probably separated very widely from each other. The globe may now be equally divided, so that one hemisphere shall be entirely covered with water, with the exception of some promontories and islands, while the other shall contain less water than land; and what is still more extraordinary, on comparing the extra-tropical lands in the northern and southern hemispheres, the former are found to be to the latter in the proportion of thirteen to one*! To imagine all the lands, therefore, in high, and all the sea in low latitudes, would scarcely be a more anomalous state of the surface.

Let us now turn from the contemplation of the winter of the "great year," and consider the opposite train of circumstances, which would bring on the spring and summer. That some part of the vast ocean which forms the Atlantic and Pacific, should at certain periods occupy entirely one or both of the polar regions, and should extend, interspersed with islands, only to the parallels of 40°, and even 30°, is an event that may be supposed in the highest degree probable, in the course of many great geological revolutions. In order to estimate the degree to which the general temperature would then be elevated, we should begin by considering separately the effect of the diminution of certain portions of land, in high northern latitudes, which might cause the sea to be as open in every direction, as it is at present towards the north pole, in the meridian of Spitzbergen. By transferring the same lands to the torrid zone, we might gain farther accessions of heat, and cause the ice towards the south pole to diminish. We might first continue these geographical mutations, until we had produced as mild a climate in high latitudes as exists at those points in the same parallel where the mean annual heat is now greatest. We should then endeavour to calculate what farther alterations would be required to double the amount of change; and the great deviation of isothermal lines at present

* Humboldt, on Isothermal Lines.

seems to authorize us to infer, that without an entire revolution of the surface, we might cause the mean temperature to vary to an extent equivalent to 20° or even 30° of latitude,—in other words, we might transfer the temperature of the torrid zone, to the mean parallel, and of the latter, to the arctic regions. By additional transpositions, therefore, of land and sea, we might bring about a still greater variation, so that, throughout the year, all signs of frost should disappear from the earth.

The plane of congelation would rise in the atmosphere in all latitudes; and as our hypothesis would place all the highest mountains in the torrid zone, they would be clothed with rich vegetation to their summits. We must recollect that even now it is necessary to ascend to the height of 15,000 feet in the Andes under the line; and in the Himalaya mountains, which are without the tropic, to 17,000 feet before we reach the limit of perpetual snow. When the absorption of the solar rays was unimpeded, even in winter, by a coat of snow, the mean heat of the earth's crust would augment to considerable depths, and springs, which we know to be an index of the mean temperature of the climate, would be warmer in all latitudes. The waters of lakes, therefore, and rivers, would be much hotter in winter, and would be never chilled in summer by the melting of snow. A remarkable uniformity of climate would prevail amid the numerous archipelagos of the polar ocean, amongst which the tepid waters of equatorial currents would freely circulate. The general humidity of the atmosphere would far exceed that of the present period, for increased heat would promote evaporation in all parts of the globe. The winds would be first heated in their passage over the tropical plains, and would then gather moisture from the surface of the deep, till, charged with vapour, they would arrive at northern regions, and, encountering a cooler atmosphere, would discharge their burden in warm rain. If, during the long night of a polar winter, the snows should whiten the summit of some arctic islands, and ice collect in the bays of the remotest Thule, they would be dissolved as rapidly by the returning sun, as are the snows of Etna by the blasts of the sirocco.

We learn from those who have studied the geographical distribution of plants, that in very low latitudes, at present, the vegetation of small islands remote from continents has a peculiar character, and the ferns and allied families, in particular, bear a great proportion to the total number of other vegetables. Other circumstances being the same, the more remote the isles are from the continents, the greater does this proportion become. Thus, in the continent of India, and the tropical parts of New Holland, the proportion of ferns to the phanerogamic plants is only as one to twenty-six; whereas, in the South Sea Islands, it is as one to four, or even as one to three*. We might expect, therefore, in the summer of the "great year," which we are now considering, that there would be a great predominance of tree-ferns and plants allied to palms and arborescent grasses in the isles of the wide ocean, while the dicotyledonous plants and other forms now most common in temperate regions would almost disappear from the earth. Then might those genera of animals return, of which the memorials are preserved in the ancient rocks of our continents. The huge iguanodon might reappear in the woods, and the ichthyosaur in the sea, while the pterodactyle might flit again through umbrageous groves of tree-ferns. Coral reefs might be prolonged beyond the arctic circle, where the whale and the narwal now abound. Turtles might deposit their eggs in the sand of the sea beach, where now the walrus sleeps, and where the seal is drifted on the ice-floe.

But, not to indulge these speculations farther, we may observe, in conclusion, that however great, in the lapse of ages, may be the vicissitudes of temperature in every zone, it accords with our theory that the general climate should not experience any sensible change in the course of a few thousand years, because that period is insufficient to affect the leading features of the physical geography of the globe. Notwithstanding the apparent uncertainty of the seasons, it is found that the mean temperature of particular localities is very constant, provided we compare observations made at different periods for a series of years. Yet, there must be exceptions to this rule, and even the labours of man have, by the drainage of lakes and marshes, and the felling of extensive forests, caused such changes in the atmosphere as raise our conception of the important

* Ad. Brongniart, Consid. Générales sur la Nat. de la Végét., &c. Ann. des Sciences Nat., Nov. 1828.

influence of those forces to which even the existence in certain latitudes of land or water, hill or valley, lake or sea, must be ascribed. If we possessed accurate information of the amount of local fluctuation in climate in the course of twenty centuries, it would often, undoubtedly, be considerable. Certain tracts, for example, on the coast of Holland and of England, consisted of cultivated land in the time of the Romans, which the sea, by gradual encroachments, has at length occupied. Here an alteration has been effected; for neither the division of heat in the different seasons, nor the mean annual heat of the atmosphere investing the sea is precisely the same as that which rests on the land. In those countries also where the earthquake and volcano are in full activity, a much shorter period may produce a sensible variation. The climate of the once fertile plain of Malpais in Mexico must differ materially from that which prevailed before the middle of the last century; for, since that time, six mountains, the highest of them rising 1700 feet above the plateau, have been thrown up by volcanic eruptions. It is by the repetition of an indefinite number of local revolutions due to volcanic and various other causes, that a general change of climate is finally brought about.

CHAPTER VIII.

Geological proofs that the geographical features of the northern hemisphere, at the period of the deposition of the carboniferous strata, were such as would, according to the theory before explained, give rise to an extremely hot climate—Origin of the transition and mountain limestones, coal-sandstones, and coal—Change in the physical geography of northern latitudes, between the era of the formation of the carboniferous series and the lias—Character of organic remains, from the lias to the chalk inclusive—State of the surface when these deposits originated—Great accession of land, and elevation of mountain-chains, between the consolidation of the newer secondary and older tertiary rocks—Consequent refrigeration of climate—Abrupt transition from the organic remains of the secondary to those of the tertiary strata—Maestricht beds—Remarks on the theory of the diminution of central heat.

We stated, in the sixth chapter, our reasons for concluding that the mean annual temperature of the northern hemisphere was considerably more elevated when the old carboniferous strata were deposited; as also that the climate had been modified more than once since that epoch, and that it approximated by successive changes more and more nearly to that now prevailing in the same latitudes. Further, we endeavoured, in the last chapter, to prove that vicissitudes in climate of no less importance may be expected to recur in future, if it be admitted that causes now active in nature have power, in the lapse of ages, to vary to an unlimited extent the relative position of land and sea. It next remains for us to inquire whether the alterations, which the geologist can prove to have actually taken place at former periods, in the geographical features of the northern hemisphere, coincide in their nature, and in the time of their occurrence, with such revolutions in climate as would naturally have followed, according to the meteorological principles already explained.

We may select the great carboniferous series, including the transition and mountain limestones, and the coal, as the oldest system of rocks of which the organic remains furnish any decisive evidence as to climate. We have already insisted on the indications which they afford of great heat and uniformity of temperature, extending over a vast area, from about 45° to 60°, or perhaps, if we include Melville Island, to near 75° north latitude *.

When we attempt to restore in imagination the distribution of land and sea, as they existed at that remote epoch, we discover that our information is at present limited to latitudes north of the tropic of cancer, and we can only hope, therefore, to point out that the condition of the earth, so far as relates to our temperate and arctic zones, was such as the theory before offered would have led us to anticipate. Now there is scarcely any land hitherto examined in Europe, Northern Asia, or North America, which has not been raised from the bosom of the deep, since the origin of the carboniferous rocks, or which, if previously raised, has not subsequently acquired additional altitude. If we were to submerge again all the marine strata, from the transition limestone to the most recent shelly beds, the summits of some primary mountains alone would remain above the waters. These facts, it is true, considered singly, are not conclusive as to the universality of the ancient ocean in the northern hemisphere, because the movements of earthquakes occasion the subsidence as well as the upraising of the surface, and by the alternate rising and sinking of particular spaces, at successive periods, a great area may become entirely covered with marine deposits, although the whole has never been beneath the waters at one time, nay, even though the relative proportion of land and sea may have continued unaltered throughout the whole period. There is, however, the highest presumption against such an hypothesis, because the land in the northern hemisphere is now in great excess, and this circumstance alone should induce us to suppose that, amidst the repeated changes which the surface has undergone, the sea has usually predominated in a much greater degree. But when we study the mineral composition and fossil contents

* Our ancient coal-formation has not been found in Italy, Spain, Sicily, or any of the more southern countries of Europe. Whether any of the ammonitiferous limestones of the Southern Apennines and Sicily (Taormina for example) can be considered as of contemporaneous origin with our carboniferous series, is not yet determined; but it is conjectured, from the general character of the organic remains of the Apennine limestones, that they belong to some part of our secondary series, from the lias to the chalk inclusive.

of the older strata, we find evidence of a more positive and unequivocal kind in confirmation of the same opinion.

Calcareous rocks, containing the same class of organic remains as our transition and mountain limestones, extend over a great part of the central and northern parts of Europe, are found in the lake district of North America, and even appear to occur in great abundance as far as the border of the Arctic sea *. The organic remains of these rocks consist principally of marine shells, corals, and the teeth and bones of fish ; and their nature, as well as the continuity of the calcareous beds of homogeneous mineral composition, concur to prove that the whole series was formed in a deep and expansive ocean, in the midst of which, however, there were many isles. These isles were composed partly of primary and partly of volcanic rocks, which being exposed to the erosive action of torrents, to the undermining power of the waves beating against the cliffs, and to atmospheric decomposition, supplied materials for pebbles, sand, and shale, which, together with substances introduced by mineral springs and volcanos in frequent eruption, contributed the inorganic parts of the carboniferous strata. The disposition of the beds in that portion of this group which is of mechanical origin, and which incloses the coal, has been truly described to be such as would result from the waste of small islands placed in rows and forming the highest points of ridges of submarine mountains. The disintegration of such clusters of isles would produce around and between them detached deposits of various dimensions, which, when subsequently raised above the waters, would resemble the strata formed in a chain of lakes. The insular masses of primary rock would preserve their original relative superiority of height,

* It appears from the observations of Dr. Richardson, made during the expedition under the command of Captain Franklin to the north-west coast of America, and from the specimens presented by him to the Geological Society of London, that, between the parallels of 60° and 70° north latitude, there is a great calcareous formation, stretching towards the mouth of the Mackenzie river, in which are included corallines, productæ, terebratulites, &c., having a close affinity in generic characters to those of our mountain limestone, of which the group has been considered the equivalent. There is also in the same region a newer series of strata, in which are shales with impressions of ferns, lepidodendrons, and other vegetables, and also ammonites. These, it is supposed, may belong to the age of our oolitic series.—*Proceedings of Geological Society, March* 1828.

and would often surround the newer strata on several sides, like the boundary heights of lake basins*.

As might have been expected, the zoophytic, and shelly limestones of the same era, (as the mountain limestone,) sometimes alternate with the rocks of mechanical origin, but appear to have been, in ordinary cases, diffused far and wide over the bottom of the sea, remote from any islands, and where no grains of sand were transported by currents. The associated volcanic rocks, resemble the products of submarine eruptions, the tuffs being sometimes interstratified with calcareous shelly beds, or with sandstones, just as might be expected if the sand and ejected matter of which they are probably composed had been intermixed with the waters of the sea, and had then subsided like other sediment. The lavas also often extend in spreading sheets, and must have been poured out on a surface rendered horizontal by sedimentary depositions. There is, moreover, a compactness and general absence of porosity in these igneous rocks which distinguishes them from most of those which are produced on the sides of Etna or Vesuvius, and other land-volcanos. The modern submarine lavas of Sicily, which alternate with beds of shells specifically identical with those now living in the Mediterranean, have almost all their cavities filled with calcareous and other ingredients, and have been converted into amygdaloids, and this same change we must suppose such parts of the Etnean lava currents as enter the sea to be undergoing at present, because we know the water on the adjoining coast to be copiously charged with carbonate of lime in solution. It is, therefore, one among many reasons for inferring the submarine origin of our ancient trap rocks, that there are scarcely any instances, in which the cellular hollows, left by bubbles of elastic fluid, have not subsequently been filled by calcareous, siliceous, or other mineral ingredients, such as now abound in the hot springs of volcanic countries.

If, on the other hand, we examine the fossil remains in these strata, we find the vegetation of the coal strata declared by botanists to possess the characters of an insular, not a continental flora, and we may suppose the carbonaceous matter to

* See some ingenious remarks to this effect, in the work of M. Ad. Brongniart, Consid. Générales sur la Nat. de la Végét. &c. Ann. des Sci. Nat., Nov. 1828.

have been derived partly from trees swept from the rock by torrents into the sea, and partly from such peaty matter as often discolours and blackens the rills flowing through marshy grounds in our temperate climate, where the vegetation is probably less rank, and its decomposition less rapid than in the moist and hot climate of the era under consideration. There is only one instance yet on record of the remains of a saurian animal having been found in a member of the carboniferous series*. The larger oviparous reptiles usually inhabit rivers of considerable size in warm latitudes, and had crocodiles and other animals of that class been as abundant as in some secondary formations, we must have inferred the existence of many rivers, which could only have drained large tracts of land. Nor have the bones of any terrestrial mammalia rewarded our investigations. Had any of these, belonging to quadrupeds of large size, occurred, they would have supplied an argument against the resemblance of the ancient northern archipelagos to those of the modern Pacific, since in the latter no great indigenous quadrupeds have been met with. It is, indeed, a general character of small islands situated at a remote distance from continents, to be altogether destitute of land quadrupeds, except such as appear to have been conveyed to them by man. Kerguelen's land, which is of no inconsiderable size, placed in a latitude corresponding to that of the Scilly islands, may be cited as an example, as may all the groups of fertile islands in the Pacific ocean between the tropics, where no quadrupeds have been found, except the dog, the hog, and the rat, which have probably been brought to them by the natives, and also bats, which may have made their way along the chain of islands which extend from the shores of New Guinea far into the southern Pacific †. Even the isles of New Zealand, which may be compared to Ireland and Scotland

* Amongst other fossils collected from the mountain-limestone of Northumberland, the Rev. Charles V. Vernon has been fortunate enough

Unius sese dominum fecisse lacertæ,

having found a saurian vertebra together with patellæ and echinal spines, and an impression of a fern analogous to those of the coal-measures in the mountain limestone. In the same district, coal of a good quality and in great abundance occurs in the lower part of the limestone series. Annual Report of the Yorkshire Phil. Soc. for 1826, p. 14.

† Prichard's Physical History of Man, vol. i., p. 75.

in dimensions, appear to possess no indigenous quadrupeds, except the bat; and this is rendered the more striking, when we recollect that the northern extremity of New Zealand stretches to latitude 34°, where the warmth of the climate must greatly favour the prolific development of organic life. Lastly, no instance has yet been discovered of a pure lacustrine formation of the carboniferous era; although there are some instances of shells, apparently fresh-water, which may have been washed in by small streams, and do not by any means imply a considerable extent of dry land. All circumstances, therefore, point to one conclusion;—the subaqueous character of the igneous products—the continuity of the calcareous strata over vast spaces—the marine nature of their organic remains—the basin-shaped disposition of the mechanical rocks—the absence of large fluviatile and of land quadrupeds—the non-existence of pure lacustrine strata—the insular character of the flora,—all concur with wonderful harmony to establish the prevalence throughout the northern hemisphere of a great ocean, interspersed with small isles. If we seek for points of analogy to this state of things, we must either turn to the north Pacific, and its numerous submarine or insular volcanos between Kamtschatka and New Guinea, or, in order to obtain a more perfect counterpart to the coralline and shelly limestones, we may explore the archipelagos of the south Pacific, between Australia and South America, where volcanos are not wanting, and where coral reefs, consisting in great part of compact limestone, are spread over an area not inferior, perhaps, to that of our ancient calcareous rocks, though we suppose these to be prolonged from the lakes of North America to central Europe *.

No geologists have ever denied, that when our oldest conchiferous rocks were produced, great continents were wanting

* Captain King found a continued line of coral reef seven hundred miles in length, stretching from the N.E. coast of Australia towards New Guinea. It was interrupted only by a few intervals, not exceeding in all thirty miles. If we pass from these calcareous formations to the Friendly Isles and Society Isles, we find a succession of coral islands and submarine reefs; and Captain Beechey informs me, that in Ducie's Isle, W. long. 120°, he found the same formation in progress, and there he ascertained that the corals were growing at the depth *of one hundred and eighty feet.* He also observed that compact limestone constitutes a large portion of recent reefs.

in the temperate and arctic zones north of the equator; but they have even gone farther, and have been disposed to speculate on the universality of what they termed the primeval ocean. As well might a new Zealander, who had surveyed and measured the quantity of land between the south pole and the tropic of Capricorn, assume that the same proportion would be found to exist between the tropic of Cancer and the north pole. By this generalization, he would imagine twelve out of thirteen parts of the land of our temperate and arctic zones to be submerged. Such theorists should be reminded, that if the ocean was ever universal, its mean depth must have been inferior, and if so, the probability of deep water within the arctic circle is much lessened, and the likelihood of a preponderance of ice increased, and the heat of the ancient climate rendered more marvellous. To this objection, however, they will answer, that they do not profess to restrict themselves to existing analogies, and they may suppose the volume of water in the primeval ocean to have been greater. Besides, the high temperature, say they, was caused by heat which emanated from the interior of the new-born planet. In vain should we suggest to such reasoners, that when the ocean was in excess in high latitudes, the land in all probability predominated within the tropics, where, being exposed to the direct rays of the sun, it may have heated the winds and currents which flowed from lower to higher latitudes. In vain should we contend that a greater expanse of ocean, if general throughout the globe, would imply a comparative evenness of the superficial crust of the earth, and such an hypothesis would oblige us to conclude that the disturbances caused by subterranean movements in ancient times were inferior to those of later date. Will these arguments be met by the assumption, that earthquakes were feebler in the earlier ages, or wholly unknown,—as, according to Werner, there were no volcanos? Such a doctrine would be inconsistent with other popular prejudices respecting the extraordinary violence of the operations of nature in the olden time; and it is probable, therefore, that refuge will be taken in the old dogma of Lazzoro Moro, who imagined that the bed of the first ocean was as regular as its surface, and if so, it may be contended that sufficient time did not elapse between the creation of the world and the origin of the carboniferous strata, to allow the derangement necessary to produce great continents and Alpine chains.

But it would be idle to controvert, by reference to modern analogies, the conjectures of those who think they can ascend in their retrospect to the origin of our system. Let us, therefore, consider what changes the crust of the globe suffered after the consolidation of that ancient series of rocks to which we have adverted. Now, there is evidence that, before our secondary strata were formed, those of older date (from the old red sandstone to the coal inclusive) were fractured and contorted, and often thrown into vertical positions. We cannot enter here into the geological details by which it is demonstrable, that at an epoch extremely remote, some parts of the carboniferous series were lifted above the level of the sea, others sunk to greater depths beneath it, and the former, being no longer protected by a covering of water, were partially destroyed by torrents and the waves of the sea, and supplied matter for newer horizontal beds. These were arranged on the truncated edges of the submarine portions of the more ancient series, and the fragments included in the more modern conglomerates still retain their fossil shells and corals, so as to enable us to determine the parent rocks from whence they were derived*. By such remodelling of the surface the small islands of the first period increased in size, and new land was introduced into northern regions, consisting partly of primary and volcanic rocks and partly of the newly raised carboniferous strata. Among other proofs that earthquakes were then governed by the same laws which now regulate the subterranean forces, we find that they were restrained within limited areas, so that the site of Germany was not agitated, while that of some parts of England was convulsed. The older rocks, therefore, remained in some cases undisturbed at the bottom of the ancient ocean, and in this case the strata of

* Thus, for example, on the banks of the Avon, in the Bristol coal-field, the dolomitic conglomerate, a rock of an age intermediate between the carboniferous series and the lias, rests on the truncated edges of the coal and mountain limestone, and contains rolled and angular fragments of the latter, in which are seen the characteristic mountain-limestone fossils. For accurate sections illustrating the disturbances which rocks of the carboniferous series underwent before the newer red sandstone was formed, the reader should consult the admirable memoir on the south-western coal district of England, by Dr. Buckland and Mr. Conybeare, Geol. Trans., vol. i., second series.

the succeeding epoch were deposited upon them in conformable position. By reference to groups largely developed on the continent, but which are some of them entirely wanting, and others feebly represented in our own country, we find that the apparent interruption in the chain of events between the formation of our coal and the lias arises merely from local deficiency in the suite of geological monuments*. During the great interval which separated the formation of these groups, new species of animals and plants made their appearance, and in their turn became extinct; volcanos broke out, and were at length exhausted; rocks were destroyed in one region, and others accumulated elsewhere, while, in the mean time, the geographical condition of the northern hemisphere suffered material modifications. Yet the sea still extended over the greater part of the area now occupied by the lands which we inhabit, and was even of considerable depth in many localities where our highest mountain-chains now rise. The vegetation, during a part at least of this new period (from the lias to the chalk inclusive), appears to have approached to that of the larger islands of the equatorial zone†. These islands appear to have been drained by rivers of considerable size, which were inhabited by crocodiles and gigantic oviparous reptiles, both herbivorous and carnivorous, belonging for the most part to extinct genera. Of the contemporary inhabitants of the land we have as yet acquired but scanty information, but we know that there were flying reptiles, insects, and small insectivorous mammifera, allied to the opossum. In farther confirmation of the opinion that countries of considerable extent now rose above the sea in the temperate zone, we may mention the discovery of a large estuary formation in the south-west of England of higher antiquity than the chalk, containing terrestrial plants and fresh-

* In many parts of Germany, the newer red sandstone, and other rocks of about the same age, lie in conformable strata on the coal. In some districts, as in the Thuringerwald, among others, there is an immense series of formations intervening between the coal and the lias; one of these groups, called the *muschelkalkstein*, which seems to have no existence in England, is of great thickness and full of organic remains. See Professor Sedgwick's Memoir on the Geological relations and internal structure of the Magnesian Limestone, &c. Geol. Trans., second series, vol. iii., part 1, p. 121.

† Ad. Brongniart, Consid. Générales sur la Nat. de la Végét. &c. Ann. des Sci. Nat., Nov. 1828.

water testacea, tortoises, and large reptiles,—in a word, such an assemblage as the delta of the Ganges, or a large river in a hot climate might be expected to produce*.

In the present state of our knowledge, we cannot pretend to institute a close comparison between the climate which prevailed during the gradual deposition of our secondary formations and that of the older carboniferous rocks, for the general temperature of the surface must at both epochs have been so dissimilar to that now experienced in the same, or perhaps in any latitudes, that proofs from analogy lose much of their value, and a larger body of facts is required to support theoretical conclusions. If the signs of intense heat diminish, as some suppose, in the newer groups of this great series, there are nevertheless indications in the animal forms of the continued prevalence of a climate which we might consider as tropical in its character.

We may now turn our attention to the phenomena of the tertiary strata, which afford evidence of an abrupt transition from one description of climate to another. If this remarkable break in the regular sequence of physical events is merely apparent, arising from the present imperfect state of our knowledge, it nevertheless serves to set in a clearer point of view the intimate connexion between great changes in the physical geography of the earth, and revolutions in the mean temperature of the air and water. We have already shewn that when the climate was hottest, the northern hemisphere was for the most part occupied by the ocean, and it remains for us to point out, that the refrigeration did not become considerable, until a very large portion of that ocean was converted into land, nor even until it was in some parts replaced by high mountain chains. Nor did the cold reach its maximum until these chains attained their full height, and the lands their full extension. A glance at the best geological maps now constructed of

* We do not mean to compare the extent of the Wealden formation (from the Weald clay to the Purbeck limestone inclusive) to that of the Gangetic delta, for we shall afterwards see that the most modern addition made to the latter is equal in superficial area to North and South Wales. But, judging from the great continuity of some minor subdivisions of the Wealden group in our island, characterized as they are throughout their whole range by certain fresh-water remains, we may safely conclude that a considerable body of fresh water must have been permanently supplied by a large river.

various countries in the northern hemisphere, whether in North America or Europe, will satisfy the inquirer that the greater part of the present land has been raised from the deep, either between the period of the deposition of the chalk and that of the strata termed tertiary, or at subsequent periods, during which, various tertiary groups were formed in succession. For, as the secondary rocks from the lias to the chalk inclusive, are, with a few unimportant exceptions, marine, it follows that every district now occupied by them has been converted into land since they originated. We may prove, by reference to the relative altitudes of the secondary and tertiary groups, and several other circumstances, that a considerable part of the elevation of the older series was accomplished before the newer was formed. The Apennines, for example, as the Italian geologists hinted long before the time of Brocchi, and as that naturalist more clearly demonstrated, rose* several thousand feet above the level of the Mediterranean, before the deposition of the recent Subapennine beds which flank them on either side. What now constitutes the central calcareous chain of the Apennines, must for a long time have been a narrow ridgy peninsula, branching off at its northern extremity from the Alps near Savonna. A line of volcanos afterwards burst out in the sea, parallel to the axis of the older ridge. These igneous vents were extremely numerous, and the ruins of some of their cones and craters (as those in Tuscany, for example) indicate such a continued series of eruptions, almost all subsequent to the deposition of the Subapennine strata, that we cannot wonder at the vast changes in the relative level of land and sea which were produced. However minute the effect of each earthquake which preceded or intervened between such countless eruptions, the aggregate result of their elevating or depressing operation may well be expected to display itself in seas of great depth, and hills of considerable altitude. Accordingly, the more recent shelly beds, which often contain rounded pebbles derived from the waste of contiguous parts of the older Apennine rocks, have been raised from one to two thousand feet; but they never attain the loftier

* The greater number of Italian naturalists, and Brocchi among the rest, attributed the change of level to the lowering of the Mediterranean; rejecting the more correct theory of Moro and his followers, that the land had been upheaved.

eminences of the Apennines, nor penetrate far into the higher and more ancient valleys; for the whole peninsula was evidently subjected to the action of the same subterranean movements, and the older and newer groups of strata changed their level, in relation to the sea, but not to each other.

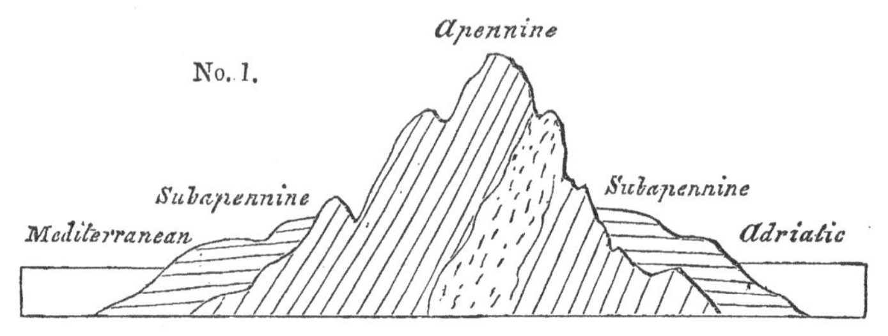

In the above diagram, exhibiting a transverse section of the Italian peninsula, the superior elevation of the more ancient group, and its unconformable stratification in relation to the more recent beds is expressed. The latter, however, are often much more disturbed at the point of contact than is here represented, and in some cases they have suffered such derangement as to dip towards, instead of from, the more ancient chain. There is usually, moreover, a valley at the junction of the Apennine and Subapennine strata, owing to the greater degradation which the newer and softer beds have undergone; but this intervening depression is not universal.

These phenomena are exhibited in the Alps on a much grander scale; those mountains being encircled by a great zone of tertiary rocks of different ages, both on their southern flank towards the plains of the Po, and on the side of Switzerland and Austria *, and at their eastern termination towards Styria and Hungary. This tertiary zone marks the position of former seas or gulfs, like the Adriatic, which were many thousand feet deep, and wherein strata accumulated, some single groups of which are not inferior in thickness to the whole of our secondary formations in England. These marine tertiary strata rise to the height of from two to four thousand feet and upwards, and consist of formations of different ages, characterized by dif-

* See a Memoir by Professor Sedgwick and Mr. Murchison, On the Tertiary Deposits of the Vale of Gosau, in the Salzburg Alps, Proceedings of Geol. Soc. No. 13, Nov. 1829.

ferent assemblages of organized fossils. The older tertiary groups generally rise to greater heights, and form interior zones nearest to the Alps. We may imagine some future convulsion once more to upraise this stupendous chain, together with the adjoining bed of the sea, so that the greatest mountains of Europe might rival the Andes in elevation, in which case the deltas of the Po, Adige, and Brenta now encroaching upon the Adriatic, might be uplifted so as to form another exterior belt, of considerable height, around the south-eastern flank of the Alps. Although we have not yet ascertained the number of different periods at which the Alps gained accessions to their height and width, yet we can affirm, that the last series of movements occurred when the seas were inhabited by *many existing species of animals* *.

There appears to be no sedimentary formations in the Alps so ancient as the rocks of our carboniferous series; while, on the other hand, secondary strata as modern as the green sand of English geologists, and perhaps the chalk, enter into some of the higher and central ridges. Down to the period, therefore, when the rocks, from our lias to the chalk inclusive, were deposited, there was sea where now the principal chain of Europe extends, and that chain attained more than half its present elevation and breadth between the eras when our newer secondary and oldest tertiary rocks originated. The remainder of its growth, if we may so speak, is of much more recent date, some of the latest changes, as we have stated, having been coeval with the existence of many animals belonging to species now contemporary with man. The Pyrenees, also, have acquired the whole of their present altitude, which in Mont Perdu exceeds eleven thousand feet, since the origin of some of

* Brocchi supposed the Subapennine beds to occur *abundantly* on both sides of the plains of the Po; but in this he was mistaken. The subalpine tertiary deposits are for the most part distinct and older formations. Professors Bonelli and Guidotti informed me, that they have recognized the Subapennine shells in one or two districts only north of the Po. They form in these cases, as might have been anticipated, the outermost belt, as at Azolo, at the foot of the Alps near the plains of Venice, and at Bassano, on the Brenta. In the section given by Mr. Murchison of the strata laid open by the Brenta, between Bassano and the Alps above Campese, it will be seen that the older chain must have partaken of the movement which raised the newest tertiary strata of the age of the Subapennines. Phil. Mag. and Annals, June, 1829.

the newer members of our secondary series. The granitic axis of that chain does not rise so high as a ridge formed by marine calcareous beds, the organic remains of which shew them to be the equivalents of our lower chalk, or a formation of about that age *. The tertiary strata at the base of this great chain are only slightly raised above the sea, and retain a horizontal position, without partaking of any of the disturbances to which the older series has been subjected, so that the great barrier between France and Spain was almost entirely upheaved in the interval between the deposition of the secondary and tertiary strata †. The Jura, also, owe the greatest part of their present elevation to subterranean convulsions which happened after the deposition of certain tertiary groups; at which time that portion which had been previously raised above the level of the sea underwent an entire alteration of form ‡. In other parts of the continent, as in France and England, where the newer rocks lie in basins surrounded by gently-rising hills, we find evidence that considerable spaces were redeemed from the original ocean and converted into dry land after the chalk was formed, and before the origin of the tertiary deposits. In these cases, the secondary strata were not raised into lofty mountain chains, like the Alps, Apennines, and Pyrenees, but the proofs are not less clear of their partial conversion into land anterior to the tertiary era. The chalk, for example, must have originated in the sea in the form of sediment from tranquil water; but before the tertiary rocks of the Paris and London basins were deposited, large portions of it had been so raised as to be exposed to the destroying power of the elements. The layers of flint had been washed out by torrents and rivers from their cretaceous matrix, rounded by attrition, and transported to the sea, where oysters attached themselves, and in some localities grew to a full size, until covered by other beds of flint-pebbles or sand. These newer derivative deposits are found abundantly along the borders, and in the inferior strata of our tertiary basins, and they are often interstratified with lignite. We may fairly infer, that the various trees

* This observation, first made by M. Boué, has been since confirmed by M. Dufrénoy.

† See a Memoir by M. Elie de Beaumont, Ann. des Sci. Nat., Nov. 1829, p. 286.

‡ M. Elie de Beaumont, ibid., Dec. 1829, p. 346.

and plants which enter into the composition of this lignite, grew on the surface of the same chalk which was then wasting away and affording to the torrents a constant supply of flint gravel.

We cannot dwell longer on the distinct periods when the secondary and various tertiary groups were upraised, without anticipating details which belong to other parts of this treatise; but we may observe, that although geologists have neglected to point out the relation of changes in the configuration of the earth's surface with fluctuations in general temperature, they do not dispute the fact, that the sea covered the regions where a great part of the land in Europe is now placed, until after the period when the newer groups of secondary rocks were formed. There is, therefore, confessedly a marked coincidence in point of time between the greatest alteration in climate and the principal revolution in the physical geography of the northern hemisphere. It is very probable that the abruptness of the transition from the organic remains of the secondary to those of the tertiary epoch, may not be wholly ascribable to the present deficiency of our information. We shall doubtless hereafter discover many intermediate gradations, (and one of these may be recognized in the calcareous beds of Maestricht,) by which a passage was effected from one state of things to another; but it is not impossible that the interval between the chalk and tertiary formations constituted an era in the earth's history, when the passage from one class of organic beings to another was, comparatively speaking, rapid. For if the doctrines explained by us in regard to vicissitudes of temperature are sound, it will follow that changes of equal magnitude in the geographical features of the globe, may at different periods produce very unequal effects on climate, and, so far as the existence of certain animals and plants depends on climate, the duration of species may often be shortened or protracted, according to the rate at which the change in temperature proceeded.

Let us suppose that the laws which regulate the subterranean forces are constant and uniform, (which we are entitled to assume, until some convincing proofs can be adduced to the contrary;) we may then infer, that a given amount of alteration in the superficial inequalities of the surface of the planet always requires for its consummation nearly equal periods of time. Let us then imagine the quantity of land between the equator and the tropic in one hemisphere to be to that in the other as thirteen to one, which, as we before stated, represents the unequal proportion of the extra-tropical lands in the two hemispheres at present. Then let the first geographical change consist in the shifting of this preponderance of land from one side of the line to the other, from the southern hemisphere, for example, to the northern. Now this would not affect the *general* temperature of the earth. But if, at another epoch, we suppose a continuance of the same agency to transfer an equal volume of land from the torrid zone to the temperate and arctic regions of the northern hemisphere, there might be so great a refrigeration of the mean temperature *in all latitudes*, that scarcely any of the pre-existing races of animals would survive, and, unless it pleased the Author of Nature that the planet should be uninhabited, new species would be substituted in the room of the extinct. We ought not, therefore, to infer, that equal periods of time are always attended by an equal amount of change in organic life, since a great fluctuation in the mean temperature of the earth, the most influential cause which can be conceived in exterminating whole races of animals and plants, must, in different epochs, require unequal portions of time for its completion.

The only geological monument yet discovered, which throws light on the period immediately succeeding the deposition of the chalk, is the series of calcareous beds in St. Peter's Mount at Maestricht. The turtles and gigantic reptiles there found, seem to indicate that the hot climate of the secondary era had not then been greatly modified; but as it seems that but a small proportion of the fossil species hitherto discovered are identical with known chalk fossils, there may perhaps have been a considerable lapse of ages between the consolidation of our upper chalk, and the completion of the Maestricht group *. During these ages, part of the gradual rise of the Alps and Pyrenees may have been accomplished; for we know that earth-

* It appears from a Memoir by Dr. Fitton, read before the Geological Society of London, Dec. 1829, that the Maestricht beds extend over a considerable area, preserving the same mineral characters and organic remains. Out of fifty species of shells and zoophytes collected by him, ten only could be identified with the copious list of chalk fossils published by Mr. Mantell, in the Geol. Trans., vol. iii. part 1., second series, p. 201.

quakes may work mighty changes during what we may call a small portion of one zoological era, since there are hills in Sicily which have gained more than three thousand feet in height, while the assemblage of testacea and zoophytes inhabiting the Mediterranean has only suffered slight alterations, and a large part of the countries bordering the Mediterranean have been remodelled since about one-third of the existing species were in being.

Before we conclude this chapter, we may be expected to offer some remarks on the gradual diminution of the supposed central heat of the globe, a doctrine which appears of late years to have increased in popularity. Baron Fourier, after making a curious series of experiments on the cooling of incandescent bodies, has endeavoured by profound mathematical calculations to prove that the actual distribution of heat in the earth's envelope is precisely that which would have taken place if the globe had been formed in a medium of a very high temperature, and had afterwards been constantly cooled *. He supposes that the matter of our planet, as Leibnitz formerly conjectured, was in an intensely heated state at the era of its creation, and that the incandescent fluid nucleus has been parting ever since with portions of its original heat, thereby contracting its dimensions,—a process which has not yet entirely ceased. But it is admitted, that there are no positive facts in support of this contraction; on the contrary, La Place has shewn, by reference to astronomical observations made in the time of Hipparchus, that in the last two thousand years there has been no sensible contraction of the globe by cooling down, for had this been the case, even to an extremely small amount, the day would have been shortened in an appreciable degree. The reader will bear in mind, that the question as to the existence of a central heat is very different from that of the gradual refrigeration of the interior of the earth. Many observations and experiments appear to countenance the idea, that in descending from the surface to those slight depths to which man can penetrate, there is a progressive increase of heat; but if this be established, and if, as some are not afraid to infer,

* See a Memoir on the Temperature of the Terrestrial Globe, and the Planetary Spaces, Ann. de Chimie et Phys., tom. xxvii. p. 136. Oct. 1824.

we dwell on a thin crust which covers a central ocean of liquid incandescent lava, we ought still to be very reluctant to concede on slight evidence that the internal heat is *variable in quantity*.

In our ignorance of the sources and nature of volcanic fire, it seems more consistent with philosophical caution, to assume that there is no instability in this part of the terrestrial system. We know that different regions have been subject in succession to a series of violent subterranean convulsions, and that fissures have opened from which hot vapours, thermal springs, and at some points red hot liquid lavas have issued to the surface. This evolution of heat often continues for ages after the extinction of volcanos and after the cessation of earthquakes, as in Central France, for example, and it seems perfectly natural, that each part of the earth's crust should, as M. Fourier states to be the fact, present the appearance of a heated body slowly cooling down. This may be owing chiefly to the shifting of the volcanic foci; but some effect may perhaps be due to that unequal absorption of the solar rays to which we have alluded, when speaking of the different temperature of the earth, according to the varying distribution of its superficial inequalities. M. Cordier announces as the result of his experiments and observations on the temperature of the interior of the earth, that the heat increases rapidly with the depth, but the increase does not follow the same law over the whole earth, being twice or three times as much in one country as in another, and these differences not being in constant relation either with the latitudes or longitudes of places. All this is precisely what we should have expected to arise from variations in the intensity of volcanic heat, and from that change of position, which the principal theatres of volcanic action have undergone at different periods, as the geologist can distinctly prove. But M. Cordier conjectures that there is a connexion between such phenomena and the secular refrigeration and contraction of the internal fluid mass, and that the changes of climate, of which there are geological proofs, favour this hypothesis *.

We cannot help suspecting that if it had appeared that *the same species of animals and plants* had continued to inhabit the

* See M. Cordier's Memoir on the Temperature of the Interior of the Earth, read to the Academy of Sciences, 4th June, 1827. Edin. New Phil. Journ., No. viii., p. 273.

seas, lakes, and continents, before and after the great physical mutations which the northern hemisphere has undergone since the secondary strata were formed, the difficulty of explaining the ancient climate of the globe would have appeared far more insurmountable than at present. It would have been so contrary to the elementary truths of meteorology to suppose no refrigeration to have followed from the rising of so many new mountain-chains in northern latitudes, that recourse would probably have been had in that case also to cosmological speculations. It might have been argued with much plausibility, that as the accession of high ridges covered with perpetual snow and glaciers had not occasioned any perceptible increase of cold, so as to affect the state of organic life, there must have been some new source of heat which counterbalanced that refrigerating cause. This, it might have been said, was the increased developement of *central fire* issuing from innumerable fissures opened in the crust of the earth, when it was shaken by convulsions which raised the Alps and other colossal chains.

But, without entering into farther discussion on the merits of the hypothesis of gradual refrigeration, let us hope that experiments will continue to be made, to ascertain whether there be internal heat in the globe, and what laws may govern its distribution. When its existence has been incontrovertibly established, it will be time to enquire whether it be subject to secular variations. Should these also be confirmed, we may begin to indulge speculations respecting the cause, but let us not hastily assume that it has reference to the original formation of the planet, with which it might be as unconnected as with its final dissolution. In the mean time we know that great changes in the external configuration of the earth's crust have at various times taken place, and we may affirm that they *must* have produced *some effect* on climate. The extent of their influence ought, therefore, to form a primary object of enquiry, more especially as there seems an obvious coincidence between the eras at which the principal accessions of land in high latitudes were made, and the successive periods when the diminution of temperature was most decided.

CHAPTER IX.

Theory of the progressive development of organic life considered—Evidence in its support wholly inconclusive—Vertebrated animals in the oldest strata—Differences between the organic remains of successive formations—Remarks on the comparatively modern origin of the human race—The popular doctrine of successive development not confirmed by the admission that man is of modern origin—In what manner the change in the system caused by the introduction of man affects the assumption of the uniformity of the past and future course of physical events.

We have considered, in the preceding chapters, many of the most popular grounds of opposition to the doctrine, that all former changes of the organic and inorganic creation are referrible to one uninterrupted succession of physical events, governed by the laws now in operation.

As the principles of the science must always remain unsettled so long as no fixed opinions are entertained on this fundamental question, we shall proceed to examine other objections which have been urged against the assumption of uniformity in the order of nature. We shall cite the words of a late distinguished writer, who has formally advanced some of the weightiest of these objections. "It is impossible," he affirms, "to defend the proposition, that the present order of things is the ancient and constant order of nature, only modified by existing laws—in those strata which are deepest, and which must, consequently, be supposed to be the earliest deposited forms, even of vegetable life, are rare; shells and vegetable remains are found in the next order; the bones of fishes and oviparous reptiles exist in the following class; the remains of birds, with those of the same genera mentioned before, in the next order; those of quadrupeds of extinct species in a still more recent class; and it is only in the loose and slightly-consolidated strata of gravel and sand, and which are usually called diluvian formations, that the remains of animals such as now people the globe are found, with others belonging to extinct species. But, in none of these formations, whether called secondary, tertiary, or diluvial, have the remains of man, or any of his works, been discovered; and whoever dwells upon this subject must be convinced, that the present order of things, and the comparatively recent existence of man as the master of the globe, is as certain as the destruction of a former and a different order, and the extinction of a number of living forms which have no types in being. In the oldest secondary strata there are no remains of such animals as now belong to the surface; and in the rocks, which may be regarded as more recently deposited, these remains occur but rarely, and with abundance of extinct species;—there seems, as it were, a gradual approach to the present system of things, and a succession of destructions and creations preparatory to the existence of man*."

In the above passages, the author deduces two important conclusions from geological data; first, that in the successive groups of strata, from the oldest to the most recent, there is a progressive development of organic life, from the simplest to the most complicated forms;—secondly, that man is of comparatively recent origin. It will be easy to shew that the first of these propositions, though very generally received, has no foundation in fact. The second, on the contrary, is indisputable, and it is important, therefore, to consider how far its admission is inconsistent with the assumption, that the system of the natural world has been uniform from the beginning, or rather from the era when the oldest rocks hitherto discovered were formed.

We shall first examine the geological proofs appealed to in support of the theory of the successive development of animal and vegetable life, and their progressive advancement to a more perfect state. No geologists, who are in possession of all the data now established respecting fossil remains, will for a moment contend for the doctrine in all its detail, as laid down by the great chemist to whose opinions we have referred. But naturalists, who are not unacquainted with recent discoveries, continue to defend the ancient doctrine in a somewhat modified form. They say that, in the first period of the world, (by which they mean the earliest of which we have yet procured

* Sir H. Davy, Consolations in Travel, Dialogue 3, "The Unknown."

any memorials,) the vegetation consisted almost entirely of cryptogamic plants, while the animals which co-existed were almost entirely confined to zoophytes, testacea, and a few fish. Plants of a less simple structure succeeded in the next epoch, when oviparous reptiles began also to abound. Lastly, the terrestrial flora became most diversified and most perfect when the highest orders of animals, the mammifera and birds, were called into existence.

Now, in the first place, we may observe, that many naturalists have been guilty of no small inconsistency in endeavouring to connect the phenomena of the earliest vegetation with a nascent condition of organic life, and at the same time to deduce, from the numerical predominance of certain types of form, the greater heat of the ancient climate. The arguments in favour of the latter conclusion are without any force, unless we can assume that the rules followed by the Author of Nature in the creation and distribution of organic beings were the same formerly as now; and that as certain families of animals and plants are now most abundant, or exclusively confined to regions where there is a certain temperature, a certain degree of humidity, intensity of light, and other conditions, so also the same phenomena were exhibited at every former era. If this postulate be denied, and the prevalence of particular families be declared to depend on a certain order of precedence in the introduction of different classes into the earth, and if it be maintained that the standard of organization was raised successively, we must then ascribe the numerical preponderance in the earlier ages of plants of simpler structure, *not to the heat*, but to those different laws which regulate organic life in newly created worlds. If, according to the laws of progressive development, cryptogamic plants always flourish for ages before the dicotyledonous order can be established, then is the small proportion of the latter fully explained; for in this case, whatever may have been the mildness or severity of the climate, they could not make their appearance. Before we can infer an elevated temperature in high latitudes, from the presence of arborescent Ferns, Lycopodiaceæ, and other allied families, we must be permitted to assume, that at all times, past and future, a heated and moist atmosphere pervading the northern hemisphere has a tendency to produce in the vegetation a predominance of analogous types of form. We grant, indeed, that there may be a connexion between an extraordinary profusion of monocotyledonous plants, and a youthful condition of the world, if the dogma of certain cosmogonists be true, that planets, like certain projectiles, are always red hot when they are first cast; but to this arbitrary hypothesis we need not again revert.

Between two and three hundred species of plants are now enumerated as belonging to the carboniferous era, and, with very few exceptions, not one of them are dicotyledonous*. But these exceptions are as fatal to the doctrine of successive development as if there were a thousand, although they do not by any means invalidate the conclusion in regard to the heat of the ancient climate, for that depends on the numerical relations of the different classes.

The animal remains in the most ancient series of European sedimentary rocks (from the graywacke to the coal inclusive), consist chiefly of corals and testacea. Some estimate may generally be formed of the comparative extent of our information concerning the fossil remains of a particular era, by reference to the number of species of shells obtained from a particular group of strata. Some of the rarest species cannot be discovered, unless the more abundant kinds have been found again and again; and if the variety brought to light be very considerable, it proves not only great diligence of research, but a good state of preservation of the organic contents of that formation. In the older rocks, many causes of destruction have operated, of which the influence has been rendered considerable by the immense lapse of ages during which they have acted. Mechanical pressure, derangement by subterranean movements, the action of chemical affinity, the percolation of acidulous waters and other agencies, have obliterated, in a greater or less degree, all traces of organization in fossil bodies. Sometimes only obscure or unintelligible impressions are left,

* Fragments of dicotyledonous wood which have evidently belonged to at least two different species of trees, have been obtained from the coal-field of Fife, by Dr. Fleming, of Flisk, and the same gentleman has shewn me a large dicotyledonous stem which he procured from the graywacke of Cork. See a memoir by Dr. Fleming on the neighbourhood of Cork. (Trans. of Wern. Soc. Edin.) I am informed also by Dr. Buckland, that he has received from the coal-field of Northumberland another specimen of dicotyledonous wood, which is now in the Oxford Museum.

and the lapidifying process has often effaced not only the characters by which the species, but even those whereby the class might be determined. The number of organic forms which have disappeared from the oldest strata, may be conjectured from the fact, that their former existence is in many cases merely revealed to us by the unequal weathering of an exposed face of rock, by which certain parts are made to stand out in relief. As the number of species of shells found in the English series, from the graywacke to the coal inclusive, after attentive examination, amounts only to between one and two hundred species, we cannot be surprised that so few examples of vertebrated animals have as yet occurred. The remains of fish, however, appear in one of the lowest members of the group *, which entirely destroys the theory of the precedence of the simplest forms of animals. The vertebra also of a saurian, as we before stated, has been met with in the mountain limestone of Northumberland †, so that the only negative fact remaining in support of the doctrine of the imperfect development of the higher orders of animals in remote ages, is the absence of birds and mammalia. The former are generally wanting in deposits of all ages, even where the highest order of animals occurs in abundance. Land mammifera could not, as we have before suggested, be looked for in strata formed in an ocean interspersed with isles, such as we must suppose to have existed in the northern hemisphere, when the carboniferous rocks were formed.

* Numerous scales of fish have been found by Dr. Fleming in quarries of the old red sandstone at Clashbinnie in Perthshire, where I have myself collected them. These beds are decidedly older than the coal and mountain limestone of Fifeshire.

† I do not insist on the abundant occurrence of the scales of a tortoise nearly allied to Trionyx, in the bituminous schists of Caithness, and in the same formation in the Orkneys in Scotland, as another example of a fossil reptile in rocks as old as the carboniferous series; because the geological position of those schists is not yet determined with precision. Professor Sedgwick and Mr. Murchison indeed infer, that they alternate with a sandstone of the age of the old red sandstone; but this opinion wants confirmation. The numerous fish, and the tortoise of Caithness, are certainly in strata older than the lias, for that rock rests upon them unconformably; but as the strata between the schists and the granite contain no organic remains, and as no fossils of the carboniferous era have yet been found in the Caithness beds, the relative date of the tortoise cannot be determined with confidence. It might possibly be of the age of our magnesian limestone. See Geol. Trans. second series, vol. iii., part 1, p. 144, and for a representation of the scales of the Trionyx, plate 16 of the same part.

As all are agreed that the ancient strata in question were subaqueous, and for the most part submarine, from what data we may ask do naturalists infer the non-existence or even the rarity of warm-blooded quadrupeds in the earlier ages? Have they dredged the bottom of the ocean throughout an area co-extensive with that now occupied by the carboniferous rocks, and have they found that with the number of between one and two hundred species of shells they always obtain the remains of at least one land quadruped? Suppose our mariners were to report that on sounding in the Indian ocean near some coral reefs, and at some distance from the land, they drew up on hooks attached to their line portions of a leopard, elephant, or tapir; should we not be sceptical as to the accuracy of their statements; and if we had no doubt of their veracity, might we not suspect them to be unskilful naturalists? or, if the fact were unquestioned, should we not be disposed to believe that some vessel had been wrecked on the spot? The casualties must be rare indeed whereby land quadrupeds are swept by rivers and torrents into the sea, and still rarer must be the contingency of such a floating body not being devoured by sharks or other predaceous fish, such as were those of which we find the teeth preserved in some of the carboniferous strata *. But if the carcase should escape and should happen to sink where sediment was in the act of accumulating, and if the numerous causes of subsequent disintegration should not efface all traces of the body included for countless ages in solid rock, is it not contrary to all calculation of chances that we should hit upon the exact spot,—that mere point in the bed of the ancient ocean, where the precious relic was entombed? Can we expect for a moment that when we have only succeeded amidst several thousand fragments of corals and shells, in finding a few bones of *aquatic* or *amphibious* animals, that we should meet with a single skeleton of an inhabitant of the land?

Clarence, in his dream, saw "in the slimy bottom of the deep,"

> —— a thousand fearful wrecks;
> A thousand men, that fishes gnaw'd upon;
> Wedges of gold, great anchors, heaps of pearl.

Had he also beheld amid "the dead bones that lay scatter'd

* I have seen in the collection of Dr. Fleming, the teeth of carnivorous fish from the mountain limestone of Fife, which alternates with the coal.

by," the carcasses of lions, deer, and the other wild tenants of the forest and the plain, the fiction would have been deemed unworthy of the genius of Shakspeare. So daring a disregard of probability, so avowed a violation of analogy, would have been condemned as unpardonable even where the poet was painting those incongruous images which present themselves to a disturbed imagination during the visions of the night. But the cosmogonist is not amenable, even in his waking hours, to these laws of criticism; for he assumes either that the order of nature was formerly distinct, or that the globe was in a condition to which it can never again be reduced by changes which the existing law of nature can bring about. This assumption being once admitted, inexplicable anomalies and violations of analogy, instead of offending his judgment, give greater consistency to his reveries.

The organic contents of the secondary strata in general consist of corals and marine shells. Of the latter, the British strata (from the inferior oolite to the chalk inclusive) have yielded about six hundred species. Vertebrated animals are very abundant, but they are almost entirely confined to fish and reptiles. But some remains of cetacea have also been met with in the oolitic series of England *, and the bones of two species of warm-blooded quadrupeds of extinct genera allied to the Opossum †. The occurrence of one individual of the higher classes of mammalia, whether marine or terrestrial, in these ancient strata, is as fatal to the theory of successive development, as if several hundreds had been discovered.

The tertiary strata, as will appear from what we have already stated, were deposited when the physical geography of the northern hemisphere had been entirely altered. Large inland lakes

* On the authority of Dr. Buckland. Trans. Geol. Soc. vol. i. part 2, second series, p. 394.

† The mammiferous remains of the Stonesfield slate, near Oxford, consist of three or perhaps four jaws, one of which, now in the Oxford Museum, has been examined by M. Cuvier, and pronounced to belong to a species of Didelphis. Another of these valuable fossils in the possession of my friend Mr. Broderip, appears to be not only specifically, but generically distinct, from that shewn to M. Cuvier. See Observations on the Jaw of a fossil Mammiferous Animal found in the Stonesfield Slate, by W. J. Broderip, Esq., Sec. G.S., F.R.S., F.L.S., &c., Zool. Journ., vol. iii., p. 408; 1827.

had become numerous as in central France and many other countries. There were gulfs of the sea into which large rivers emptied themselves, where strata were formed like those of the Paris basin. There were then also littoral formations in progress, such as are indicated by the English *Crag*, and the *Faluns* of the Loire. The state of preservation of the organic remains of this period is very different from that of fossils in the older rocks, the colours of the shells, and even the cartilaginous ligaments uniting the valves being in some cases retained. No less than twelve hundred species of testacea have been found in the beds of the Paris basin, and an equal number in the more modern formations of the Subapennine hills; and it is a most curious fact in natural history, that the zoologist has already acquired more extensive information concerning the testacea which inhabited the ancient seas of northern latitudes at that era, than of those now living in the same parallels in Europe. The strata of the Paris basin are partly of fresh-water origin, and filled with the spoils of the land. They have afforded a great number of skeletons of land quadrupeds, but these relics are confined almost entirely to one small member of the group, and their conservation may be considered as having arisen from some local and accidental combination of circumstances. On the other hand, the scarcity of terrestrial mammalia in submarine sediment is elucidated, in a striking manner, by the extremely small number of such remains hitherto procured from the Subapennine hills. The facilities of investigation in these strata, which undergo rapid disintegration, are perhaps unexampled in the rest of Europe, and they have been examined by collectors for three hundred years. But, although they have already yielded twelve hundred species of testacea, the authenticated examples of associated remains of terrestrial mammalia are extremely scanty; and several of those which have been cited by earlier writers as belonging to the elephant or rhinoceros, have since been declared, by able anatomists, to be the bones of whales and other cetacea. In about five or ten instances, perhaps, bones of the mastodon, rhinoceros, and some other animals, have been observed in this formation with marine shells attached. These must have been washed into the bed of the ancient sea when the strata were forming, and they serve to attest the contiguity of land inhabited by large herbivora, which renders the rarity of such exceptions more worthy of attention. On the contrary, the number of skeletons of existing animals in the upper Val d'Arno, which are usually considered to be referrible to the same age as the Subapennine beds, occur in a deposit which was formed entirely in an inland lake, surrounded by lofty mountains.

The inferior member of our oldest tertiary formations in England, usually termed the plastic clay, has hitherto proved as destitute of mammiferous remains, as our ancient coal strata; and this point of resemblance between these deposits is the more worthy of observation, because the lignite, in the one case, and the coal in the other, are exclusively composed of terrestrial plants. From the London clay we have procured three or four hundred species of testacea, but the only bones of vertebrated animals are those of reptiles and fish. On comparing, therefore, the contents of these strata with those of our oolitic series, we find the supposed order of precedence inverted. In the more ancient system of rocks, mammalia, both of the land and sea, have been recognized, whereas in the newer, if negative evidence is to be our criterion, nature has made a retrograde, instead of a progressive, movement, and no animals more exalted in the scale of organization than reptiles are discoverable.

Not a single bone of a quadrumanous animal has ever yet been discovered in a fossil state, and their absence has appeared, to some geologists, to countenance the idea that the type of organization most nearly resembling the human came last in the order of creation, and was scarcely perhaps anterior to that of man. But the evidence on this point is quite inconclusive, for we know nothing, as yet, of the details of the various classes of the animal kingdom which inhabited the land up to the consolidation of the newest of the secondary strata; and when a large part of the tertiary formations were in progress, the climate does not appear to have been of such a tropical character as seems necessary for the development of the tribe of apes, monkeys, and allied genera. Besides, it must not be forgotten, that almost all the animals which occur in subaqueous deposits are such as frequent marshes, rivers, or the borders of lakes, as the rhinoceros, tapir, hippopotamus, ox, deer, pig, and others. On the other hand, species which live in trees are extremely rare in a fossil state, and we have no data as yet for determining how great a number of the one kind we ought to find, before we have a right to expect a single individual of the other. If, therefore, we are led to infer, from the presence of crocodiles and turtles in the London clay, and from the cocoa-nuts and spices found in the isle of Sheppey, that at the period when our older tertiary strata were formed, the climate was hot enough for the quadrumanous tribe, we nevertheless could not hope to discover any of their skeletons until we had made considerable progress in ascertaining what were the contemporary Pachydermata; and not one of these, as we have already remarked, has been discovered as yet in any strata of this epoch in England*.

It is, therefore, clear, that there is no foundation in geological facts, for the popular theory of the successive development of the animal and vegetable world, from the simplest to the most perfect forms; and we shall now proceed to consider another question, whether the recent origin of man lends any support to the same doctrine, or how far the influence of man may be considered as such a deviation from the analogy of the order of things previously established, as to weaken our confidence in the uniformity of the course of nature. We need not dwell on the proofs of the low antiquity of our species, for it is not controverted by any geologist; indeed, the real difficulty which we experience consists in tracing back the signs of man's existence on the earth to that comparatively modern period when species, now his contemporaries, began to predominate. If there be a difference of opinion respecting the occurrence in certain deposits of the remains of man and his works, it is always in reference to strata confessedly of the most modern order; and

* The only exception of which I have heard is the tooth of an Anoplotherium, mentioned by Dr. Buckland as having been found in the collection of Mr. Allan, labelled "Binstead, Isle of Wight." The quarries of Binstead are entirely in the lower fresh-water formation, and such is undoubtedly the geological position in which we might look for the bones of such an animal. My friend Mr. Allan has shewn me this tooth, to which, unfortunately, none of the matrix is attached, so that it is still open to a captious sceptic to suspect that a Parisian fossil was so ticketed by mistake.

it is never pretended that our race co-existed with assemblages of animals and plants, of which *all the species* are extinct. From the concurrent testimony of history and tradition, we learn that parts of Europe, now the most fertile and most completely subjected to the dominion of man, were, within less than three thousand years, covered with forests, and the abode of wild beasts. The archives of nature are in perfect accordance with historical records; and when we lay open the most superficial covering of peat, we sometimes find therein the canoes of the savage, together with huge antlers of the wild stag, or horns of the wild bull. Of caves now open to the day in various parts of Europe, the bones of large beasts of prey occur in abundance; and they indicate, that at periods extremely modern in the history of the globe, the ascendancy of man, if he existed at all, had scarcely been felt by the brutes*.

No inhabitant of the land exposes himself to so many dangers on the waters as man, whether in a savage or a civilized state, and there is no animal, therefore, whose skeleton is so liable to become imbedded in lacustrine or submarine deposits; nor can it be said, that his remains are more perishable than those of other animals, for in ancient fields of battle, as Cuvier has observed, the bones of men have suffered as little decomposition as those of horses which were buried in the same grave. But even if the more solid parts of our species had disappeared, the impression would have remained engraven on the rocks as have the traces of the tenderest leaves of plants, and the integuments of many animals. Works of art, moreover, composed of the most indestructible materials, would have outlasted almost all the organic contents of sedimentary rocks; edifices, and even entire cities have, within the times of history, been buried under volcanic ejections, or submerged beneath the sea, or engulphed by earthquakes; and had these catastrophes been repeated throughout an indefinite lapse of ages, the high antiquity of man would have been inscribed in far more legible characters on the frame-work of the globe, than are the forms of the ancient vegetation which once covered the isles of the

* We shall discuss in a subsequent chapter, when treating of animal remains in caves, the probable antiquity assignable to certain human bones and works of art found intermixed with remains of extinct animals in the cavern of Bize, and in several localities in the department of Herault, in France.

northern ocean, or of those gigantic reptiles, which at later periods peopled the seas and rivers of the northern hemisphere.

Assuming, then, that man is, comparatively speaking, of modern origin, can his introduction be considered as one step in a progressive system by which, as some suppose, the organic world advanced slowly from a more simple to a more perfect state? To this question we may reply, that the superiority of man depends not on those faculties and attributes which he shares in common with the inferior animals, but on his reason by which he is distinguished from them.

If the organization of man were such as would confer a decided pre-eminence upon him, even if he were deprived of his reasoning powers, and provided only with such instincts as are possessed by the lower animals, he might then be supposed to be a link in a progressive chain, especially if it could be shewn that the successive development of the animal creation had always proceeded from the more simple to the more compound, from species most remote from the human type to those most nearly approaching to it. But this is an hypothesis which, as we have seen, is wholly unsupported by geological evidence. On the other hand, we may admit, that man is of higher dignity than were any pre-existing beings on the earth, and yet question whether his coming was a step in the gradual advancement of the organic world: for the most highly civilized people may sometimes degenerate in strength and stature, and become inferior in their physical attributes to the stock of rude hunters from which they descended. If then the physical organization of man may remain stationary, or even become deteriorated, while the race makes the greatest progress to higher rank and power in the scale of rational being, the animal creation also may be supposed to have made no progress by the addition to it of the human species, regarded merely as a part of the organic world. But, if this reasoning appear too metaphysical, let us waive the argument altogether, and grant that the animal nature of man, even considered apart from the intellectual, is of higher dignity than that of any other species; still the introduction at a certain period of our race upon the earth, raises no presumption whatever that each former exertion of creative power was characterized by the successive development of *irrational* animals of higher orders. The comparison here instituted is between things so dissimilar, that when we attempt to draw such inferences, we strain analogy beyond all reasonable bounds. We may easily conceive that there was a considerable departure from the succession of phenomena previously exhibited in the organic world, when so new and extraordinary a circumstance arose, as the union, for the first time, of moral and intellectual faculties capable of indefinite improvement, with the animal nature. But we have no right to expect that there were any similar deviations from analogy —any corresponding steps in a progressive scheme, at former periods, when no similar circumstances occurred.

But another, and a far more difficult question may arise out of the admission that man is comparatively of modern origin. Is not the interference of the human species, it may be asked, such a deviation from the antecedent course of physical events, that the knowledge of such a fact tends to destroy all our confidence in the uniformity of the order of nature, both in regard to time past and future? If such an innovation could take place after the earth had been exclusively inhabited for thousands of ages by inferior animals, why should not other changes as extraordinary and unprecedented happen from time to time? If one new cause was permitted to supervene, differing in kind and energy from any before in operation, why may not others have come into action at different epochs? Or what security have we that they may not arise hereafter? If such be the case, how can the experience of one period, even though we are acquainted with all the possible effects of the then existing causes, be a standard to which we can refer all natural phenomena of other periods?

Now these objections would be unanswerable, if adduced against one, who was contending for the absolute uniformity throughout all time of the succession of sublunary events—if, for example, he was disposed to indulge in the philosophical reveries of some Egyptian and Greek sects, who represented all the changes both of the moral and material world as repeated at distant intervals, so as to follow each other in their former connexion of place and time. For they compared the course of events on our globe to astronomical cycles, and not only did they consider all sublunary affairs to be under the influence of the celestial bodies, but they taught that on the earth, as well as in the heavens, the same identical phenomena recurred again and again in a perpetual vicissitude. The same individual men were doomed to be re-born, and to perform the same actions as before; the same arts were to be invented, and the same cities built and destroyed. The Argonautic expedition was destined to sail again with the same heroes, and Achilles with his Myrmidons, to renew the combat before the walls of Troy.

> Alter erit tum Tiphys et altera quæ vehat Argo
> Dilectos heroas; erunt etiam altera bella,
> Atque iterum ad Trojam magnus mittetur Achilles *.

The geologist, however, may condemn these tenets as absurd, without running into the opposite extreme, and denying that the order of nature has, from the earliest periods, been uniform in the same sense in which we believe it to be uniform at present. We have no reason to suppose, that when man first became master of a small part of the globe, a greater change took place in its physical condition than is now experienced when districts, never before inhabited, become successively occupied by new settlers. When a powerful European colony lands on the shores of Australia, and introduces at once those arts which it has required many centuries to mature; when it imports a multitude of plants and large animals from the opposite extremity of the earth, and begins rapidly to extirpate many of the indigenous species, a mightier revolution is effected in a brief period, than the first entrance of a savage horde, or their continued occupation of the country for many centuries, can possibly be imagined to have produced. If there be no impropriety in assuming that the system is uniform when disturbances so unprecedented occur in certain localities, we can with much greater confidence apply the same language to those primeval ages when the aggregate number and power of the human race, or the rate of their advancement in civilization, must be supposed to have been far inferior.

If the barren soil around Sidney had at once become fertile upon the landing of our first settlers; if, like the happy isles

* Virgil, Eclog. 4. For an account of these doctrines, see Dugald Stewart's Elements of the Philosophy of the Human Mind, vol.. ii. chap. 2, sect. 4, and Prichard's Egypt. Mythol., p. 177.

whereof the poets have given us such glowing descriptions, those sandy tracts had begun to yield spontaneously an annual supply of grain, we might then, indeed, have fancied alterations still more remarkable in the economy of nature to have attended the first coming of our species into the planet. Or if, when a volcanic island like Ischia was, for the first time brought under cultivation by the enterprise and industry of a Greek colony, the internal fire had become dormant, and the earthquake had remitted its destructive violence, there would then have been some ground for speculating on the debilitation of the subterranean forces, when the earth was first placed under the dominion of man. But after a long interval of rest, the volcano bursts forth again with renewed energy, annihilates one-half of the inhabitants, and compels the remainder to emigrate. Such exiles, like the modern natives of Cumana, Calabria, Sumbawa, and other districts, habitually convulsed by earthquakes, would probably form no very exalted estimate of the sagacity of those geological theorists, who, contrasting the *human* with antecedent epochs, have characterized it as *the period of repose.*

In reasoning on the state of the globe immediately before our species was called into existence, we may assume that all the present causes were in operation, with the exception of man, until some geological arguments can be adduced to the contrary. We must be guided by the same rules of induction as when we speculate on the state of America in the interval that elapsed between the period of the introduction of man into Asia, the cradle of our race, and that of the arrival of the first adventurers on the shores of the New World. In that interval, we imagine the state of things to have gone on according to the order now observed in regions unoccupied by man. Even now, the waters of lakes, seas, and the great ocean, which teem with life, may be said to have no immediate relation to the human race—to be portions of the terrestrial system of which man has never taken, nor ever can take, possession, so that the greater part of the inhabited surface of the planet remains still as insensible to our presence, as before any isle or continent was appointed to be our residence.

The variations in the external configuration of the earth, and the successive changes in the races of animals and plants inhabiting the land and sea, which the geologist beholds when he restores in imagination the scenes presented by certain regions at former periods, are not more full of wonderful or inexplicable phenomena, than are those which a traveller would witness who traversed the globe from pole to pole. Or if there be more to astonish and perplex us in searching the records of the past, it is because one district may, in an indefinite lapse of ages, become the theatre of a greater number of extraordinary events, than the whole face of the globe can exhibit at one time. However great the multiplicity of new appearances, and however unexpected the aspect of things in different parts of the present surface, the observer would never imagine that he was transported from one system of things to another, because there would always be too many points of resemblance, and too much connexion between the characteristic features of each country visited in succession, to permit any doubt to arise as to the continuity and identity of the whole plan.

"In our globe," says Paley, "new countries are continually "discovered, but the old laws of nature are always found "in them: new plants perhaps, or animals, but always in "company with plants and animals which we already know, "and always possessing many of the same general properties. "We never get amongst such original, or totally different modes "of existence, as to indicate that we are come into the province "of a different Creator, or under the direction of a different "will. In truth, the same order of things attends us wherever "we go*." But the geologist is in danger of drawing a contrary inference, because he has the power of passing rapidly from the events of one period to those of another—of beholding, at one glance, the effects of causes which may have happened at intervals of time incalculably remote, and during which, nevertheless, no local circumstances may have occurred to mark that there is a great chasm in the chronological series of nature's archives. In the vast interval of time which may really have elapsed between the results of operations thus compared, the physical condition of the earth may, by slow and insensible modifications, have become entirely altered, one or more races

* Natural Theology, Chap. xxv.

of organic beings may have passed away, and yet have left behind, in the particular region under contemplation, no trace of their existence. To a mind unconscious of these intermediate links in the chain of events, the passage from one state of things to another must appear so violent, that the idea of revolutions in the system inevitably suggests itself. The imagination is as much perplexed by such errors as to time, as it would be if we could annihilate space, and by some power, such as we read of in tales of enchantment, could transfer a person who had laid himself down to sleep in a snowy arctic wilderness, to a valley in a tropical region, where on awaking he would find himself surrounded by birds of brilliant plumage, and all the luxuriance of animal and vegetable forms of which nature is there so prodigal. The most reasonable supposition, perhaps, which a philosopher could make, if by the necromancer's art he was placed in such a situation, would be, that he was dreaming; and if a geologist forms theories under a similar delusion, we should not expect him to preserve more consistency in his speculations, than in the train of ideas in an ordinary dream.

But if, instead of inverting the natural order of inquiry, we cautiously proceed in our investigations, from the known to the unknown, and begin by studying the most modern periods of the earth's history, attempting afterwards to decipher the monuments of more ancient changes, we can never so far lose sight of analogy, as to suspect that we have arrived at a new system, governed by different physical laws. In more recent formations, consisting often of strata of great thickness, the shells of the present seas and lakes, and the remains of animals and plants now living on the land, are imbedded in great numbers. In those of more ancient date, many of the same species are found associated with others now extinct. These unknown kinds again are observed in strata of still higher antiquity, connected with a great number of others which have also no living representatives, till at length we arrive at periods of which the monuments contain exclusively the remains of species with many genera foreign to the present creation. But even in the oldest rocks which contain organic remains, some genera of marine animals are recognized, of which species still exist in our seas, and these are repeated at different intervals in all the intermediate groups of strata, attesting that, amidst the great variety of revolutions of which the earth's surface has been the theatre, there has never been a departure from the conditions necessary for the existence of certain unaltered types of organization. The uniformity of animal instinct, observes Mr. Stewart *, pre-supposes a corresponding regularity in the physical laws of the universe, "insomuch that if the established order of the material world were to be essentially disturbed, (the instincts of the brutes remaining the same,) all their various tribes would inevitably perish." Now, any naturalist will be convinced, on slight reflection, of the justice of this remark. He will also admit that the same species have always retained the same instincts, and therefore that all the strata wherein any of *their* remains occur, must have been formed when the phenomena of inanimate matter were the same as they are in the actual condition of the earth. The same conclusion must also be extended to the extinct animals with which the remains of these living species are associated; and by these means we are enabled to establish the permanence of the existing physical laws, throughout the whole period when the tertiary deposits were formed. We have already stated that, during that vast period, a large proportion of all the lands in the northern hemisphere were raised above the level of the sea.

The modifications in the system of which man is the instrument, do not, in all probability, constitute so great a deviation from analogy as we usually imagine; we often, for example, form an exaggerated estimate of the extent of the power displayed by man in extirpating some of the inferior animals, and causing others to multiply; a power which is circumscribed within certain limits, and which, in all likelihood, is by no means exclusively exerted by our species. The growth of human population cannot take place without diminishing the numbers, or causing the entire destruction of many animals. The larger carnivorous species give way before us, but other quadrupeds of smaller size, and innumerable birds, insects, and plants, which are inimical to our interests, increase in spite of us, some attacking our food, others our raiment and persons, and others interfering with our agricultural and horticultural

* Phil. of the Human Mind, vol. ii., p. 230.

labours. We force the ox and the horse to labour for our advantage, and we deprive the bee of his store; but, on the other hand, we raise the rich harvest with the sweat of our brow, and behold it devoured by myriads of insects, and we are often as incapable of arresting their depredations as of staying the shock of an earthquake, or the course of a stream of burning lava. The changes caused by other species, as they gradually diffuse themselves over the globe, are inferior probably in magnitude, but are yet extremely analogous to those which we occasion. The lion, for example, and the migratory locust, must necessarily, when they first made their way into districts now occupied by them, have committed immense havoc amongst the animals and plants which became their prey. They may have caused many species to diminish, perhaps wholly to disappear; but they must also have enabled some others greatly to augment in number, by removing the natural enemies by which they had been previously kept down. It is probable from these, and many other considerations, that as we enlarge our knowledge of the system, we shall become more and more convinced, that the alterations caused by the interference of man deviate far less from the analogy of those effected by other animals than we usually suppose. We are often misled, when we institute such comparisons, by our knowledge of the wide distinction between the instincts of animals and the reasoning power of man; and we are apt hastily to infer, that the effects of a rational and an irrational species, considered merely *as physical agents*, will differ almost as much as the faculties by which their actions are directed. A great philosopher has observed, that we can only command nature by obeying her laws, and this principle is true even in regard to the astonishing changes which are superinduced in the qualities of certain animals and plants by domestication and garden culture. We can only effect such surprising alterations by assisting the development of certain instincts, or by availing ourselves of that mysterious law of their organization, by which individual peculiarities are transmissible from one generation to another.

We are not, however, contending that a real departure from the antecedent course of physical events cannot be traced in the introduction of man. If that latitude of action which enables the brutes to accommodate themselves in some measure to accidental circumstances, could be imagined to have been at any former period so great, that the operations of instinct were as much diversified as are those of human reason, it might perhaps be contended, that the agency of man did not constitute an anomalous deviation from the previously established order of things. It might then have been said, that the earth's becoming at a particular period the residence of human beings, was an era in the moral, not in the physical world—that our study and contemplation of the earth, and the laws which govern its animate productions, ought no more to be considered in the light of a disturbance or deviation from the system, than the discovery of the satellites of Jupiter should be regarded as a physical event in the history of those heavenly bodies, however influential they may have become from that time in advancing the progress of sound philosophy among men, and in augmenting human resources by aiding navigation and commerce. The distinctness, however, of the human, from all other species, considered merely as an efficient cause in the physical world, is real, for we stand in a relation to contemporary species of animals and plants, widely different from that which other irrational animals can ever be supposed to have held to each other. We modify their instincts, relative numbers, and geographical distribution, in a manner superior in degree, and in some respects very different in kind from that in which any other species can affect the rest. Besides, the progressive movement of each successive generation of men causes the human species to differ more from itself in power at two distant periods, than any one species of the higher order of animals differs from another. The establishment, therefore, by geological evidence of the first intervention of such a peculiar and unprecedented agency, long after other parts of the animate and inanimate world existed, affords ground for concluding that the experience during thousands of ages of all the events which may happen on this globe would not enable a philosopher to speculate with confidence concerning future contingencies. If an intelligent being, therefore, after observing the order of events for an indefinite series of ages had witnessed at last so wonderful an innovation as this, to what extent would his belief in the regularity of the system be weakened?—would he cease to assume that there was permanency in the laws of nature?—would he no longer be guided in his speculations by the strictest rules of induction? To this question we may reply, that had he previously presumed to dogmatize respecting the absolute uniformity of the order of nature, he would undoubtedly be checked by witnessing this new and unexpected event, and would form a more just estimate of the limited range of his own knowledge, and the unbounded extent of the scheme of the universe. But he would soon perceive that no one of the fixed and constant laws of the animate or inanimate world was subverted by human agency, and that the modifications produced were on the occurrence of new and extraordinary circumstances, and those not of a *physical*, but a *moral* nature. The deviation permitted, would also appear to be as slight as was consistent with the accomplishment of the new *moral* ends proposed, and to be in a great degree temporary in its nature, so that whenever the power of the new agent was withheld, even for a brief period, a relapse would take place to the ancient state of things; the domesticated animal, for example, recovering in a few generations its wild instinct, and the garden-flower and fruit-tree reverting to the likeness of the parent stock.

Now, if it would be reasonable to draw such inferences with respect to the future, we cannot but apply the same rules of induction to the past. It will scarcely be disputed that we have no right to anticipate any modifications in the results of existing causes in time to come, which are not conformable to analogy, unless they be produced by the progressive development of human power, or perhaps from some other new relations between the moral and material worlds. In the same manner we must concede, that when we speculate on the vicissitudes of the animate and inanimate creation in former ages, we have no ground for expecting any anomalous results, unless where man has interfered, or unless clear indications appear of some other *moral* source of temporary derangement. When we are unable to explain the monuments of past changes, it is always more probable that the difficulty arises from our ignorance of all the existing agents, or all their possible effects in an indefinite lapse of time, than that some cause was formerly in operation which has ceased to act; and if in any part of the globe the energy of a cause appears to have decreased, it is always probable, that the diminution of intensity in its action is merely local, and that its force is unimpaired, when the whole globe is considered. But should we ever establish by unequivocal proofs, that certain agents have, at particular periods of past time, been more potent instruments of change over the entire surface of the earth than they now are, it will be more consistent with philosophical caution to presume, that after an interval of quiescence they will recover their pristine vigour, than to regard them as worn out.

The geologist who yields implicit assent to the truth of these principles, will deem it incumbent on him to examine with minute attention all the changes now in progress on the earth, and will regard every fact collected respecting the causes in diurnal action, as affording him a key to the interpretation of some mystery in the archives of remote ages. Our estimate, indeed, of the value of all geological evidence, and the interest derived from the investigation of the earth's history, must depend entirely on the degree of confidence which we feel in regard to the permanency of the laws of nature. Their immutable constancy alone can enable us to reason from analogy, by the strict rules of induction, respecting the events of former ages, or, by a comparison of the state of things at two distinct geological epochs, to arrive at the knowledge of general principles in the economy of our terrestrial system.

The uniformity of the plan being once assumed, events which have occurred at the most distant periods in the animate and inanimate world will be acknowledged to throw light on each other, and the deficiency of our information respecting some of the most obscure parts of the present creation will be removed. For as by studying the external configuration of the existing land and its inhabitants, we may restore in imagination the appearance of the ancient continents which have passed away, so may we obtain from the deposits of ancient seas and lakes an insight into the nature of the subaqueous processes now in operation, and of many forms of organic life, which, though now existing, are veiled from our sight. Rocks, also produced by subterranean fire in former ages at great depths in the bowels of the earth, present us, when upraised by gradual movements, and exposed to the light of heaven, with an image of those changes which the deep-seated volcano may now occasion in the nether regions. Thus, although we are mere sojourners on the surface of the planet, chained to a mere point in space, enduring but for a moment of time, the human mind is not only enabled to number worlds beyond the unassisted ken of mortal eye, but to trace the events of indefinite ages before the creation of our race, and is not even withheld from penetrating into the dark secrets of the ocean, or the interior of the solid globe; free, like the spirit which the poet described as animating the universe,

——— ire per omnes
Terrasque tractusque maris, cœlumque profundum.

CHAPTER X.

Division of the subject into changes of the organic and inorganic world—Inorganic causes of change divided into the aqueous and igneous—Aqueous causes—Destroying and transporting power of running water—Sinuosities of rivers—Two streams when united do not occupy a bed of double surface—Heavy matter removed by torrents and floods—Recent inundations in Scotland—Effects of ice in removing stones—Erosion of chasms through hard rocks—Excavations in the lavas of Etna by Sicilian rivers—Gorge of the Simeto—Gradual recession of the cataracts of Niagara—Speculations as to the time required for their reaching Lake Erie.

We defined geology to be the science which investigates the former changes that have taken place in the organic, as well as in the inorganic kingdoms of nature; and we now proceed to inquire what changes are now in progress in both these departments. Vicissitudes in the inorganic world are most apparent, and as on them all fluctuations in the animate creation must in a great measure depend, they may claim our first consideration. We may divide the great agents of change in the inorganic world into two principal classes, the aqueous and the igneous. To the former belong Rivers, Torrents, Springs, Currents, and Tides; to the latter, Volcanos and Earthquakes. Both these classes are instruments of decay as well as of reproduction; but they may be also regarded as antagonist forces. The *aqueous* agents are incessantly labouring to reduce the inequalities of the earth's surface to a level, while the *igneous*, on the other hand, are equally active in restoring the unevenness of the external crust, partly by heaping up new matter in certain localities, and partly by depressing one portion, and forcing out another of the earth's envelope. It is difficult, in a scientific arrangement, to give an accurate view of the combined effects of so many forces in simultaneous operation; because, when we consider them separately, we cannot easily estimate either the extent of their efficacy, or the kind of results which they produce. We are in danger, therefore, when we attempt to examine the influence exerted singly by each, of overlooking the modifications which they produce on one another; and these are so complicated, that sometimes the igneous and aqueous forces co-operate to produce a joint effect, to which neither of them unaided by the other could give rise,—as when repeated earthquakes unite with running water to widen a valley. Sometimes the organic combine with the inorganic causes; as when a reef, composed of shells and corals, protects one line of coast from the destroying power of tides or currents, and turns them against some other point; or when drift timber floated into a lake, fills a hollow to which the stream would not have had sufficient velocity to convey earthy sediment.

It is necessary, however, to divide our observations on these various causes, and to classify them systematically, endeavouring as much as possible to keep in view that the effects in nature are mixed, and not simple, as they may appear in an artificial arrangement.

In treating, first, of the aqueous causes, we may consider them under two divisions: first, those which are connected with the circulation of water from the land to the sea, under which are included all the phenomena of rivers and springs; secondly, those which arise from the movements of water in lakes, seas, and the ocean, wherein are comprised the phenomena of tides and currents. In turning our first attention to the former division, we find that the effects of rivers may be subdivided into those of a destroying and those of a renovating nature. In the former are included the erosion of rocks and the transportation of matter to lower levels; in the latter, the formation of sand-bars and deltas, the shallowing of seas, &c.

Action of Running Water.—We shall begin, then, by describing the destroying and transporting power of running water, as exhibited by torrents and rivers. It is well known that the lands elevated above the sea attract in proportion to their volume and density a larger quantity of that aqueous vapour which the heated atmosphere continually absorbs from the surface of lakes and seas. By this means, the higher regions become perpetual reservoirs of water, which descend and irrigate the lower valleys and plains. In consequence of this provision, almost all the water is first carried to the highest regions, and is then made to descend by steep declivities towards the sea; so that it acquires superior velocity, and removes a greater quantity of soil than it would do if the rain had been distributed over the low plains and high mountains equally in proportion to their relative areas. Almost all the water is also made by these means to pass over the greatest distances which each region affords, before it can regain the sea. The rocks in the higher regions are particularly exposed to atmospheric influences, to frost, rain, and vapour, and to great annual alternations, of moisture and desiccation—of cold and heat. Among the most powerful agents of decay may be mentioned the mechanical action of water, which possesses the remarkable property of expanding during congelation. When water has penetrated into crevices and cavities, it rends open, on freezing, the most solid rocks with the force of a lever, and for this reason, although in cold climates the comparative quantity of rain which falls is very inferior, and although it descends more gradually than in tropical regions, yet the severity of frost, and the greater inequalities of temperature, compensate for this diminished power of degradation, and cause it to proceed with equal, if not greater rapidity than in high latitudes. The solvent power of water also is very great, and acts particularly on the calcareous and alkaline elements of stone, especially when it holds carbonic acid in solution, which is abundantly supplied to almost every large river by springs, and is collected by rain from the atmosphere. The oxygen of the atmosphere is also gradually absorbed by all animal and vegetable productions, and by almost all mineral masses exposed to the open air. It gradually destroys the equilibrium of the elements of rocks, and tends to reduce into powder, and to render fit for soils, even the hardest aggregates belonging to our globe *. And as it is well known that almost every thing affected by rapid combustion may also be affected gradually by the slow absorption of oxygen, the surface of the hardest rocks exposed to the air may be said to be slowly burning away.

When earthy matter has once been intermixed with running

* Sir H. Davy, Consolations in Travel, p. 271.

water, a new mechanical power is obtained by the attrition of sand and pebbles, borne along with violence by a stream. Running water charged with foreign ingredients being thrown against a rock, excavates it by mechanical force, sapping and undermining till the superincumbent portion is at length precipitated into the stream. The obstruction causes a temporary increase of the water, which then sweeps down the barrier. By a repetition of these land-slips, the ravine is widened into a small, narrow valley, in which sinuosities are caused by the deflexion of the stream first to one side and then to the other. The unequal hardness of the materials through which the channel is eroded, tends also to give new directions to the lateral force of excavation. When by these, or by accidental shiftings of the alluvial matter in the channel, and numerous other causes, the current is made to cross its general line of descent, it eats out a curve in the opposite bank, or the side of the hill bounding the valley, from which curve it is turned back again at an equal angle, and recrossing the line of descent, it gradually hollows out another curve lower down, in the opposite bank, till the whole sides of the valley, or river-bed, present a succession of salient and retiring angles.

Among the causes of deviation from a straight course by which torrents and rivers tend to widen the valleys through which they flow, may be mentioned the confluence of lateral torrents, swoln irregularly at different seasons in mountainous regions by partial storms, and discharging at different times unequal quantities of debris into the main channel.

When the tortuous flexures of a river are extremely great, the aberration from the direct line of descent is often restored by the river cutting through the isthmus which separates two neighbouring curves. Thus, in the annexed diagram, the

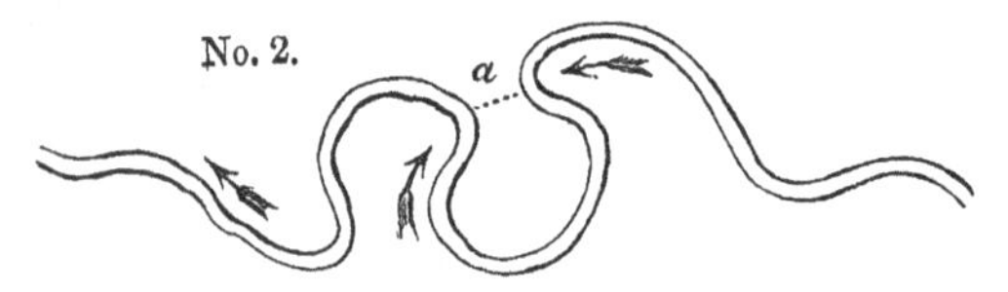

extreme sinuosity of the river has caused it to return for a brief space in a contrary direction to its main course, so that a peninsula is formed, and the isthmus (at *a*) is consumed on both sides by currents flowing in opposite directions. In this case an island is soon formed,—on either side of which a portion of the stream usually remains *. These windings occur not only in the channels of rivers flowing through flat alluvial plains, but large valleys also are excavated to a great depth through solid rocks in this serpentine form. In the valley of the Moselle, between Berncastle and Roarn, which is sunk to a depth of from six to eight hundred feet through an elevated platform of transition rocks, the curves are so considerable that the river returns, after a course of seventeen miles in one instance, and nearly as much in two others, to within a distance of a few hundred yards of the spot it passed before †. The valley of the Meuse, near Givet, and many others in different countries, offer similar windings. Mr. Scrope has remarked, that these tortuous flexures are decisively opposed to the hypothesis, that any violent and transient rush of water suddenly swept out such valleys; for great floods would produce straight channels in the direction of the current, not sinuous excavations, wherein rivers flow back again in an opposite direction to their general line of descent.

Our present purpose, however, relates to the force of aqueous erosion, and the transportation of materials by running water, considered separately, and not to the question so much controverted respecting the formation of valleys in general. This subject cannot be fully discussed without referring to all the powers to which the inequalities of the earth's surface, and the very existence of land above the level of the sea, are due. Nor even when we have described the influence of all the chemical and mechanical agents which operate at one period in effecting changes in the external form of the land, shall we be enabled to present the reader with a comprehensive theory of the origin of the present valleys. It will be necessary to consider the complicated effects of all these causes at distinct geological epochs, and to inquire how particular regions, after having remained for ages in a state of comparative tranquillity, and under the influence of a certain state of the atmosphere, may be subsequently remodelled by another series of subterranean movements,—how

* See a paper on the Excavation of Valleys, &c., by G. Poulett Scrope, Esq. Proceedings of Geol. Soc., No. 14, 1830.

† Ibid.

the new direction, volume, and velocity acquired by rivers and torrents may modify the former surface,—what effects an important difference in the mean temperature of the climate, or the greater intensity of heat and cold at different seasons, may produce,—what pre-existing valleys, under a new configuration of the land, may cease to give passage to large bodies of water, or may become entirely dried up,—how far the relative level of certain districts in the more modern period may become precisely the reverse of those which prevailed at the more ancient era. When these and other essential elements of the problem are all duly appreciated, the reader will not be surprised to learn, that amongst geologists who have neglected them there has prevailed a great contrariety of opinion on these topics. Some writers of distinguished talent have gone so far as to contend, that the origin of the greater number of existing valleys was simply due to the agency of one cause, and that it was consummated in a brief period of time. But without discussing the merits of the general question, we may observe that we agree with the author before cited, that the sinuosity of deep valleys is one among many proofs that they have been shaped out progressively, and not by the simultaneous action of one or many causes; and when we consider other agents of change, we shall have opportunities of pointing out a multitude of striking facts in confirmation of the gradual nature of the process to which the inequalities of hill and valley owe their origin.

In regard to the transporting power of water, we are often surprised at the facility wherewith streams of a small size, and which descend a slight declivity, bear along coarse sand and gravel; for we usually estimate the weight of rocks in air, and do not reflect sufficiently on their comparative buoyancy when submerged in a denser fluid. The specific gravity of many rocks is not more than twice that of water, and very rarely more than thrice, so that almost all the fragments propelled by a stream have lost a third, and many of them half of what we usually term their weight.

It has been proved by experiment, in contradiction to the theories of the early writers on hydrostatics, to be a universal law, regulating the motion of running water, that the velocity at the bottom of the stream is everywhere less than in any part above it, and is greatest at the surface. Also that the superficial particles in the middle of the stream move swifter than those at the sides. This retardation of the lowest and lateral currents is produced by friction, and when the velocity is sufficiently great, the soil composing the sides and bottom gives way. A velocity of three inches per second is ascertained to be sufficient to tear up fine clay,—six inches per second, fine sand,—twelve inches per second, fine gravel,—and three feet per second, stones of the size of an egg *.

When this mechanical power of running water is considered, we are prepared for the transportation of large quantities of gravel, sand, and mud, by the torrents and rivers which descend with great velocity from the mountainous regions. But a question naturally arises, how the more tranquil rivers of the valleys and plains, flowing on comparatively level ground, can remove the prodigious burden which is discharged into them by their numerous tributaries, and by what means they are enabled to convey the whole mass to the sea. If they had not this power, their channels would be annually choked up, and the lower valleys and districts adjoining mountain-chains would be continually strewed over with fragments of rock and sterile sand. But this evil is prevented by a general law regulating the conduct of running water, that two equal streams do not occupy a bed of double surface. In proportion, therefore, as the whole fluid mass increases, the space which it occupies decreases relatively to the volume of water; and hence there is a smaller proportion of the whole retarded by friction against the bottom and sides of the channel. The portion thus unimpeded moves with great velocity, so that the main current is often accelerated in the lower country, notwithstanding that the slope of the channel is lessened. It not unfrequently happens, as we shall afterwards demonstrate by examples, that two large rivers, after their junction, have only the surface which one of them had previously; and even in some cases their united waters are confined in a narrower bed than each of them filled before. By this beautiful adjustment, the water which drains the interior country is made continually to occupy less room as it approaches the sea; and thus the most valuable part of our continents, the rich deltas,

* Encycl. Brit.—Art. Rivers.

and great alluvial plains, are prevented from being constantly under water *.

Many remarkable illustrations of the power of running water in moving stones and heavy materials were afforded by the late storm and flood which occurred on the 3rd and 4th of August 1829, in Aberdeenshire and other counties, in Scotland. The floods extended almost simultaneously and in equal violence over a space of about five thousand square miles, being that part of the north-east of Scotland which would be cut off by two lines drawn from the head of Lochrannoch, one towards Inverness, and another to Stonehaven. All the rivers within that space were flooded, and the destruction of roads, lands, buildings, and crops along the courses of the streams was very great. The elements during this storm assumed all the characters which mark the tropical hurricanes: the wind blowing in sudden gusts and whirlwinds, the lightning and thunder being such as is rarely witnessed in that climate, and heavy rain falling without intermission. The bridge over the Dee at Ballatu consisted of five arches, having upon the whole a water-way of two hundred and sixty feet. The bed of the river on which the piers rested, was composed of rolled pieces of granite and gneiss. The bridge was built of granite, and had stood uninjured for twenty years, but the different parts were swept away in succession by the flood, and the whole mass of masonry disappeared in the bed of the river †. "The river Don," observes Mr. Farquharson, in his account of the inundations, "has upon my own premises forced a mass of four or five hundred tons of stones, many of them two or three hundred pounds weight, up an inclined plane, rising six feet in eight or ten yards; and left them in a rectangular heap, about three feet deep on a flat ground; and, singularly enough, the heap ends abruptly at its lower extremity. A large stone, of three or four tons which I have known for many years in a deep pool of the river, has been moved about one hundred yards from its place ‡."

The power even of a small rivulet, when swoln by rain, in removing heavy bodies, was lately exemplified in the College, a

* See Article Rivers, Ency. Brit.

† From the account given by the Rev. James Farquharson, in the Quarterly Journ. of Sci., &c., No. 12, new series, p. 328.

‡ Ibid., p. 331.

small stream which flows at a moderate declivity from the eastern water-shed of the Cheviot-Hills. Several thousand tons weight of gravel and sand were transported to the plain of the Till, and a bridge then in progress of building was carried away, some of the arch-stones of which, weighing from half to three-quarters of a ton each, were propelled two miles down the rivulet. On the same occasion, the current tore away from the abutment of a mill-dam a large block of greenstone-porphyry, weighing nearly two tons, and transported the same to the distance of a quarter of a mile. Instances are related as occurring repeatedly, in which from one to three thousand tons of gravel are in like manner removed to great distances in one day *.

In the cases above adverted to, the waters of the river and torrent were dammed back by the bridges which acted as partial barriers, and illustrate the irresistible force of a current when obstructed. Bridges are also liable to be destroyed by the tendency of rivers to shift their course, whereby the pier, or the rock on which the foundation stands, is undermined.

When we consider how insignificant are the volume and velocity of the rivers and streams in our island, when compared to those of the Alps and other lofty chains, and how, during the various changes which the levels of different districts have undergone, the various contingencies which give rise to floods, must in the lapse of ages be multiplied, we may easily conceive that the quantity of loose superficial matter distributed over Europe must be very considerable. That the position also of a great portion of these travelled materials should now appear most irregular, and should often bear no relation to the existing water-drainage of the country, is a necessary consequence, as we shall afterwards see, of the combined operations of running water and subterranean movements.

In mountainous regions and high northern latitudes, the moving of heavy stones by water is greatly assisted by the ice which adheres to them, and which forming together with the rock a mass of less specific gravity †, is readily borne along. The glaciers also of alpine regions, formed of consolidated snow,

* See a paper by Mr. Culley, F.G.S., Proceedings of Geol. Soc., No. 12, 1829.

† Silliman's Journal, No. 30, p. 303.

bear down upon their surface a prodigious burden of rock and sand mixed with ice. These materials are generally arranged in long ridges, which sometimes in the Alps are thirty or forty feet high, running parallel to the borders of the glacier, like so many lines of intrenchment. These mounds of debris are sometimes three or more deep, and have generally been brought in by lateral glaciers: the whole accumulation is slowly conveyed to the lower valleys, where, on the melting of the glacier, it is swept away by rivers *.

The rapidity with which even the smallest streams hollow out deep channels in soft and destructible soils is remarkably exemplified in volcanic countries, where the sand and half-consolidated tuffs oppose but a slight resistance to the torrents which descend the mountain side. After the heavy rains which followed the eruption of Vesuvius is 1822, the water flowing from the Atrio del Cavallo, cut in three days a new chasm through strata of tuff and volcanic ejected matter to the depth of twenty-five feet. The old mule road was seen, in 1828, intersected by this new ravine. But such facts are trifling when compared to the great gorges which are excavated in somewhat similar materials in the great plateau of Mexico, where an ancient system of valleys, originally worn out of granite and secondary rocks, has been subsequently filled with strata of tuff, pumice, lava, and trachytic conglomerate, to the thickness of several thousand feet. The rivers and torrents annually swoln by tropical rains, are now actively employed in removing these more recent deposits, and in re-excavating the ancient water-courses †.

The gradual erosion of deep chasms through some of the hardest rocks, by the constant passage of running water charged with foreign matter, is another phenomenon of which striking examples may be adduced. Some of the clearest illustrations of this excavating power are presented by many valleys in Central France, where the channels of rivers have been barred up by solid currents of lava, through which the streams have re-excavated a passage to the depth of from twenty to seventy feet and upwards, and often of great width. In these cases

* Saussure, Voyage dans les Alpes, tom. i.

† I am indebted to Captain Vetch for this information, whose researches in Mexico will, it is hoped, be soon communicated to the scientific world.

there are decisive proofs that neither the sea nor any denuding wave, or extraordinary body of water, have passed over the spot since the melted lava was consolidated. Every hypothesis of the intervention of sudden and violent agency is entirely excluded, because the cones of *loose* scoriæ, out of which the lavas flowed, are oftentimes at no great elevation above the rivers, and have remained undisturbed during the whole period which has been sufficient for the hollowing out of such enormous ravines. But we shall reserve a more detailed account of the volcanic district of Central France for another part of this work, and at present confine ourselves to examples derived from events which have happened since the time of history.

Some lavas of Etna, produced by eruptions of which the date is known, have flowed across two of the principal rivers in Sicily; and in both cases the streams, dispossessed of their ancient beds, have opened for themselves new channels. An eruption from Mount Mojo, an insulated cone at the northern base of Etna, sent forth, in the year 396, B.C., in the reign of Dionysius I., a great lava-stream, which crossed the river Caltabianca in two places. The lowermost point of obstruction is seen on the eastern side of Etna, on the high road from Giardini to Catania, where one pier of the bridge on either bank is based upon a remnant of the solid lava, which has been breached by the river to the depth of fourteen feet. But the Caltabianca, although it has been at work for more than two and twenty centuries, has not worn through the igneous rock so as to lay open the gravel of its ancient bed. The declivity, however, of the alluvial plain is very slight; and as the extent of excavation in a given time depends on the volume and velocity of the stream, and the destructibility of the rock, we must carefully ascertain all these circumstances before we attempt to deduce from such examples a measure of the force of running water in a given period *.

* I omitted to visit the higher point near the village of Mojo, where the Caltabianca has cut through the lava. Some future traveller would probably derive much instruction from inspecting that spot, which is laid down in Gemmellaro's Quadro Istorico, &c. dell' Etna, 1824.

Recent Excavation of the Simeto.—The power of running water to hollow out compact rock is exhibited, on a larger scale, at the western base of Etna, where a great current of lava (A A, diagram 3), descending from near the summit

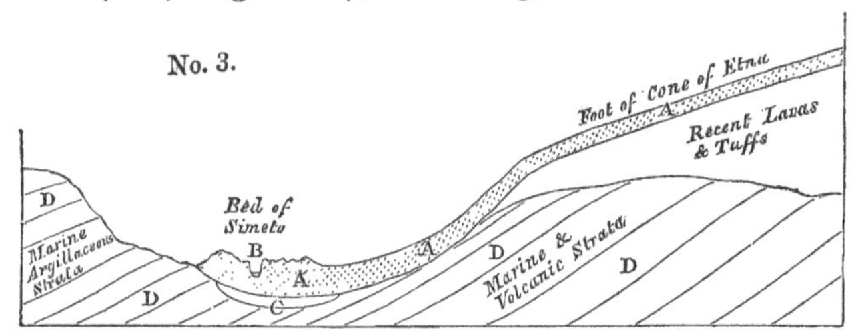

of the great volcano, has flowed to the distance of five or six miles, and then reached the alluvial plain of the Simeto, the largest of the Sicilian rivers which skirts the base of Etna, and falls into the sea a few miles south of Catania. The lava entered the river about three miles above the town of Adernò, and not only occupied its channel for some distance, but, crossing to the opposite side of the valley, accumulated there in a rocky mass. Gemmellaro gives the year 1603 as the date of the eruption *. The appearance of the current clearly proves that it is one of the most modern of those of Etna, for it has not been covered or crossed by subsequent streams or ejections, and the olives on its surface are all of small size, yet older than the natural wood on the same lava. In the course, therefore, of about two centuries the Simeto has eroded a passage from fifty to several hundred feet wide, and in some parts from forty to fifty feet deep.

The portion of lava cut through is in no part porous or scoriaceous, but consists of a compact homogeneous mass of hard blue rock, somewhat lighter than ordinary basalt, containing crystals of olivine and glassy felspar. The general declivity of this part of the bed of the Simeto is not considerable, but, in consequence of the unequal waste of the lava, two waterfalls occur at Passo Manzanelli, each about six feet in height. Here the chasm (B, diagram No. 3.) is about forty feet deep, and only fifty broad.

* Quadro Istorico dell' Etna, 1824. Some doubts are entertained as to the exact date of this current by others, but all agree that it is not one of the older streams even of the historical era.

The sand and pebbles in the river bed consist chiefly of a brown quartzose sandstone, derived from the upper country; but the matter derived from the volcanic rock itself must have greatly assisted the attrition. This river, like the Caltabianca, has not yet cut down to the ancient bed of which it was dispossessed, and of which we have indicated the probable position in the annexed diagram (c, No. 3.)

On entering the narrow ravine where the water foams down the two cataracts, we are entirely shut out from all view of the surrounding country; and a geologist who is accustomed to associate the characteristic features of the landscape with the relative age of certain rocks, can scarcely dissuade himself from the belief that he is contemplating a scene in some rocky gorge of a primary district. The external forms of the hard blue lava are as massive as any of the most ancient trap-rocks of Scotland. The solid surface is in some parts smoothed and almost polished by attrition, and covered in others with a white lichen, which imparts to it an air of extreme antiquity, so as greatly to heighten the delusion. But the moment we reascend the cliff, the spell is broken; for we scarcely recede a few paces, before the ravine and river disappear, and we stand on the black and rugged surface of a vast current of lava, which seems unbroken, and which we can trace up nearly to the distant summit of that majestic cone which Pindar called "the pillar of heaven," and which still continues to send forth a fleecy wreath of vapour, reminding us that its fires are not extinct, and that it may again give out a rocky stream, wherein other scenes like that now described may present themselves to future observers.

Falls of Niagara.—The falls of Niagara afford a magnificent example of the progressive excavation of a deep valley in solid rock. That river flows from Lake Erie to Lake Ontario, the former lake being three hundred and thirty feet above the latter, and the distance between them being thirty-two miles. On flowing out of the upper lake, the river is almost on a level with its banks; so that, if it should rise perpendicularly eight or ten feet, it would lay under water the adjacent flat country of Upper Canada on the West, and of the State of New York on the

East*. The river, where it issues, is about three quarters of a mile in width. Before reaching the falls, it is propelled with great rapidity, being a mile broad, about twenty-five feet deep, and having a descent of fifty feet in half a mile. An island at the very verge of the cataract divides it into two sheets of water; one of these, called the Horse-shoe Fall, is six hundred yards wide, and one hundred and fifty-eight feet perpendicular; the other, called the American Falls, is about two hundred yards in width, and one hundred and sixty-four feet in height. The breadth of the island is about five hundred yards. This great sheet of water is precipitated over a ledge of hard limestone, in horizontal strata, below which is a somewhat greater thickness of soft shale, which decays and crumbles away more rapidly, so that the calcareous rock forms an overhanging mass, projecting forty feet or more above the hollow space below. The blasts of wind, charged with spray, which rise out of the pool into which this enormous cascade is projected, strike against the shale beds, so that their disintegration is constant; and the superincumbent limestone, being left without a foundation, falls from time to time in rocky masses. When these enormous fragments descend, a shock is felt at some distance, accompanied by a noise like a distant clap of thunder. After the river has passed over the falls, its character, observes Captain Hall, is immediately and completely changed. It then runs furiously along the bottom of a deep wall-sided valley, or huge trench, which has been cut into the horizontal strata by the continued action of the stream during the lapse of ages. The cliffs on both sides are in most places perpendicular, and the ravine is only perceived on approaching the edge of the precipice†.

The waters which expand at the falls, where they are divided by the island, are contracted again, after their union, into a stream not more than one hundred and sixty yards broad. In the narrow channel, immediately below this immense rush of water, a boat can pass across the stream with ease. The pool, it is said, into which the cataract is precipitated, being one hundred and seventy feet deep, the descending water sinks down

* Captain Hall's Travels in North America, vol. i., p. 179.

† Ibid., pp. 195, 196, 216.

and forms an under-current, while a superficial eddy carries the upper stratum back *towards* the main fall *. This is not improbable; and we must also suppose, that the confluence of two streams, which meet at a considerable angle, tends mutually to neutralize their forces. The bed of the river below the falls is strewed over with huge fragments which have been hurled down into the abyss. By the continued destruction of the rocks, the falls have, within the last forty years, receded nearly fifty yards, or, in other words, the ravine has been prolonged to that extent. Through this deep chasm the Niagara flows for about seven miles; and then the table-land, which is almost on a level with Lake Erie, suddenly sinks down at a town called Queenstown, and the river emerges from the ravine into a plain which continues to the shores of Lake Ontario †.

There seems good foundation for the general opinion, that the falls were once at Queenstown, and that they have gradually retrograded from that place to their present position, about seven miles distant. If the ratio of recession had never exceeded fifty yards in forty years, it must have required nearly ten thousand years for the excavation of the whole ravine; but no probable conjecture can be offered as to the quantity of time consumed in such an operation, because the retrograde movement may have been much more rapid when the whole current was confined within a space not exceeding a fourth or fifth of that which the falls now occupy. Should the erosive action not be accelerated in future, it will require upwards of thirty thousand years for the falls to reach Lake Erie (twenty-five miles distant), to which they seem destined to arrive in the course of time, unless some earthquake changes the relative levels of the district. The table-land, extending from Lake Erie, consists uniformly of the same geological formations as are now exposed to view at the falls. The upper stratum is an ancient alluvial sand, varying in thickness from ten to one hundred and forty feet; below which is a bed

* See Mr. Bakewell, jun., on Falls of Niagara. Loudon's Magazine, No. 12, March, 1830.

† The Memoir of Mr. Bakewell, jun. above referred to, contains two very illustrative sketches of the physical geography of the country between Lakes Erie and Ontario, including the Falls.

of hard limestone, about ninety feet in thickness, stretching nearly in a horizontal direction over the whole country, and forming the bed of the river *above* the falls, as do the inferior shales *below*. The lower shale is nearly of the same thickness as the limestone. Should Lake Erie remain in its present state until the period when the ravine recedes to its shores, the sudden escape of that great body of water would cause a tremendous deluge; for the ravine would be much more than sufficient to drain the whole lake, of which the average depth was found, during the late survey, to be only ten or twelve fathoms. But, in consequence of its shallowness, Lake Erie is fast filling up with sediment, and the annual growth of the deltas of many rivers and torrents which flow into it is remarkable. Long Point, for example, near the influx of Big Creek River, was observed, during the late survey, to advance three miles in as many years. A question therefore arises, whether Lake Erie may not be converted into dry land before the Falls of Niagara recede so far. In speculating on this contingency, we must not omit one important condition of the problem. As the surface of the lake is contracted in size, the loss of water by evaporation will diminish; and unless the supply shall decrease in the same ratio (which seems scarcely probable), Niagara must augment continually in volume, and by this means its retrograde movement may hereafter be much accelerated.

CHAPTER XI.

Action of running water, *continued*—Course of the Po—Desertion of its old channel —Articial embankments of the Po, Adige, and other Italian rivers—Basin of the Mississippi—Its meanders—Islands—Shifting of its course—Raft of the Atchafalaya—Drift wood—New-formed lakes in Louisiana—Earthquakes in the valley of the Mississippi—Floods caused by land-slips in the White mountains—Bursting of a lake in Switzerland—Devastations caused by the Anio at Tivoli.

Course of the Po.—THE Po affords a grand example of the manner in which a great river bears down to the sea the matter poured into it by a multitude of tributaries descending from lofty chains of mountains. The changes gradually effected in the great plain of Northern Italy, since the time of the Roman republic, are very considerable. Extensive lakes and marshes have been gradually filled up, as those near Placentia, Parma, and Cremona, and many have been drained naturally by the deepening of the beds of rivers. Deserted river-courses are not unfrequent, as that of the Serio Morto, which formerly fell into the Adda, in Lombardy; and the Po itself has often deviated from its course. Subsequently to the year 1390, it deserted part of the territory of Cremona, and invaded that of Parma; its old channel being still recognizable, and bearing the name of Po Morto. Bressello is one of the towns of which the site was formerly on the left of the Po, but which is now on the right bank. There is also an old channel of the Po in the territory of Parma, called Po Vecchio, which was abandoned in the twelfth century, when a great number of towns were destroyed. There are records of parish-churches, as those of Vicobellignano, Agojolo, and Martignana, having been pulled down and afterwards rebuilt at a greater distance from the devouring stream. In the fifteenth century the main branch again resumed its deserted channel, and carried away a great island opposite Casalmaggiore. At the end of the same century it abandoned, a second time, the bed called "Po Vecchio," carrying away three streets of Casalmaggiore. The friars in the monastery de' Serviti took the alarm in 1471, demolished their buildings, and reconstructed them at Fontana, whither they had transported the materials. In like manner, the church of S. Rocco was demolished in 1511. In the seventeenth century also the Po shifted its course for a mile in the same district, causing great devastations *.

To check these and similar aberrations, a general system of embankment has been adopted; and the Po, Adige, and almost all their tributaries, are now confined between high artificial banks. The increased velocity acquired by streams thus closed in, enables them to convey a much larger portion of foreign matter to the sea; and consequently the deltas of the Po and Adige have gained far more rapidly on the Adriatic since the practice of embankment became almost universal. But although more sediment is borne to the sea, part of the sand, and mud, which in the natural state of things would be spread out by annual inundations over the plain, now subsides in the bottom of the river-channels, and their capacity being thereby diminished, it is necessary, in order to prevent inundations in the following spring, to extract matter from the bed, and to add it to the banks, of the river. Hence it has arisen that these streams now traverse the plain on the top of high mounds, like the waters of aqueducts, and the surface of the Po has become more elevated than the roofs of the houses of the city of Ferrara †. The magnitude of these barriers is a subject of increasing expense and anxiety, it having sometimes of late years been found necessary to give an additional height of nearly one foot to the banks of the Adige and Po in a single season. The practice of embankment was adopted on some of the Italian rivers as early as the thirteenth century; and Dante, writing in the beginning of the fourteenth, describes, in the seventh circle of hell, a rivulet of tears separated from a burning sandy desert by embankments "like those which, between Ghent and Bruges, were raised against the ocean, or those which the Paduans had

* Dell' Antico Corso de' Fiumi Po, Oglio, ed Adda dell' Giovanni Romani, Milan, 1828.

† Prony, see Cuvier, Disc. Prelim., p. 146.

erected along the Brenta to defend their villas on the melting of the Alpine snows."

> Quale i Fiamminghi tra Guzzante e Bruggia,
> Temendo il fiotto che in ver lor s' avventa,
> Fanno lo schermo, perchè il mar si fuggia,
> E quale i Padovan lungo la Brenta,
> Per difender lor ville e lor castelli,
> Anzi che Chiarentana il caldo senta—
>
> *Inferno,* Canto xv.

Basin of the Mississippi.—The hydrographical basin of the Mississippi displays, on the grandest scale, the action of running water on the surface of a vast continent. This magnificent river rises nearly in the forty-ninth parallel of north latitude, and flows to the Gulf of Mexico in the twenty-ninth—a course, including its meanders, of nearly five thousand miles. It passes from a cold arctic climate, traverses the temperate regions, and discharges its waters into the sea, in the region of the olive, the fig, and the sugar-cane *. No river affords a more striking illustration of the law before mentioned, that an augmentation of volume does not occasion a proportional increase of surface, nay, is even sometimes attended with a narrowing of the channel. The Mississippi is a mile and a half wide at its junction with the Missouri, the latter being half a mile wide; yet the united waters have only, from their confluence to the mouth of the Ohio, a medial width of about three quarters of a mile. The junction of the Ohio seems also to produce no increase, but rather a decrease of surface †. The St. Francis, White, Arkansas, and Red rivers, are also absorbed by the main stream with scarcely any apparent increase of its width; and, on arriving near the sea at New Orleans, it is somewhat less than half a mile wide. Its depth there is very variable, the greatest at high water being one hundred and sixty-eight feet. The mean rate at which the whole body of water flows, is variously estimated. According to some, it does not exceed one mile an hour ‡. The alluvial plain of this great river is bounded on the east and west by great ranges of mountains stretching along their respective oceans. Below the junction

* Flint's Geography, vol. i., p. 21. † Ibid., p. 140.

‡ Hall's Travels in North America, vol. iii., p. 330, who cites Darby.

of the Ohio, the plain is from thirty to fifty miles broad, and after that point it goes on increasing in width till the expanse is perhaps three times as great! On the borders of this vast alluvial tract are perpendicular cliffs, or "bluffs," as they are called, composed of limestone and other rocks. For a great distance the Mississippi washes the eastern "bluffs;" and below the mouth of the Ohio, never once comes in contact with the western. The waters are thrown to the eastern side, because all the large tributary rivers enter from the west, and have filled that side of the great valley with a sloping mass of clay and sand. For this reason, the eastern bluffs are continually undermined, and the Mississippi is slowly but incessantly progressing eastward *.

The river traverses the plain in a meandering course, describing immense and uniform curves. After sweeping round the half of a circle, it is precipitated from the point in a current diagonally across its own channel, to another curve of the same uniformity upon the opposite shore †. These curves are so regular, that the boatmen and Indians calculate distances by them. Opposite to each of them, there is always a sand-bar, answering, in the convexity of its form, to the concavity of "the bend," as it is called ‡. The river, by continually wearing these curves deeper, returns, like many other streams before described, on its own tract, so that a vessel in some places, after sailing for twenty-five or thirty miles, is brought round again to within a mile of the place whence it started. When the waters approach so near to each other, it often happens at high floods that they burst through the small tongue of land; and, having insulated a portion, rush through what is called the "cut off" with great velocity. At one spot called the "grand cut off," vessels now pass from one point to another in half a mile, to a distance which it formerly required twenty miles to reach §. After the flood season, when the river subsides within its channel, it acts with destructive force upon the alluvial banks, softened and diluted by the recent overflow. Several acres at a time, thickly covered with wood, are precipitated into the stream; and the islands formed by the pro-

* Geograph. Descrip. of the State of Louisiana, by W. Darby, Philadelphia, 1816, p. 102.

† Flint's Geog., vol. i., p. 152. ‡ Ibid. § Ibid., vol. i., p. 154.

cess before described, lose large portions of their outer circumference.

"Some years ago," observes Captain Hall, "when the Mississippi was regularly surveyed, all its islands were numbered, from the confluence of the Missouri to the sea; but every season makes such revolutions, not only in the number but in the magnitude and situation of these islands, that this enumeration is now almost obsolete. Sometimes large islands are entirely melted away—at other places they have attached themselves to the main shore, or, which is the more correct statement, the interval has been filled up by myriads of logs cemented together by mud and rubbish *." When the Mississippi and many of its great tributaries overflow their banks, the waters, being no longer borne down by the main current, and becoming impeded amongst the trees and bushes, deposit the sediment of mud and sand with which they are abundantly charged. Islands arrest the progress of floating trees, and they become in this manner reunited to the land; the rafts of trees, together with mud, constituting at length a solid mass. The coarser portion subsides first, and the most copious deposition is found near the banks where the soil is most sandy. Finer particles are found at the farthest distances from the river, where an impalpable mixture is deposited, forming a stiff unctuous black soil. Hence the alluvions of these rivers are highest directly on the banks, and slope back like a natural "glacis" towards the rocky cliffs bounding the great valley †. The Mississippi, therefore, by the continual shifting of its course, sweeps away, during a great portion of the year, considerable tracts of alluvium which were gradually accumulated by the overflow of former years, and the matter now left during the spring-floods will be at some future time removed.

One of the most interesting features in this basin is "the raft." The dimensions of this mass of timber were given by Darby, in 1816, as ten miles in length, about two hundred and twenty yards wide, and eight feet deep, the whole of which had accumulated, in consequence of some obstruction, during

* Travels in North America, vol. iii., p. 361.

† Flint's Geography, vol. i., p. 151.

about thirty-eight years, in an arm of the Mississippi called the Atchafalaya, which is supposed to have been at some past time a channel of the Red River, before it intermingled its waters with the main stream. This arm is in a direct line with the direction of the Mississippi, and it catches a large portion of the drift wood annually brought down. The mass of timber in the raft is continually increasing, and the whole rises and falls with the water. Although floating, it is covered with green bushes, like a tract of solid land, and its surface is enlivened in the autumn by a variety of beautiful flowers. Notwithstanding the astonishing number of cubic feet of timber collected here in so short a time, greater deposits have been in progress at the extremity of the delta in the Bay of Mexico. "Unfortunately for the navigation of the Mississippi," observes Captain Hall, "some of the largest trunks, after being cast down from the position on which they grew, get their roots entangled with the bottom of the river, where they remain anchored, as it were, in the mud. The force of the current naturally gives their tops a tendency downwards, and by its flowing past, soon strips them of their leaves and branches. These fixtures, called snags or planters, are extremely dangerous to the steam-vessels proceeding up the stream, in which they lie like a lance in rest, concealed beneath the water, with their sharp ends pointed directly against the bow of vessels coming up. For the most part these formidable snags remain so still, that they can be detected only by a slight ripple above them, not perceptible to inexperienced eyes. Sometimes, however, they vibrate up and down, alternately showing their heads above the surface and bathing them beneath it*." So imminent is the danger caused by these obstructions, that almost all the boats on the Mississippi are constructed on a particular plan, to guard against fatal accidents †.

* Travels in North America, vol. iii., p. 362.

† "The boats are fitted," says Captain Hall, "with what is called a snag-chamber; a singular device, and highly characteristic of this peculiar navigation. At the distance of from twelve or fourteen feet from the stem of the vessel, a strong bulk-head is carried across the hold from side to side, as high as the deck, and reaching to the kelson. This partition, which is formed of stout planks, is caulked, and made so effectually water-tight, that the foremost end of the vessel is cut off as entirely from the rest of the hold as if it belonged to another boat. If the steam-vessel happen to run against a snag, and

The prodigious quantity of wood annually drifted down by the Mississippi and its tributaries, is a subject of geological interest, not merely as illustrating the manner in which abundance of vegetable matter becomes, in the ordinary course of Nature, imbedded in submarine and estuary deposits, but as attesting the constant destruction of soil and transportation of matter to lower levels by the tendency of rivers to shift their courses. Each of these trees must have required many years, some of them many centuries, to attain their full size; the soil, therefore, whereon they grew, after remaining undisturbed for long periods, is ultimately torn up and swept away. Yet notwithstanding this incessant destruction of land and up-rooting of trees, the region which yields this never-failing supply of drift wood is densely clothed with noble forests, and is almost unrivalled in its power of supporting animal and vegetable life.

Innumerable herds of wild deer and bisons feed on the luxuriant pastures of the plains. The jaguar, the wolf, and the fox, are amongst the beasts of prey. The waters teem with alligators and tortoises, and their surface is covered with millions of migratory water-fowl, which perform their annual voyage between the Canadian lakes and the shores of the Mexican gulf. The power of man begins to be sensibly felt, and the wilderness to be replaced by towns, orchards, and gardens. The gilded steam-boat, like a moving city, now stems the current with a steady pace—now shoots rapidly down the descending stream through the solitudes of the forests and prairies. Already does the flourishing population of the great valley exceed that of the thirteen United States when first they declared their independence, and after a sanguinary struggle were severed from the parent country *. Such is the state of a continent where rocks and trees are hurried annually, by a thousand torrents, from the mountains to the plains, and where sand and finer matter are swept down by a vast current to the sea, together with the wreck of countless forests and the bones of animals which perish in the inundations. When these materials reach the Gulf, they do not render the waters unfit for

that a hole is made in her bow, under the surface, this chamber merely fills with water; for the communication being cut off from the rest of the vessel, no further mischief need ensue." Travels in North America, vol. iii., p. 363.

* Flint's Geography, vol. i.

aquatic animals; but, on the contrary, the ocean here swarms with life, as it generally does where the influx of a great river furnishes a copious supply of organic and mineral matter. Yet many geologists, when they behold the spoils of the land heaped in successive strata, and blended confusedly with the remains of fishes, or interspersed with broken shells and corals, imagine that they are viewing the signs of a turbulent, instead of a tranquil and settled state of the planet. They read in such phenomena the proof of chaotic disorder, and reiterated catastrophes, instead of indications of a surface as habitable as the most delicious and fertile districts now tenanted by man. They are not content with disregarding the analogy of the present course of Nature, when they speculate on the revolutions of past times, but they often draw conclusions concerning the former state of things directly the reverse of those to which a fair induction from facts would infallibly lead them.

There is another striking feature in the basin of the Mississippi, illustrative of the changes now in progress, which we must not omit to mention—the formation by natural causes of great lakes, and the drainage of others. These are especially frequent in the basin of the Red River in Louisiana, where the largest of them, called Bistineau, is more than *thirty miles* long, and has a medium depth of from *fifteen* to *twenty* feet. In the deepest parts are seen numerous cypress-trees, of all sizes, now dead, and most of them with their tops broken by the wind, yet standing erect under water. This tree resists the action of air and water longer than any other, and, if not submerged throughout the whole year, will retain life for an extraordinary period *. Lake Bistineau, as well as Black Lake, Cado Lake, Spanish Lake, Natchitoches Lake, and many others, have been formed, according to Darby, by the gradual elevation of the bed of Red River, in which the alluvial accumulations have been so great as to raise its channel, and cause its waters, during the flood season, to flow up the mouths of many tributaries, and to convert parts of their courses into lakes. In

* Captains Clark and Lewis found a forest of pines standing erect under water in the body of the Columbia River in North America, which they supposed, from the appearance of the trees, to have been only submerged about twenty years.—Vol. ii., p. 241.

the autumn, when the level of Red River is again depressed, the waters rush back again, and some lakes become grassy meadows, with streams meandering through them *. Thus, there is a periodical flux and reflux between Red River and some of these basins, which are merely reservoirs, alternately emptied and filled like our tide estuaries—with this difference, that in the one case the land is submerged for several months continuously, and, in the other, twice in every twenty-four hours. It has happened, in several cases, that a bar has been thrown by Red River across some of the openings of these channels, and then the lakes become, like Bistineau, constant repositories of water. But even in these cases, their level is liable to annual elevation and depression, because the flood when at its height passes over the bar; just as, where sand-hills close the entrance of an estuary on the Norfolk or Suffolk coast, the sea, during some high tide or storm, has often breached the barrier and inundated again the interior country.

The frequent fluctuations in the direction of river-courses, and the activity exerted by running water in various parts of the basin of the Mississippi, are partly, perhaps, to be ascribed to the co-operation of subterranean movements, which alter from time to time the relative levels of various parts of the surface. So late as the year 1812, the whole valley, from the mouth of the Ohio to that of the St. Francis, including a front of three hundred miles, was convulsed to such a degree, as to create new islands in the river, and lakes in the alluvial plain, some of which were *twenty miles in extent*. We shall allude to this event when we treat of earthquakes, but may state here that they happened exactly at the same time as the fatal convulsions at Caraccas; and the district shaken was nearly five degrees of latitude farther removed from the great centre of volcanic disturbance, than the basin of the Red River, to which we before alluded †. When countries are liable to be so extensively and permanently affected by earthquakes, speculations concerning changes in their hydrographical features must not be made with-

* Darby's Louisiana, p. 33.

† Darby mentions beds of marine shells on the banks of Red River, which seem to indicate that Lower Louisiana is of recent formation: its elevation, perhaps, above the sea, may have been due to the same series of earthquakes which continues to agitate equatorial America.

out regard to the igneous as well as the aqueous causes of change. It is scarcely necessary to observe, that the inequalities produced even by one shock, might render the study of the alluvial plain of the Mississippi, at some future period, most perplexing to a geologist who should reason on the distribution of transported materials, without being aware that the configuration of the country had varied materially during the time when the excavating or removing power of the river was greatest. The region convulsed in 1812, of which New Madrid was the centre, exceeded in length the whole basin of the Thames, and the shocks were connected with active volcanos more distant from New Madrid than are the extinct craters of the Eyfel or of Auvergne from London. If, therefore, during the innumerable eruptions which formerly broke forth in succession in the parts of Europe last alluded to, the basin of the principal river of our island was frequently agitated, and the relative levels of its several parts altered (an hypothesis in perfect accordance with modern analogy), the difficulties of some theorists might, perhaps, be removed; and they might no longer feel themselves under the necessity of resorting to catastrophes out of the ordinary course of Nature, when they endeavour to explain the alluvial phenomena of that district.

FLOODS, BURSTING OF LAKES, &c.

The power which running water may exert, in the lapse of ages, in widening and deepening a valley, does not so much depend on the volume and velocity of the stream usually flowing in it, as on the number and magnitude of the obstructions which have, at different periods, opposed its free passage. If a torrent, however small, be effectually dammed up, the size of the valley above the barrier, and its declivity below, will determine the violence of the debacle, and not the dimensions of the torrent. The most universal source of local deluges are land-slips, slides, or avalanches, as they are sometimes called, when great masses of rock and soil, or sometimes ice and snow, are precipitated into the bed of a river, either by the undermining of a cliff, by the loosening of a sub-stratum by springs, by the shock of an earthquake, or other causes. volumes might be filled were we to enumerate all the instances which are on record of these terrific catastrophes: we may therefore select a few examples of recent occurrence, the facts of which are well authenticated.

Two dry seasons in the White Mountains, in New Hampshire, were followed by heavy rains on the 28th August, 1826, when from the steep and lofty declivities which rise abruptly on both sides of the river Saco, innumerable rocks and stones, many of sufficient size to fill a common apartment, were detached, and in their descent swept down before them, in one promiscuous and frightful ruin, forests, shrubs, and the earth which sustained them. No tradition existed of any similar slides at former times, and the growth of the forest on the flanks of the hills clearly shewed that for a long interval nothing similar had occurred. One of these moving masses was afterwards found to have slid three miles, with an average breadth of a quarter of a mile. The excavations commenced generally in a trench a few yards in depth and a few rods in width, and descended the mountains, widening and deepening till they became vast chasms. At the base of such hollow ravines was seen a wide and deep mass of ruins, consisting of transported earth, gravel, rocks, and trees. Forests of spruce-fir and hemlock were prostrated with as much ease as if they had been fields of grain; for, where they disputed the ground, the torrent of mud and rock accumulated behind till it gathered sufficient force to burst the temporary barrier.

The valleys of the Amonoosuck and Saco presented, for many miles, an uninterrupted scene of desolation, all the bridges being carried away, as well as those over their tributary streams. In some places, the road was excavated to the depth of from fifteen to twenty feet; in others, it was covered with earth, rocks, and trees, to as great a height. The water flowed for many weeks after the flood, as densely charged with earth as it could be without being changed into mud, and marks were seen in various localities of its having risen on either side of the valley to more than twenty-five feet above its ordinary level. Many sheep and cattle were swept away, and the Willey family, nine in number, who in alarm had deserted their house, were destroyed on the banks of the Saco: seven of their mangled bodies were afterwards found near the river, buried beneath drift-wood and mountain-ruins *. It is almost superfluous to point out to the reader that the lower alluvial plains are most exposed to such violent floods, and are at the same time best fitted for the sustenance of herbivorous animals. If, therefore, any organic remains are found amidst the superficial heaps of transported matter, resulting from those catastrophes, at whatever periods they may have happened, and whatever may have been the former configuration and relative levels of the country, we may expect the imbedded fossil relics to be principally referrible to this class of mammalia.

But these catastrophes are insignificant, when compared to those which are occasioned by earthquakes, when the boundary hills, for miles in length, are thrown down into the hollow of a valley. We shall have an opportunity of alluding to inundations of this kind when treating of earthquakes, and shall content ourselves at present with selecting an example, of modern date, of a flood caused by the bursting of a lake; the facts having been described, with more than usual accuracy, by scientific observers.

Flood in the Valley of Bagnes, 1818.—The valley of Bagnes is one of the largest of the lateral embranchments of the main valley of the Rhone, above the Lake of Geneva. Its upper portion was, in 1818, converted into a lake by the damming up of a narrow pass, in consequence of the fall of avalanches of snow and ice, precipitated from an elevated glacier into the bed of the river Dranse. In the winter season, during continued frost, scarcely any water flows in the bed of this river to preserve an open channel, so that the ice-barrier remained entire until the melting of the snows in spring, when a lake was formed above, about half a league in length, which finally attained a depth of about two hundred feet in parts, and a width of about seven hundred feet. To prevent or lessen the mischief apprehended from the sudden bursting of the barrier, an artificial gallery, seven hundred feet in length, was cut through the ice, before the waters had risen to a great height. When at length they accumulated and flowed through

* Silliman's Journal of Science, vol. xv., No. 2, p. 216, Jan. 1829.

this tunnel, they dissolved the ice, and thus deepened their channel, until nearly half of the whole contents of the lake were slowly drained off. But, at length, on the approach of the hot season, the central portion of the remaining mass of ice gave way with a tremendous crash, and the residue of the lake was emptied in half an hour. In the course of its descent, the waters encountered several narrow gorges, and at each of these they rose to a great height, and then burst, with new violence, into the next basin, sweeping along rocks, forests, houses, bridges, and cultivated land. For the greater part of its course the flood resembled a moving mass of rock and mud, rather than of water. Some fragments of primary rock, of enormous magnitude, and which, from their dimensions, might be compared without exaggeration to houses, were torn out of a more ancient alluvion, and borne down for a quarter of a mile. The velocity of the water, in the first part of its course, was thirty-three feet per second, which diminished to six feet before it reached the Lake of Geneva, where it arrived in six hours and a half, the distance being forty-five miles *. This flood left behind it, on the plains of Martigny, thousands of trees torn up by the roots, together with the ruins of buildings. Some of the houses in that town were filled with mud up to the second story. After expanding in the plain of Martigny, it entered the Rhone, and did no further damage; but some bodies of men, who had been drowned above Martigny, were afterwards found at the distance of about thirty miles, floating on the further side of the Lake of Geneva, near Vevey. Inundations, precisely similar, are recorded to have occurred at former periods in this district, and from the same cause. In 1595, for example, a lake burst, and the waters, descending with irresistible fury, destroyed the town of Martigny, where from sixty to eighty persons perished. In a similar flood, fifty years before, one hundred and forty persons were drowned. For several months after the débâcle of 1818, the Dranse, having no settled channel, shifted its position continually from one side to the other of the valley, carrying away newly-erected bridges, under-

* See an account of the inundation of the Val de Bagnes, in 1818, in Ed. Phil. Journ., vol. i., p. 187. Drawn up from the Memoir of M. Escher, with a section, &c.

mining houses, and continuing to be charged with as large a quantity of earthy matter as the fluid could hold in suspension *.

The waters, on escaping from the lake, intermixed with mud and rock, swept along, for the first four miles, at the rate of above twenty miles an hour; and Mr. Escher, the engineer, calculated that the flood furnished three hundred thousand cubic feet of water every second,—an efflux which is five times greater than that of the Rhine below Basle. Now, if part of the lake had not been gradually drained off, the flood would have been nearly double, approaching in volume to some of the largest rivers in the world. It is evident, therefore, that when we are speculating on the excavating force which running water may have exerted in any particular valley, the most important question is not the volume of the existing stream, nor the present levels of the river-channel, nor the size of the gravel, but the probability of a succession of floods, at some period since the time when some of the land in question may have been first elevated above the bosom of the sea.

Flood at Tivoli, 1826.—We shall conclude with one more example, derived from a land of classic recollections, the ancient Tibur, and which, like all the other inundations to which we have alluded, occurred within the present century. The younger Pliny, it will be remembered, describes a flood on the Anio, which destroyed woods, rocks, and houses, with the most sumptuous villas and works of art †. For four or five centuries consecutively, this headlong stream, as Horace truly called it, has often remained within its bounds, and then, after such long intervals of rest, at different periods inundated its banks again, and widened its channel. The last of these catastrophes happened 15th Nov. 1826, after heavy rains, such as produced the floods before alluded to in Scotland. The waters appear also to have been impeded by an artificial

* I visited this valley four months after the flood, and was witness to the sweeping away of a bridge, and the undermining of part of a house. The greater part of the ice-barrier was then standing, presenting a vertical cliff, one hundred and fifty feet high, like the lava-currents of Etna or Auvergne, intersected by a river.

† Lib. viii., Epist. 17.

dike, by which they were separated into two parts, a short distance above Tivoli. They broke through this dike, and, leaving the left trench dry, precipitated themselves, with their whole weight, on the right side. Here they undermined, in the course of a few hours, a high cliff, and widened the river's channel about fifteen paces. On this height stood the church of St. Lucia, and about thirty-six houses of the town of Tivoli, which were all carried away, presenting, as they sank into the roaring flood, a terrific scene of destruction to the spectators on the opposite bank *. As the foundations were gradually removed, each building, some of them edifices of considerable height, was first traversed with numerous rents, which soon widened into large fissures, until at length the roofs fell in with a crash, and then the walls sank into the river, and were hurled down the cataract below.

The destroying agency of the flood came within two hundred yards of the precipice on which the beautiful temple of Vesta stands; but fortunately this precious relic of antiquity was spared, while the wreck of modern structures was hurled down the abyss. Vesta, it will be remembered, in the heathen mythology, personified the stability of the earth; and when the Samian astronomer, Aristarchus, first taught that the earth revolved on its axis, and round the sun, he was publicly accused of impiety, "for moving the everlasting Vesta from her place." Playfair † observed, that when Hutton ascribed instability to the earth's surface, and represented the continents which we inhabit as the theatre of incessant change and movement, his antagonists, who regarded them as unalterable, assailed him, in a similar manner, with accusations founded on religious prejudices. We might appeal to the excavating power of the Anio as corroborative of one of the most controverted parts of the Huttonian theory; and if the days of omens had not gone by, the geologists who now worship Vesta might regard the late catastrophe as portentous. We may, at least, recommend the modern votaries of the goddess to lose no time in making a pilgrimage to her shrine, for the next flood may not respect the temple.

* When at Tivoli, in 1829, I received this account from eye-witnesses of the event.

† Illustr. of Hutt. Theory, § 3, p. 147.

CHAPTER XII.

Difference between the transporting power of springs and rivers—Many springs carry matter from below upwards—Mineral ingredients most abundant in springs—Connexion of mineral waters with volcanic phenomena—Calcareous springs—Travertin of the Elsa—Baths of San Vignone and of San Filippo, near Radicofani—Spheroidal structure in travertin, as in English magnesian limestone—Bulicami of Viterbo—Lake of the Solfatara, near Rome—Travertin at Cascade of Tivoli—Ferruginous Springs—Cementing and colouring property of iron—Brine Springs—Carbonated Springs—Disintegration of Auvergne granite—Caverns in limestone—Petroleum Springs—Pitch Lake of Trinidad.

We have hitherto considered the destroying and transporting power of those atmospheric waters which circulate on the surface of the land; but another portion which sink deep into the earth, present phenomena essentially different in character. Rivers, as we have seen, remove earthy matter from higher to lower levels, but springs not only effect this purpose, but sometimes, like volcanos, carry matter from below upwards. Almost all springs are impregnated with some foreign ingredients, which render them more agreeable to our taste, and more nutritive than pure rain water; but as their mineral contents are in a state of chemical solution, they rarely, even when in great abundance, affect the clearness of the water, and for this reason, we are usually unconscious of the great instrumentality of these agents in the transfer of solid matter from one part of the globe to another. The specific gravity of spring water being greater than that of rain, it augments the carrying power of rivers, enabling them to bear down a greater quantity of matter in mechanical suspension towards the sea. Springs, both cold and thermal, rise up beneath the waters of lakes and seas, as well as in different parts of the land, and must often greatly modify the mineral character of subaqueous deposits.

The number of metals, earths, acids, and alkalies, held in solution by different springs, comprehends a considerable portion of all known substances, and recent observations have tended continually to augment the list; but those alone which are most abundant, need be regarded as of geological importance. These are lime, iron, magnesia, silica, alumine, soda, and the carbonic and sulphuric acids. Besides these, springs of petroleum, or liquid bitumen, and its various modifications, such as mineral pitch, naptha, and asphaltum; are largely distributed over the surface of the earth, but usually in close connexion with volcanos. The relation, indeed, of almost all springs impregnated copiously with mineral matter, to the sources of subterranean heat, seems placed beyond all reasonable doubt by modern research. Mineral waters, as they have been termed, are most abundant in regions of active volcanos, or where earthquakes are most frequent and violent. Their temperature is often very high, and has been known to be permanently heightened or lowered by the shock of an earthquake. The volume of water also given out has been sometimes affected by the same cause. With the exception of silica, the minerals entering most abundantly into *thermal* waters do not seem to differ from those in cold springs. There is, moreover, a striking analogy between the earthy matters evolved in a gaseous state by volcanos, and those wherewith springs in the same region are impregnated; and when we proceed from the site of active to that of extinct volcanos, we find the latter abounding in precisely the same kind of springs. Where thermal and mineral waters occur far from active or extinct volcanos, some great internal derangement in the strata almost invariably marks the site to have been at some period, however remote, the theatre of violent earthquakes.

Springs, are in general, ascribable to the percolation of rain-water through porous rocks, which, meeting at last with argillaceous strata, is thrown out to the surface. But, in all likelihood, they sometimes descend by fissures, even to the regions of subterranean heat. Michell, in 1760, suggested that those pent-up volcanic vapours which cause earthquakes, penetrate also through rents and cavities, and drive up water impregnated with sulphurous and other matters, whereby springs are charged with their mineral ingredients. Nor is it by any means improbable, that the same power which when intense is able to lift up a column of lava many thousand feet in height, should even in its more languid state be capable of raising to the surface considerable quantities of water from the interior. But as the geographical limits of mineral waters are not confined to volcanic regions, being coextensive with the whole globe, as far as is hitherto known, we must consider them apart, and in their connexion with rivers rather than volcanos. We might divide the consideration of springs, like that of rivers, into their destroying and reproductive agency; but the former class of effects being chiefly subterranean, are beyond the reach of our observation; while their reproductive power consists chiefly in augmenting the quantity of matter deposited by rivers in deltas, or at the bottom of the sea. We shall, therefore, arrange the facts of geological interest, respecting mineral springs, under the head of the different ingredients which predominate in their waters.

CALCAREOUS SPRINGS.

Our first attention is naturally directed to springs which are highly charged with calcareous matter; for these produce a variety of phenomena of much interest to the geologist. It is well known that rain-water has the property of dissolving the calcareous rocks over which it flows, and by these means, matter is often supplied for the earthy secretions of testacea, and certain plants on which they feed, in the smallest ponds and rivulets. But many springs hold so much carbonic acid in solution, that they are enabled to dissolve a much larger quantity of calcareous matter than rain-water; and when the acid is dissipated in the atmosphere, the mineral ingredients are slowly thrown down in the form of tufa or travertin. Calcareous springs, although most abundant in limestone districts, are by no means confined to them, but flow out indiscriminately from all rock formations. In Central France, a district where the primary rocks are unusually destitute of limestone, springs copiously charged with carbonate of lime rise up through the granite and gneiss. Some of these are thermal, and probably derive their origin from the deep source of volcanic heat, once so active in that region. One of these springs, at the northern base of the hill upon which Clermont is built, issues from volcanic peperino, which rests on granite. It has formed, by its incrustations, an elevated mound of solid travertin, or calc-sinter, as it s sometimes called, two hundred and forty feet in length, and, at its termination, sixteen feet high, and twelve wide. Another incrusting spring in the same department, situated at Chaluzet, near Pont Gibaud, rises in a gneiss country, at the foot of a regular volcanic cone, at least twenty miles from any calcareous rock. Some masses of tufaceous deposit, produced by this spring, have an oolitic texture.

Valley of the Elsa.—If we pass from the volcanic district of France to that which skirts the Apennines in the Italian peninsula, we meet with innumerable springs, which have precipitated so much calcareous matter, that the whole ground in some parts of Tuscany is coated over with travertin, and sounds hollow beneath the foot.

In other places in the same country, compact rocks are seen descending the slanting sides of hills, very much in the manner of lava-currents, except that they are of a white colour, and terminate abruptly when they reach the course of a river. These consist of the calcareous precipitate of springs, some of them still flowing, while others have disappeared or changed their position. Such masses are frequent on the slope of the hills which bound the valley of the Elsa, one of the tributaries of the Arno, which flows near Colle, through a valley several hundred feet deep, shaped out of a lacustrine formation, containing fossil shells of existing species. The travertin is unconformable to the lacustrine beds, and its inclination accords with the slope of the sides of the valley*. The Sena, and several other small rivulets which feed the Elsa, have the property of lapidifying wood and herbs; and, in the bed of the Elsa itself, aquatic plants, such as charæ, which absorb large quantities of carbonate of lime, are very abundant. Carbonic acid is also seen in the same valley, bubbling up from many springs, where no precipitate of tufa is observable. Targioni, who in his travels has mentioned a great number of mineral waters in Tuscany, found no difference between the deposits of cold and thermal springs. They issue sometimes from the older Apennine limestone, shale, and sandstone, while, in other places, they flow from more modern deposits; but, even in the latter case, their source may probably be in, or below the older series of strata.

* One of the finest examples of these which I saw, was at the Molino delle Caldane, near Colle.

Baths of San Vignone.—Those persons who have merely seen the action of petrifying waters in our own country, will not easily form an adequate conception of the scale on which the same process is exhibited in those regions which lie nearer to the modern centres of volcanic disturbance. One of the most striking examples of the rapid precipitation of carbonate of lime from thermal waters occurs in the hill of San Vignone in Tuscany, at a short distance from Radicofani, and only a few hundred yards from the high-road between Sienna and Rome. The spring issues from near the summit of a rocky hill, about one hundred feet in height. The top of the hill is flat, and stretches in a gently-inclined plateau to the foot of Mount Amiata, a lofty eminence, which consists in great part of volcanic products. The fundamental rock, from which the spring issues, is a black slate, with serpentine (*b. b. b*, diagram 4) belonging to the older Apennine for-

Baths of San Vignone.

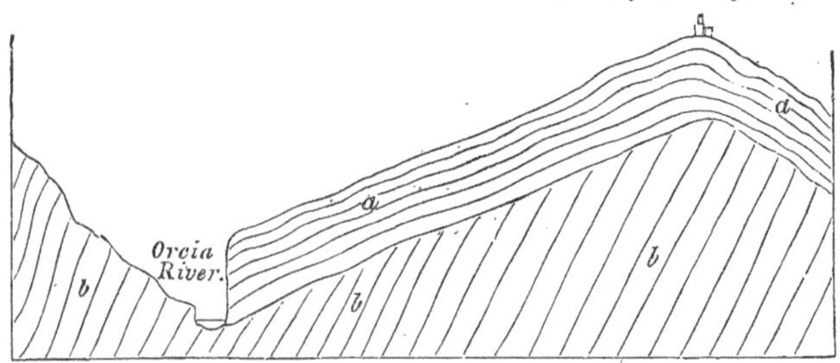

mation. The water is hot, has a strong taste, and, when not in very small quantity, is of a bright green colour. So rapid is the deposition near the source, that in the bottom of a conduit pipe for carrying off the water to the baths, inclined at an angle of 30°, half a foot of solid travertin is formed every year. A more compact rock is produced where the water flows slowly, and the precipitation in winter is said to be more solid and less in quantity by one-fourth than in summer. The rock is generally white: some parts of it are compact, and ring to the hammer; others are cellular, and with such cavities as are seen in the carious part of bone or the siliceous meulière of the Paris basin. A portion of it also below the village consists of long vegetable tubes. Sometimes the travertin assumes precisely the botroidal and mammillary forms, common to similar deposits; in Auvergne, of a much older date, hereafter to be mentioned; and like them it often scales off in thin, slightly undulating layers.

A large mass of travertin descends the hill from the point where the spring issues, and reaches to the distance of about half a mile east of San Vignone. The beds take the slope of the hill at about an angle of 6°, and the planes of stratification are perfectly parallel. One stratum, composed of many layers, is of a compact nature and fifteen feet thick; it serves as an excellent building stone, and a mass of fifteen feet in length was, in 1828, cut out for the new bridge over the Orcia. Another branch of it (*a. a.*, diagram 4,) descends to the west, for two hundred and fifty feet in length, of varying thickness, but sometimes two hundred feet deep; it is then cut off by the small river Orcia, precisely as some glaciers in Switzerland descend into a valley till their progress is suddenly arrested by a transverse stream of water. The abrupt termination of the mass of rock at the river, when its thickness is undiminished, clearly shews that it would proceed much farther if not arrested by the stream, over which it impends slightly. But it cannot encroach upon the channel of the Orcia, being constantly undermined, so that its solid fragments are seen strewed amongst the alluvial gravel. However enormous, therefore, the mass of solid rock may appear which has been given out by this single spring, we may feel assured that it is insignificant in volume, when compared to that which has been carried to the sea since the time when it began to flow. What may have been the length of that period of time, we have no data for conjecturing. In quarrying the travertin, Roman tiles have been sometimes found at the depth of five or six feet.

Baths of San Filippo.—On another hill, not many miles from that last mentioned, and also connected with Mount Amiata, the summit of which is about three miles distant, are the celebrated baths of San Filippo. The subjacent rocks consist of alternations of black slate, limestone, and serpentine, of highly inclined strata, belonging to the Apennine formation; and, as at San Vignone, near the boundary of a tertiary basin of marine origin, consisting chiefly of blue argillaceous marl. There are three warm springs here, containing carbonate and sulphate of lime, and sulphâte of magnesia. The water which supplies the bath falls into a pond, where it has been known to deposit a solid mass *thirty feet thick*, in about *twenty years**. A manufactory of medallions in basso-relievo is carried on at these baths. The water is conducted by canals into several pits, in which it deposits travertin and crystals of sulphate of lime. After being thus freed from its grosser parts, it is conveyed by a tube to the summit of a small chamber, and made to fall through a space of ten or twelve feet. The current is broken in its descent by numerous crossed sticks, by which the spray is dispersed around upon certain moulds, which are rubbed lightly over with a solution of soap, and a deposition of solid matter like marble is the result, yielding a beautiful cast of the figures formed in the mould†. The geologist may derive from these experiments considerable light, in regard to the high inclination at which some semicrystalline precipitations can be formed; for some of the moulds are disposed almost perpendicularly, yet the deposition is nearly equal in all parts.

A hard stratum of stone, about a foot in thickness, is obtained from the waters of San Filippo in four months; and, as the springs are powerful, and almost uniform in the quantity given out, we are at no loss to comprehend the magnitude of the mass which descends the hill, which is a mile and quarter in length and the third of a mile in breadth, in some places attaining a thickness of two hundred and fifty feet at least. To what length it might have reached, it is impossible to conjecture, as it is cut off, like the travertin of San Vignone, by a small stream, where it terminates abruptly. The remainder of the matter held in solution is carried on probably to the sea. But what renders this recent calcareo-magnesian limestone of peculiar interest to the geologist, is the spheroidal forms which it assumes, offering so striking an analogy, on the one hand, to the concentric structure displayed in the calcareous travertin of the cascade of Tivoli, and, on the other, to the spheroidal forms of the English magnesian limestone of Sunderland. Between this latter and many of the

* Dr. Grosse, on the Baths of San Filippo. Ed. Phil. Journ., v. 2, p. 292.

† Ibid., p. 297.

appearances exhibited at San Filippo, and several other recent deposits of the same kind in Italy, there is every feature of resemblance; the same combination of concentric and radiated structure, with small undulations in each concentric ring, occasional interferences of one circle with another, and a small globular structure subordinate to the large spheroidal, with frequent examples of laminæ passing off from the external coating of a spheroid into layers parallel to the general plane of stratification. There are also cellular cavities and vacuities in the rock, constituting what has been termed a honeycombed texture. The lamination of some of the concentric masses of San Filippo is so minute, that sixty may be counted in the thickness of an inch. Yet, notwithstanding these marks of gradual and successive deposition, the symmetry and magnitude of many of the spheroidal forms might convey the idea, that the whole was the result of chemical action, simultaneously operating on a great mass of matter. The concretionary forms of our magnesian limestone have been supposed by some to have been superinduced after the component parts of the rock had been brought together in stratiform masses; but a careful comparison of those older rocks with the numerous travertins now in progress of formation in Italy, leads the observer to a different conclusion. Such a structure seems to be the result of gradual precipitation, and not of subsequent re-arrangement of the particles*. Each minute particle of foreign matter, a reed, or the fragment of a shell, forms a nucleus, around which accessions of new laminæ are formed, until spheroids and elongated cones, from a few inches to several feet in diameter, are produced; for, as the precipitate is arranged by the force of chemical affinity, and not of

* The structure of the English magnesian limestone has been described, in an elaborate and profound paper on that formation, by Professor Sedgwick. Geol. Trans., vol. 3, second series, part i., p. 37. Examples of almost all the modifications of concretionary arrangement, together with the brecciated and honeycombed structure to which he alludes, may be found either in the deposits of travertin springs in various parts of Italy, or in the subaqueous travertins of Auvergne and Sicily,—the former of lacustrine, the latter of submarine origin. These will be alluded to in their proper places, and I shall merely observe here, that, after examining these more recent deposits, I visited Sunderland, and recognized a degree of identity in the various and complicated forms there assumed by the magnesian limestone, which satisfied me that the circumstances under which they were formed must have been perfectly analagous to those under which the mineral springs of volcanic countries are now giving birth to calcareous, calcareo-magnesian, and calcareo-siliceous rocks.

gravity, the different layers continue of the same thickness, and preserve the original form of the nucleus.

Bulicami of Viterbo.—We must not attempt to describe all the localities in Italy where the constant formation of limestone may be seen, as on the Silaro, near Pæstum, on the Velino at Terni, and near the Bulicami, or hot baths in the vicinity of Viterbo. About a mile and a half north of the latter town, in the midst of a sterile plain of volcanic sand and ashes, a monticule is seen, about twenty feet high and five hundred yards in circumference, entirely composed of concretionary travertin. The laminæ are extremely thin, and their minute undulations are so arranged, that the whole mass has at once a concentric and radiated structure. This mammillon has been largely quarried for lime, and much of it appears to have been removed. It seems to have been formed by a small jet or fountain of calcareous water, which continued to rise through the mound of travertin, which it gradually raised by overflowing from the summit. A spring of hot water still issues in the neighbourhood, which is conveyed to an open tank, used as a bath, the bottom and sides of which, as well as the open conduit which conveys the water, are encrusted with travertin.

Campagna di Roma.—The country around Rome, like many parts of the Tuscan States already referred to, has been at some former period the site of numerous volcanic eruptions; and the springs are still copiously impregnated with lime, carbonic acid, and sulphuretted hydrogen. A hot spring has lately been discovered near Civita Vecchia, by Riccioli, which deposits alternate beds of a yellowish travertin, and a white granular rock, not distinguishable, in hand specimens, either in grain, colour, or composition, from statuary marble. There is a passage between this and ordinary travertin. The mass accumulated near the spring is in some places about six feet thick *.

In the Campagna, between Rome and Tivoli, is the lake of

* I did not visit this spring myself, but Signor Riccioli, whose acquaintance with the geology of the environs of Rome is well known, favoured me with an inspection of a suite of specimens collected from the spot. Brocchi, a few years before his death, visited the locality in company with Signor Riccioli, and was much struck with the phenomenon, of which he had intended to publish a description.

the Solfatara, called also Lago di Zolfo, (lacus albula,) into which flows continually a stream of tepid water, from a smaller lake situated a few yards above it. The water is a saturated solution of carbonic acid gas, which escapes from it in such quantities in some parts of its surface, that it has the appearance of being actually in ebullition. "I have found by experiment," says Sir Humphry Davy, "that the water taken from the most tranquil part of the lake, even after being agitated and exposed to the air, contained in solution more than its own volume of carbonic acid gas, with a very small quantity of sulphuretted hydrogen. Its high temperature, which is pretty constant at 80° of Fahr., and the quantity of carbonic acid that it contains, render it peculiarly fitted to afford nourishment to vegetable life. The banks of travertin are every where covered with reeds, lichen, confervæ, and various kinds of aquatic vegetables; and at the same time that the process of vegetable life is going on, the crystallizations of the calcareous matter, which is every where deposited in consequence of the escape of carbonic acid, likewise proceed.—There is, I believe, no place in the world where there is a more striking example of the opposition or contrast of the laws of animate and inanimate nature, of the forces of inorganic chemical affinity, and those of the powers of life *."

The same observer informs us, that he fixed a stick on a mass of travertin covered by the water in May, and in the April following he had some difficulty in breaking, with a sharp-pointed hammer, the mass which adhered to the stick, and which was several inches in thickness. The upper part was a mixture of light tufa, and the leaves of confervæ: below this was a darker and more solid travertin, containing black and decomposed masses of confervæ; in the inferior part, the travertin was more solid, and of a grey colour, but with cavities probably produced by the decomposition of vegetable matter †. The stream which flows out of this lake fills a canal about nine feet broad, and four deep, and is conspicuous in the landscape by a line of vapour which rises from it. It deposits tufa in this channel, and the Tibur probably receives from it, as well as from numerous other streams, much carbonate of lime in solu-

* Consolations in Travel, pp. 123—125. † Ibid., p. 127.

tion, which contributes to the rapid growth of its delta. A large portion of the most splendid edifices of ancient and modern Rome are built of travertin, derived from the quarries of Ponte Leucano, where there has evidently been a lake at a remote period, on the same plain as that already described. But, as the consideration of these would carry us beyond the times of history, we shall conclude with one more example of the calcareous deposits of this neighbourhood,—those on the Anio.

Travertin of Tivoli.—The waters of the Anio incrust the reeds which grow on its banks, and the foam of the cataract of Tivoli forms beautiful pendant stalactites; but, on the sides of the deep chasm into which the cascade throws itself, there is seen an extraordinary accumulation of horizontal beds of tufa and travertin, from four to five hundred feet in thickness. The following seems the most probable explanation of their formation in this singular position. The Anio flows through a deep, irregular fissure or gorge in the Apennine limestone, which may have originated from subterranean movements, like many others of which we shall speak when treating of earthquakes. In this deep narrow channel there existed many small lakes, three of which have been destroyed since the time of history, by the erosive action of the torrent, the last of them having remained down to the sixth century of our era. We may suppose a similar lake of great depth to have existed at some remote period at Tivoli, and that, into this, the waters, charged with carbonate of lime, fell from a height inferior to that of the present cascade. Having, in their passage through the upper lakes, parted with their sand, pebbles, and coarse sediment, they only introduced into this lower pool, drift-wood, leaves, and other buoyant substances. In seasons when the water was low, a deposit of ordinary tufa, or travertin, formed along the bottom; but, at other times, when the torrent was swollen, the pool must have been greatly agitated, and every small particle of carbonate of lime which was precipitated, must have been whirled round again and again in various eddies, until it acquired many concentric coats, so as to resemble oolitic grains. If the violence of the motion be sufficient to cause the globule to be suspended for a sufficient length of time, it would grow to the size of a pea, or much larger. Small fragments of vegetable stems being incrusted on the sides of the stream, and then washed in, would form the nucleus of oval globules, and others of irregular shapes would be produced by the resting of fragments for a time on the bottom of the basin, where, after acquiring an unequal thickness of travertin on one side, they would again be set in motion. Sometimes globules, projecting above the general level of a stratum, would attract, by chemical affinity, other matter in the act of precipitation, and thus growing on all sides, with the exception of the point of contact, might at length form spheroids nearly perfect and many feet in diameter. Masses might increase above and below, so that a vertical section might afterwards present the phenomenon so common at Tivoli, where the nucleus of some of the concentric circles

has the appearance of having been suspended, without support, in the water, until it became a spheroidal mass of great dimensions. The section obtained of these deposits, about four hundred feet thick, immediately under the temples of Vesta and the Sibyl, displays some spheroids which are from *six to eight feet in diameter*, each concentric layer being about the eighth of an inch in thickness. The annexed diagram exhibits about fourteen feet of this immense mass as seen in the path cut out of the rock in descending from the temple of Vesta to the Grotto di Nettuno*. The beds (*a a*, diagram No. 5) are of hard travertin and soft tufa; below them is a pisolite (*b*), the globules being of different sizes; underneath this appears a mass of concretionary travertin (*c c*), some of the spheroids being of the above-mentioned extraordinary size. In some places (as at *d*), there is a mass of amorphous limestone, or tufa, surrounded by concentric layers. At the bottom is another bed of pisolite (*b*), in which the small nodules are about the size and shape of beans, and some of them of filberts, intermixed with some smaller oolitic grains. In the tufaceous strata, wood is seen converted into a light tufa. It is probable that the date of the greater portion of this calcareous formation may be anterior to the era of history, for we know that there was a great cascade at Tivoli in very ancient times; but, in the upper part of the travertin, is shewn the hollow left by a wheel, in which the outer circle and the spokes have been decomposed, and the spaces which they filled have been left void. It seems impossible to explain the position of this mould, without supposing that the wheel was imbedded before the lake was drained.

Our limits do not permit us to enter into minute details respecting the various limestones to which springs in different countries are continually giving birth. Pallas, in his journey along the Caucasus, a country now subject, from time to time, to be rent and fissured by violent earthquakes, enumerates a great many hot springs, which have deposited monticules of travertin precisely analogous in composition and

* I have not attempted to express, in this drawing, the innumerable thin layers of which these magnificent spheroids are composed, but the lines given mark some of the natural divisions into which they are separated by minute variations in the size or colour of the laminæ. The undulations also are much smaller, in proportion to the whole circumference, than is expressed in the diagram.

structure to those of the baths of San Filippo, and other localities in Italy. When speaking of the tophus-stone, as he terms these limestones, he often observes that it is *snow-white*, a description which is very applicable to the newer part of the deposit at San Filippo, where it has not become darkened by weathering. In many localities in the regions between the Caspian and Black seas, where subterranean convulsions are frequent, travellers mention calc-sinter as an abundant product of hot springs. Near the shores of the Lake Urmia (or Maragha), for example, a marble is rapidly deposited from a thermal spring, which is much used in ornamental architecture*. We might mention springs of the same kind in Calabria and Sicily, and indeed in almost all regions of volcanos and earthquakes which have been carefully investigated. In the limestone districts of England, as on Ingleborough Hill, in Yorkshire, we often see walls entirely constructed of calcareous tufa, enclosing terrestrial shells and vegetables, and similar tufa still continues to be formed in that district. The growth of stalactites, also, and stalagmites in caverns and grottos, is another familiar example of calcareous precipitates. To the solvent power of water, surcharged with carbonic acid, and percolating various winding rents and fissures, we may ascribe those innumerable subterranean cavities and winding passages which traverse the limestone in our own and many other countries.

In the marshes of the great plains of Hungary, horizontal beds of travertin, including recent fresh-water shells, are continually deposited, and are sufficiently solid to serve for building-stones, all the houses of Czeled being constructed of this material †. To analogous deposits in the lakes of Forfarshire, in Scotland, we shall refer more particularly when speaking of the imbedding of plants and animals in recent deposits. The quantity of calcareous rock which results from mineral waters in volcanic regions, conspicuous as it is, must be considered as insignificant, in comparison to that which is conveyed by rivers to the sea; and our inability to observe subaqueous accumulations resulting from this source, is one of many

* Hoff, Geschichte, &c., vol. ii., p. 114.

† Beudant, Voyage en Hongrie, tom. ii., p. 353.

causes of our inadequate conception of the changes now in progress on the earth's surface. It has often been supposed, that the greater part of the coral reefs in the Indian and Pacific oceans were based on submarine volcanos,—which seems indicated by the circular shape so frequently assumed by them; but perhaps a still stronger argument in favour of this theory might be deduced from the great abundance of carbonate of lime required for the rapid growth of zoophytic and shelly limestones,—an abundance which could only be looked for where there are active volcanos and frequent earthquakes, as amongst the isles of the South Pacific. We may confidently infer, that the development of organic life would be promoted in corals, sponges, and testaceous mollusca, by the heat, carbonic acid, lime, silica, and other mineral ingredients in a state of solution, given out by submarine springs, in the same manner as the vegetation is quickened in the lake of the Solfatara, in the Campagna di Roma, before described.

Gypseous springs.—All other mineral ingredients wherewith springs in general are impregnated, are insignificant in quantity in comparison to lime, and this earth is most frequently combined with carbonic acid. But as sulphuric acid and sulphuretted hydrogen are very frequently supplied by springs, we must presume that gypsum is now deposited largely in many seas and lakes. The gypseous precipitates, however, hitherto known on the land, appear to be confined to a very few springs. Those at Baden, near Vienna, which feed the public bath, may be cited as examples. Some of these supply, singly, from six hundred to one thousand cubic feet of water per hour, and deposit a fine powder, composed of a mixture of sulphate of lime, with sulphur and muriate of lime *.

SILICEOUS SPRINGS.

Azores.—In order that water should hold a very large quantity of silica in solution, it seems necessary that it should be raised to a high temperature †; and as it may retain a greater heat under the pressure of the sea than in the atmosphere, submarine springs may perhaps be more charged

* Prevost, Essai sur la Constitution Physique du Bassin de Vienne, p. 10.

† Daubeny, on Volcanos, p. 222.

with silex than any to which we have access. The hot springs of the Valle das Furnas, in the Island of St. Michael, rising through volcanic rocks, precipitate vast quantities of siliceous sinter, as it is usually termed. Around the circular basin of the largest spring, called "the Caldeira," which is between twenty and thirty feet in diameter, alternate layers are seen of a coarser variety of sinter mixed with clay, including grass, ferns, and reeds, in different states of petrifaction. Wherever the water has flowed, sinter is found rising in some places eight or ten inches above the ordinary level of the stream. The herbage and leaves, more or less incrusted with silex, exhibit all the successive steps of petrifaction, from the soft state to a complete conversion into stone; but, in some instances, alumina, which is likewise deposited from the hot waters, is the mineralizing material. Branches of the same ferns which now flourish in the island, are found completely petrified, preserving the same appearance as when vegetating, except that they acquire an ash-grey colour. Fragments of wood, and one entire bed from three to five feet in depth, composed of reeds now common in the island, have become completely mineralized. The most abundant variety of siliceous sinter occurs in layers from a quarter to half an inch in thickness, accumulated on each other often to the height of a foot and upwards, and constituting parallel, and for the most part horizontal, strata many yards in extent. This sinter has often a beautiful semi-opalescent lustre. One of the varieties differs from that of Iceland and Ischia in the larger proportion of water it contains, and in the absence of alumina and lime. A recent breccia is also in the act of forming, composed of obsidian, pumice, and scoriæ, cemented by siliceous sinter*.

Geysers of Iceland.—But the hot springs in various parts of Iceland, particularly the celebrated geysers, afford the most remarkable example of the deposition of silex. The circular reservoirs into which the geysers fall, are filled in the middle with a variety of opal, and round the edges with sinter. The plants, encrusted with the latter substance, have much the same appearance as those encrusted with calcareous tufa in our own

* Dr. Webster, on the Hot Springs of Furnas, Ed. Phil. Journ., vol. vi., p. 306.

country. The solution of the silex is supposed to be promoted by the presence of some mineral alkali. In some of the thermal waters of Iceland a vesicular rock is formed, containing portions of vegetables, more or less completely silicified. Amongst the various products also of springs in this island, is that admixture of clay and silica, called tripoli.

It has been found, by recent analysis, that several of the thermal waters of Ischia are impregnated with a certain proportion of silica. Some of the hot vapours of that island are above the temperature of boiling water; and many fissures, near Monte Vico, through which the hot steam passes, are coated with a siliceous incrustation, first noticed by Dr. Thompson under the name of fiorite.

In some places where silicification is in progress, the sources from whence the mineral matter is derived are as yet unknown. Thus the Danube has converted the external part of the piles of Trajan's bridge into silex; and the Irawadi, in Ava, has been supposed, ever since the time of the Jesuit Padre Duchatz, to have the same petrifying power, as has also Lough Neagh, in Ireland. Modern researches, however, in the Burman empire, have not confirmed, but have rather thrown doubt upon the lapidifying property of the Ava river *. The constant flow of mineral waters, even when charged with a small proportion of silica, as those of Ischia, may supply certain species of corals and sponges with matter for their siliceous secretions; but when in a volcanic archipelago, or a region of submarine volcanos, there are springs so saturated with silica, as those of Iceland and the Azores, we may expect beds of chert or layers and nodules of silex, to be spread out far and wide over the bed of the sea, and interstratified with shelly and calcareous deposits, which may be forming there, or with matter derived from the wasting cliffs or volcanic ejections.

Ferruginous Springs.—The waters of almost all springs contain some iron in solution; and it is a fact familiar to all, that many of them are so copiously impregnated with this metal, as to stain the rocks or herbage through which they pass, and to bind together sand and gravel into solid masses. We may

* Dr. Buckland, Geol. Trans., second series, vol. ii., part 3, p. 384.

naturally, therefore, conclude that this iron, which is constantly conveyed into lakes and seas from the interior of the earth, and not returned again to the land by evaporation in the atmospheric waters, must act as a colouring and cementing principle in the subaqueous deposits now in progress. When we find, therefore, that so many sandstones and other rocks in the sedimentary strata of ancient lakes and seas are bound together or coloured by iron, it presents us with a striking point of analogy between the state of things at very different epochs. In the older formations we meet with great abundance of carbonate and sulphate of iron; and in chalybeate waters at present, this metal is most frequently in the state of a carbonate, as in those of Tunbridge, for example. Sulphuric acid, however, is often the solvent, which is in many cases derived from the decomposition of pyrites.

Brine Springs.—So great is the quantity of muriate of soda in some springs, that they yield one-fourth of their weight in salt. They are rarely, however, so saturated, and generally contain, intermixed with salt, carbonate and sulphate of lime, magnesia, and other mineral ingredients. The brine springs of Cheshire are the richest in our country; those of Barton and Northwich being almost fully saturated. These brine springs rise up through strata of sandstone and red marl, which contain large beds of rock-salt. The origin of the brine, therefore, may be derived in this and many other instances from beds of fossil salt; but as muriate of soda is one of the products of volcanic emanations and of springs in volcanic regions, the original source of salt may be as deep seated as that of lava.

The waters of the Dead Sea contain scarcely anything, except muriatic salts, which lends countenance, observes Dr. Daubeny, to the volcanic origin of the surrounding country, these salts being frequent products of volcanic eruptions. Many springs in Sicily contain muriate of soda, and the "fiume salso," in particular, is impregnated with so large a quantity, that cattle refuse to drink of it. If rivers or springs, thus impregnated, enter a lake or estuary, it is evident that they may give rise to partial precipitates of salt.

A hot spring, rising through granite, at Saint Nectaire, in Auvergne, may be mentioned as one of many, containing a large proportion of muriate of soda, together with magnesia and other ingredients *.

Carbonated Springs.—Carbonic acid gas is very plentifully disengaged from springs in almost all countries, but particularly near active or extinct volcanos. This elastic fluid has the property of decomposing many of the hardest rocks with which it comes in contact, particularly that numerous class in whose composition felspar is an ingredient. It renders the oxide of iron soluble in water, and contributes, as was before stated, to the solution of calcareous matter. In volcanic districts, these gaseous emanations are not confined to springs, but rise up in the state of pure gas from the soil in various places. The Grotto delle Cane, near Naples, affords an example, and prodigious quantities are now annually disengaged from every part of the Limagne d'Auvergne, where it appears to have been developed in equal quantity from time immemorial. As the acid is invisible, it is not observed, except an excavation be made, wherein it immediately accumulates so that it will extinguish a candle. There are some springs in this district, where the water is seen bubbling and boiling up with much noise, in consequence of the abundant disengagement of this gas. The whole vegetation is affected, and many trees, such as the walnut, flourish more luxuriantly than they would otherwise do in the same soil and climate,—the leaves probably absorbing carbonic acid. This gas is found in springs rising through the granite near Clermont, as well as in the tertiary limestones of the Limagne †. In the environs of Pont-Gibaud, not far from Clermont, a rock belonging to the gneiss formation, in which lead-mines are worked, has been found to be quite saturated with carbonic acid gas, which is constantly disengaged. The carbonates of iron, lime, and manganese are so dissolved, that the rock is rendered soft, and the quartz alone remains unattacked ‡. Not far off is the small volcanic cone of Chaluzet, which once broke up through the gneiss, and sent forth a lava stream.

The disintegration of granite is a striking feature of large districts in Auvergne, especially in the neighbourhood of Cler-

* Annales de l'Auvergne, tom. i., p. 234.

† Le Coq, Annales de l'Auvergne, tom. i., p. 217. May, 1828,

‡ Ann. Scient. de l'Auvergne, tom. ii., June, 1829.

mont. This decay was called, by Dolomieu, "la maladie du granite;" and the rock may with propriety be said to have *the rot*, for it crumbles to pieces in the hand. The phenomenon may, without doubt, be ascribed to the continual disengagement of carbonic acid gas from numerous fissures. In the plains of the Po, between Verona and Parma, especially at Villa Franca, south of Mantua, I observed great beds of alluvium, consisting chiefly of primary pebbles percolated by spring water, charged with carbonate of lime and carbonic acid in great abundance. They are, for the most part, encrusted with calc-sinter; and the rounded blocks of gneiss, which have all the appearance of solidity, have been so disintegrated by the carbonic acid as readily to fall to pieces. The Po and other rivers, in winding through this plain, might now remove with ease those masses which, at a more remote period, the stream was unable to carry farther towards the sea; and in this example we may perceive how necessary it is, in reasoning on the transporting power of running water, to consider all the numerous agents which may co-operate, in the lapse of ages, in conveying the wreck of mountains to the sea. A granite block might remain stationary for ages, and defy the power of a large river; till at length a small spring may break out, surcharged with carbonic acid,—the rock may be decomposed, and a streamlet may transport the whole mass to the ocean.

The subtraction of many of the elements of rocks by the solvent power of carbonic acid, ascending both in a gaseous state and mixed with spring-water in the crevices of rocks, must be one of the most powerful sources of those internal changes and re-arrangements of particles so often observed in strata of every age. The calcareous matter, for example, of shells is often entirely removed and replaced by carbonate of iron, pyrites, or silex, or some other ingredient, such as mineral waters usually contain in solution. It rarely happens, except in limestone rocks, that the carbonic acid can dissolve all the constituent parts of the mass; and for this reason, probably, calcareous rocks are almost the only ones in which great caverns and long winding passages are found. The grottos and subterranean passages, in certain lava-currents, are due to a different cause, and will be spoken of in another place.

Petróleum Springs.—Springs impregnated with petroleum, and the various minerals allied to it, as bitumen, naphtha, asphaltum, and pitch, are very numerous, and are, in many cases, undoubtedly connected with subterranean fires, which raise or sublime the more subtle parts of the bituminous matters contained in rocks. Many springs in the territory of Modena and Parma, in Sicily, produce petroleum in abundance; but the most powerful, perhaps, yet known, are those on the Irawadi, in the Burman empire. In one locality there are said to be five hundred and twenty wells, which yield annually four hundred thousand hogsheads of petroleum *.

Fluid bitumen is seen to ooze from the bottom of the sea, on both sides of the island of Trinidad, and to rise up to the surface of the water. Near Cape La Braye there is a vortex which, in stormy weather, according to Captain Mallet, gushes out, raising the water five or six feet, and covers the surface for a considerable space with petroleum, or tar; and the same author quotes Gumilla, as stating in his "Description of the Orinoco," that, about seventy years ago, a spot of land on the western coast of Trinidad, near half-way between the capital and Indian village, sank suddenly, and was immediately replaced by a small lake of pitch, to the great terror of the inhabitants †. It is probable, that the great pitch-lake of Trinidad owes its origin to a similar cause, and Dr. Nugent has justly remarked, that in that district all the circumstances are now combined from which deposits of pitch may have originated. The Orinoco has, for ages, been rolling down great quantities of woody and vegetable bodies into the surrounding sea, where, by the influence of currents and eddies, they may be arrested and accumulated in particular places. The frequent occurrence of earthquakes and other indications of volcanic action in those parts, lend countenance to the opinion, that these vegetable substances may have undergone, by the agency of subterranean fire, those transformations and chemical changes which produce petroleum, and may, by the same causes, be forced up to the surface, where, by exposure

* Symes, Embassy to Ava, vol. ii.—Geol. Trans., second series, vol. ii., part 3, p. 388.

† Dr. Nugent, Geol. Trans., vol. i., p. 69.

to the air, it becomes inspissated, and forms the different varieties of pure and earthy pitch, or asphaltum, so abundant in the island *.

The bituminous shales, so common in geological formations of different ages, as well as many stratified deposits of bitumen and pitch, seem clearly to attest that, at former periods, springs, in various parts of the world, were as commonly impregnated as now with bituminous matter, which was carried down by rivers into lakes and seas. We may indeed remark generally, that a large portion of the finer particles and the more crystalline substances found in sedimentary rocks of different ages, are composed of the same elements as are now held in solution by springs, while the coarser materials bear an equally strong resemblance to the alluvial matter in the beds of existing torrents and rivers.

* Dr. Nugent, Geol. Trans., vol. i., p. 67.

CHAPTER XIII.

Reproductive effects of running water—Division of deltas into lacustrine, mediterranean, and oceanic—Lake deltas—Growth of the delta of the Rhone in the Lake of Geneva—Chronological computations of the age of deltas—Recent deposits in Lake Superior—Deltas of inland seas—Rapid shallowing of the Baltic—Arguments for and against the hypothesis of Celsius—Elevated beaches on the coast of Sweden—Marine delta of the Rhone—Various proofs of its increase—Stony nature of its deposits—Delta of the Po, Adige, Isonzo, and other rivers entering the Adriatic—Rapid conversion of that gulf into land—Mineral characters of the new deposits—Delta of the Nile—Its increase since the time of Homer—Its growth why checked at present.

We have hitherto considered the destroying agency of running water, as exhibited in the disintegration of rocks and transportation of matter from higher to lower levels. It remains for us to examine the reproductive effects of the same cause. The aggregate amount of matter accumulated in a given time at the mouths of rivers, where they enter lakes or seas, affords clearer data for estimating the energy of the excavating power of running water on the land, than the separate study of the operations of the same cause in the countless ramifications into which every great system of valleys is divided. We shall proceed to select some of the leading facts at present ascertained respecting the growth of deltas, and shall then offer some general observations on the quantity of subaqueous sediment transported by rivers, and on the manner of its distribution. Deltas may be divided into, first, those which are formed in lakes; secondly, those formed in inland seas; and thirdly, those formed on the borders of the ocean. The most characteristic distinction between the lacustrine and marine deltas consists in the nature of the organic remains, which become imbedded in their deposits; for, in the case of a lake, it is obvious that these must consist exclusively of such genera of animals as inhabit the land or the waters of a river or lake; whereas, in the other case, there will be an admixture and most frequently a predominance of animals which inhabit salt water. In regard, however, to the distribution of inorganic matter, the deposits of lakes and inland seas are formed under very analogous circumstances, and may be contra-distinguished from those on the shores of the great ocean, where the tides co-operating with currents give rise to a distinct class of phenomena. In lakes and inland seas, even of the largest dimensions, the tides are almost insensible, and the currents are, for the most part, inconsiderable, although some striking exceptions to this rule will be mentioned when we treat of tides and currents.

DELTAS IN LAKES.

Lake of Geneva.—It is natural to begin our examination with an inquiry into the new deposits in lakes, as they exemplify the first reproductive operations in which rivers are engaged when they convey the detritus of rocks and the ingredients of mineral springs from mountainous regions. The accession of new land at the mouth of the Rhone, at the upper end of the Lake of Geneva, or the Leman Lake, presents us with an example of a considerable thickness of strata, which have accumulated since the historical era. This sheet o. water is about thirty-seven miles long, and its breadth is from two to eight miles. The shape of the bottom is very irregular, the depth having been found, by late measurements, to vary from twenty to one hundred and sixty fathoms *. The Rhone, where it enters at the upper end, is turbid and discoloured; but its waters, where it issues at the town of Geneva, are beautifully clear and transparent. An ancient town, called Port Vallais, (Portus Valesiæ of the Romans,) once situated at the water's edge, at the upper end, is now more than a mile and a half inland,—this intervening alluvial tract having been acquired in about eight centuries. The remainder of the delta consists of a flat alluvial plain, about five or six miles in length, composed of sand and mud, a little raised above the level of the river, and full of marshes.

Mr. De la Beche found, after numerous soundings in all parts of the lake, that there was a pretty uniform depth of from one hundred and twenty to one hundred and sixty fathoms throughout the central region, and, on approaching the delta,

* De la Beche, Ed. Phil. Journ., vol. ii., p. 107, Jan. 1820

the shallowing of the bottom began to be very sensible at a distance of about a mile and three-quarters from the mouth of the Rhone; for a line drawn from St. Gingoulph to Vevey, gives a mean depth of somewhat less than six hundred feet, and from that part to the Rhone, the fluviatile mud is always found along the bottom *. We may state, therefore, that the strata annually produced are about two miles in length: so that, notwithstanding the great depth of the lake, the new deposits are not inclined at a high angle; the dip of the beds, indeed, is so slight, that they would be termed, in ordinary geological language, horizontal. The strata probably consists of alternations of finer and coarser particles, for during the hotter months from April to August, when the snows melt, the volume and velocity of the river are greatest, and large quantities of sand, mud, vegetable matter, and drift wood are introduced; but, during the rest of the year, the influx is comparatively feeble, so much so, that the whole lake, according to Saussure, stands six feet lower. If then, we could obtain a section of the accumulation formed in the last eight centuries, we should see a great series of strata, probably from six to nine hundred feet thick, and nearly two miles in length, inclined at a very slight angle. In the mean time, a great number of smaller deltas are growing around the borders of the lake, at the mouths of rapid torrents, which pour in large masses of sand and pebbles. The body of water in these torrents is too small to enable them to spread out the transported matter over so extensive an area as the Rhone. Thus, for example, there is a depth of eighty fathoms within half a mile of the shore, immediately opposite the great torrent which enters east of Ripaille, so that the dip of the strata in that delta is about four times as great as those deposited by the main river at the upper extremity of the lake †.

The capacity of this basin being now ascertained, it would be an interesting subject of inquiry, to determine in what number of years the Leman Lake will be converted into dry land. It would not be very difficult to obtain the elements for such a calculation, so as to approximate at least to the quantity of time required for the accomplishment of this result. The number of cubic feet of

* De la Beche, M.S.

† Ibid.

water annually discharged by the river into the lake being estimated, experiments might be made in the winter and summer months, to determine the proportion of matter held in suspension or in chemical solution by the Rhone. It would be also necessary to allow for the heavier matter drifted along at the bottom, which might be estimated on hydrostatical principles, when the average size of the gravel and the volume and velocity of the stream at different seasons were known. Supposing all these observations to have been made, it would be more easy to calculate the future than the former progress of the delta, because it would be a laborious task to ascertain, with any degree of precision, the original depth and extent of that part of the lake which is already filled up. Even if this information were accurately obtained by borings, it would only enable us to approximate within a certain number of centuries to the time when the Rhone began to form its present delta; but this would not give us the date of the origin of the Leman Lake in its present form, because the river may have flowed into it for thousands of years, without importing any sediment whatever. Such would have been the case, if the waters had first passed through a chain of upper lakes; and that this was actually the fact, is indicated both by the course of the Rhone between Martigny and the Lake of Geneva, and still more decidedly, by the channels of many of its principal feeders.

If we ascend, for example, the valley through which the Dranse flows, we find that it consists of a succession of basins, one above the other, in each of which there is a wide expanse of flat alluvial lands, separated from the next basin by a rocky gorge, once evidently the barrier of a lake. The river has filled the lake, and partially cut through the barrier, which it is still gradually eroding to a greater depth. The examination of almost all valleys in mountainous districts affords abundant proofs of the obliteration of a series of lakes, by the filling up of hollows and the cutting through of rocky barriers—a process by which running water ever labours to produce a more uniform declivity. Before, therefore, we can pretend even to hazard a conjecture as to the era at which any particular delta commenced, we must be thoroughly acquainted with the geological history of the whole system of higher valleys which communicate with the main stream, and all the changes which they have undergone since the last series of convulsions which agitated and altered the face of the country. The probability, therefore, of error in our chronological computations, where we omit to pay due attention to these circumstances, increases in proportion to the time that may have elapsed since the last disturbance of the country by subterranean movements, and in proportion to the extent of the hydrographical basin on which we may happen to speculate. The Alpine rivers of Vallais are prevented at present from contributing their sedimentary contingent to the delta of the Rhone in the Mediterranean, because they are intercepted by the Leman Lake; but when this is filled, they will transport as much, or nearly as much, matter to the sea as they now pour into that lake. They will then flow through a long, flat, alluvial plain, between Villeneuve and Geneva, from two to eight miles in breadth, which will present no superficial marks of the existence of a thickness of more than one thousand feet of recent sediment below. Many hundred alluvial tracts of equal, and some of much greater area, may be seen if we follow up the Rhone from its mouth, or explore the valleys of many of its principal tributaries.

What, then, shall we think of the presumption of De Luc, Kirwan, and their followers, who confidently deduced from the phenomena of modern deltas the recent origin of the present form of our continents, without pretending to have collected any one of the numerous data by which so complicated a problem can be solved? Had they, after making all the necessary investigations, succeeded in proving, as they desired, that the delta of the Rhone, and the new deposits at the mouths of all other rivers, whether in lakes or seas, had required about four thousand years to attain their present dimensions, the conclusion would have been fatal to the chronological theories, which they were anxious to confirm. The popular reception of these, and similar sophisms, respecting the effects of causes in diurnal action, has hitherto thrown stumbling-blocks in the way of those geologists who desire to pursue the science according to the rules of inductive philosophy. If speculations so vague and visionary can be proposed concerning natural operations now passing before our eyes—if authors may thus dogmatize, with impunity, on subjects capable of being determined with considerable degree of precision, can we be surprised that they who reason on the more obscure phenomena of remote ages, should wander in a maze of error and inconsistency *?

The Leman Lake fills a great cavity in rocky strata, composed of a tertiary conglomerate and sand, which constitutes its bottom, almost all its northern banks, and a great part of its southern or Alpine side. It has often been asked, why this cavity has not been filled up by the detritus of rocks, removed from the numerous valleys now drained by the waters which enter the lake? In order to remove this difficulty, it would be necessary to enter into a description of the strata of different ages composing the Alps and the Subalpine districts; to point out the distinct periods of their elevation above the sea, and the pre-existence of many mountain valleys, even to the formation of those deposits wherein the Lake of Geneva is contained. It would be premature, therefore, to enter upon this subject at present, to which we shall revert when we have described the phenomena of some of the ancient strata.

Lake Superior.—Lake Superior is the largest body of fresh water in the world, being about one thousand five hundred geographical miles in circumference, if we follow the sinuosities of its coasts, its length, on a curved line through its centre, being about three hundred and sixty, and its extreme breadth one hundred and forty geographical miles. Its average depth varies from eighty to one hundred and fifty fathoms; but, according to Captain Bayfield, there is reason to think that its greatest depth would not be overrated at two hundred fathoms†, so that its bottom is, in some parts, nearly six hundred feet below the level of the Atlantic, as its surface is about as much above it. There are appearances in different parts of this, as of the

* It is an encouraging circumstance, that the cultivators of the science in our own country have begun to appreciate the true value of the principles of reasoning most usually applied to geological questions. While writing this chapter (April, 1830), I happened to attend a meeting of the Geological Society of London, where the president, in his address, made use of the expression, *a geological logician*. A smile was seen on the countenances of some of the audience, while many of the members, like Cicero's augurs, could not resist laughing; so ludicrous appeared the association of Geology and Logic.

† Trans. of Lit. and Hist. Soc. of Quebec, vol. i., p. 5, 1829.

other Canadian lakes, leading us to infer that its waters have formerly occupied a much higher level than they reach at present; for at a considerable distance from the present shores, parallel lines of rolled stones and shells are seen rising one above the other, like the seats of an amphitheatre. These ancient lines of shingle are exactly similar to the present beaches in most bays, and they often attain an elevation of forty or fifty feet above the present level. As the heaviest gales of wind do not raise the waters more than three or four feet *, the elevated beaches must either be referred to the subsidence of the lake at former periods, in consequence of the wearing down of its barrier, or to the upraising of the shores by earthquakes, like those which have produced similar phenomena on the coast of Chili. But there seem to be no facts which lend countenance to the latter hypothesis, in reference to the North American lakes. The streams which discharge their waters into Lake Superior are several hundred in number, without reckoning those of smaller size; and the quantity of water supplied by them is many times greater than that discharged at the Falls of St. Mary, the only outlet. The evaporation, therefore, is very great, and such as might be expected from so vast an extent of surface.

On the northern side, which is encircled by primary mountains, the rivers sweep in many large boulders with smaller gravel and sand, chiefly composed of granitic and trap rocks. There are also currents in the lake, in various directions, caused by the continued prevalence of strong winds, and to their influence we may attribute the diffusion of finer mud far and wide over great areas; for, by numerous soundings made during the late survey, it was ascertained that the bottom consists generally of a very adhesive clay, containing shells of the species at present existing in the lake. When exposed to the air, this clay immediately becomes indurated in so great a degree, as to require a smart blow to break it. It effervesces slightly with diluted nitric acid, and is of different colours in different parts of the lake; in one district blue, in another red, and in a third

* Captain Bayfield remarks, that Dr. Bigsby, to whom we are indebted for several communications respecting the geology of the Canadian lakes, was misinformed by the fur traders in regard to the extraordinary height (twenty or thirty feet) to which he asserts that the autumnal gales will raise the water of Lake Superior. Trans. of Lit. and Hist. Soc. of Quebec, vol. i., p. 7, 1829.

white, hardening into a substance resembling pipe-clay *. From these statements, the geologist will not fail to remark how closely these recent lacustrine formations in America resemble the tertiary argillaceous and calcareous marls of lacustrine origin in Central France. In both cases, many of the genera of shells most abundant, as Lymnea and Planorbis, are the same; and in regard to other classes of organic remains, there must be the closest analogy, as we shall endeavour more fully to explain when speaking of the imbedding of plants and animals in recent deposits.

DELTAS OF INLAND SEAS.

Deltas of the Baltic.—Having offered these few remarks on lacustrine deltas now in progress, we may next turn our attention to those of inland seas.

The shallowing and conversion into land of many parts of the Baltic, especially the Gulfs of Bothnia and Finland, have been demonstrated by a series of accurate observations, for which we are in a great measure indebted to the animated controversy which has been kept up, since the middle of the last century, concerning the gradual lowering of the level of the Baltic. Celsius, the Swedish astronomer, first originated the idea that from the earliest times there had been a progressive fall of about forty-five inches in a century, in the mean level of the waters of that sea. He contended that this change rested not only on modern observations, but on the authority of the ancient geographers, who stated that Scandinavia was formerly an island. By the gradual depression of the sea, he said, that great island became connected with the continent; and that this event happened after the time of Pliny, and before the ninth century of our era. To the arguments urged in support of these positions, his opponents objected that the ancients were so ignorant of the geography of the most northern parts of Europe, that their authority was entitled to no weight; and that their representation of Scandinavia as an island, might with more propriety be adduced to prove the scantiness of their information, than to confirm so bold an hypothesis. It was also remarked that if the land which connected Scandinavia with the main continent

* Trans. of Lit. and Hist. Soc. of Quebec, vol. i., p. 5, 1829.

was laid dry between the time of Pliny and the ninth century, to the extent to which it is known to have risen above the sea at the latter period, the rate of depression could not have been uniform, as was pretended, for it ought to have fallen much more rapidly between the ninth and eighteenth century.

Many of the physical proofs relied on by Celsius and his followers, show clearly that they did not distinguish between the shallowing of the water by deposition of fresh sediment, and the diminution of depth caused by subsidence of the sea. By their own statements, it appeared that the accessions of new land, and the loss of depth, were at the mouths of rivers, or in certain deep bays, into which it is well known that sand and mud are carried by currents. As illustrating, however, the gradual conversion of the Gulf of Bothnia into land, their observations deserve great attention. Thus, for example, they pointed out the fact, that at Pitea half a mile was gained in forty-five years, and at Lulea no less than a mile in twenty-eight years. Ancient ports, on the same coast, had become inland cities. Considerable tracts of the gulf were rendered three feet shallower in the course of fifty years—many old fishing-grounds had been changed into dry land—small islands had been joined to the continent. According to Linnæus, the increase of land on the eastern side of Gothland near Hoburg, was about two or three toises annually for ninety years*. Besides these changes, it was asserted that along the southern shore, also, of the Baltic, particularly in West Prussia and Pomerania, anchors and sunk ships had been discovered far inland; and although these occurrences were partly accounted for by the silting up of river-beds, yet the tradition seems worthy of credit, that a bay of the sea penetrated, at a remote period, much farther to the south in that direction. These, and many other facts, are of geological interest, although they afford no confirmation to the theory of Celsius.

His most plausible arguments were derived from the alleged exposure of certain insular rocks in the Bothnian and other bays, which were declared to have been once entirely covered with water, but which had gradually protruded themselves more and more above the waves, until, in the course of about

* Linn. de tell. habit. increm.

a century and a half, they grew to be eight feet high. Of this phenomenon, the following explanation was offered by his opponents. The islands in question consisted of sand and drift-stones, and the waves, during great tempests, threw up new matter upon them, or converted shoals into islands. Sometimes, also, icebergs, heavily laden with rock, were stranded on a shoal or driven up on a low island; and when they melted away, they left a mass of debris, many feet in height. Browallius, and other Swedish naturalists, pointed out that some of these islands were lower than formerly; so that, by reference to this kind of evidence, there was equally good reason for contending that the level of the Baltic was gradually rising. They also added another curious and very conclusive proof of the permanency of the water-level for many centuries. On the Finland coast were some large pines, growing close to the water's edge; these were cut down, and, by counting the concentric rings of annual growth, as seen in a transverse section of the trunk, it was demonstrated that they had stood there for four hundred years. Now, according to the Celsian hypothesis, the sea had sunk fifteen feet during that period, so that the germination and early growth of these pines must have been for many seasons below the level of the water. In like manner it was shewn, that the lower walls of many ancient castles, such as those of Sonderburg and Abo, reached then to the water's edge, and must, therefore, according to the theory of Celsius, have been originally constructed below the level of the sea. Another unanswerable argument in proof of the stability of the level of the Baltic, was drawn from the island of Saltholm, not far from Copenhagen. This isle is so low that, in autumn and winter, it is permanently overflowed; and is only dry in summer, when it serves for pasturing cattle. It appears, from documents of the year 1280, that this island was then also in the same state, and exactly on a level with the mean height of the sea, instead of being twenty feet under water, as it ought to have been according to the computation of Celsius. Several towns, also, on the shores of the Baltic, as Lubeck, Wismar, Rostock, Stralsund, and others, after six and even eight hundred years, are as little elevated above the sea as at the era of their foundation, being now close to the water's edge. The lowest part of Dantzic was no higher than the mean level of the sea in the year 1000; and after eight centuries, its relative position remains exactly the same *.

Notwithstanding these convincing proofs that the supposed change in the relative level of land and sea arose from some local appearances, there are still many who contend for a lowering of the Baltic; and many Swedish officers of the pilotage establishment declared, in the year 1821, in favour of this opinion, after measuring the height of landmarks placed at certain heights above the sea, half a century before, as objects of comparison for the express purpose of settling the point at issue. Before we attach any weight to these assertions, which only relate to slight differences of elevation, we ought to be assured that the observers were on their guard against every imaginable cause of deception arising from local circumstances. Thus, for example, if the height of an alluvial plain was taken during the last century, it might have been subsequently raised by fresh deposits, and thus the sea would appear to have sunk; or, if a mark was cut in the rocks, the sea may have been several inches or even feet higher at one period than another, in consequence of the setting in of a current urged by particular winds into a long narrow gulf, which cause is well known to raise the Baltic, at some seasons, two feet above its ordinary level.

Von Buch, in his travels, discovered in Norway, and at Uddevalla, in Sweden, beds of shells of existing species at considerable heights above the level of the water. Since that time, several other naturalists have confirmed his observation; and, according to Ström, some deposits occur at an elevation of more than four hundred feet above the sea in the northern part of Norway. M. Alex. Brongniart, who has lately visited Uddevalla in Gotheborg, a port at the entrance of the Baltic, informs us that the principal mass of shells in the creek of Uddevalla rises about two hundred feet above the level of the sea, resting on rocks of gneiss. All the species are identical with those now inhabiting the contiguous sea, and are for the most part entire, although some of them are broken, as happens on a sea-beach. They are nearly free from any admixture of earthy matter. The reader need scarcely be reminded that, at the height of a

* For a full account of the Celsian controversy, we may refer our readers to Von Hoff, Geschichte, &c., vol. i., p. 439.

few feet above the beach, on our coasts, the rocks, where they are alternately submerged and laid dry by the ebbing and flowing tide, are frequently covered with barnacles or balani, which are firmly attached. On examining, with care, the smooth surface of the gneiss, immediately above the ancient shelly beach at Uddevalla, M. Brongniart found, in a similar manner, balani adhering to the rocks, so that there can be no doubt that the sea had for a long period sojourned on the spot *. Now, this interesting fact is precisely analogous to one well known to all who are acquainted with the geology of the borders of the Mediterranean. Perforating shells (Venus lithophaga, Lam.) excavate funnel-shaped hollows in the hardest limestone and marble, along the present sea-shores; and lines of these perforations, sometimes containing the same species of shells, have been discovered at various heights above the sea near Naples, in Calabria, at Monte Pelegrino, in the Bay of Palermo, and other localities. As many of these districts have been violently shaken by earthquakes within the historical era, and as the land has been sometimes raised and sometimes depressed, as we shall afterwards show by examples, there is no difficulty in explaining the phenomena, provided time be allowed. But no argument can be derived, from such observations, in support of great upheavings of the coast, whether by slow or sudden operations in modern times, unless we use the term *modern* in a geological sense. On the contrary, we know that the physical outline of the coast and heights in the bay of Palermo, when it was a Greek port more than two thousand years ago, was so nearly the same as it is at present, that the beds of recent shells, and the perforations in the rocks, must have stood nearly in the same relation to the level of the Mediterranean as they stand now. The high beaches on the Norwegian and Swedish coast establish the important and certainly very unexpected fact, that those parts of Europe have been the theatres of considerable subterranean movements within the present zoological era, or since the seas were inhabited by species now our contemporaries. But the phenomena do not lend the slightest support to the Celsian hypothesis, nor to that extraordinary notion proposed in our own times by Von Buch, who imagines that the whole of the land

* Tableau des Terrains, &c., p. 89. 1829.

along the northern and western shores of the Baltic is slowly and insensibly rising! No countries have been more entirely free from earthquakes since the times of authentic history than Norway, Sweden, and Denmark. In common with our own island, and, indeed with almost every spot on the globe, they have experienced some slight shocks at certain periods, as during the earthquake of Lisbon, and on a few other occasions, but these may rather be considered as prolonged vibrations in the crust of the earth, extending in the manner of sounds through the air to almost indefinite distances, than as those violent movements which in the great regions of active volcanos change, from time to time, the relative level of land and sea.

Delta of the Rhone.—We may now turn our attention to some of the principal deltas of the Mediterranean, for no other inland sea affords so many examples of accessions of new lands at the mouths of rivers within the records of authentic history. We have already considered the lacustrine delta of the Rhone in Switzerland, and we shall now describe its contemporaneous marine delta. Scarcely has the river passed out of the Leman Lake, before its pure waters are again filled with sand and sediment by the impetuous Arve descending from the highest Alps, and bearing along in its current the granitic detritus annually carried down by the glaciers of Mont Blanc. The Rhone afterwards receives vast contributions of transported matter from the Alps of Dauphiny, and the primary and volcanic mountains of Central France; and when at length it enters the Mediterranean, it discolours its blue waters with a whitish sediment for the distance of between six and seven miles from its mouth, throughout which space the current of fresh-water is perceptible. Strabo's description of the delta is so inapplicable to its present configuration, as to attest a complete alteration in the physical features of the country since the Augustan age. It appears, however, that the head of the delta, or the point at which it begins to ramify, has remained unaltered since the time of Pliny, for he states that the Rhone divided itself at Arles into two arms. This is the case at present; one of the branches being now called Le petit Rhône, which is again subdivided before entering the Mediterranean. The advance of the base of the delta, in the last eighteen centuries, is demonstrated by many curious antiquarian monuments. The most striking of these is the great detour made by the old Roman road from Ugernum to Beziers (part of the high road between Aix, Aquæ Sextiæ, and Nismes, Nemausus). It is clear that, when this was first constructed, it was impossible to pass in a direct line, as now, across the delta, and that either the sea or marshes intervened in a tract now consisting of terra firma *. Astruc also remarks, that all the places on the low lands, lying to the north of the old Roman road between Nismes and Beziers, have names of Celtic origin evidently given to them by the first inhabitants of the country; whereas the places lying south of that road, towards the sea, have names of Latin derivation, and were clearly founded after the Roman language had been introduced. Another proof, also, of the great extent of land which has come into existence since the Romans conquered and colonized Gaul, is derived from the fact, that the Roman writers never mention the thermal waters of Balaruc in the delta, although they were well acquainted with those of Aix and others, still more distant, and attached great importance to them, as they invariably did to all hot springs. The waters of Balaruc, therefore, must have formerly issued under the sea—a common phenomenon on the borders of the Mediterranean; and on the advance of the delta they continued to flow out through the new deposits. Among the more direct proofs of the increase of land, we find that Mese, described under the appellation of Mesua Collis by Pomponius Mela †, and stated by him to be nearly an island, is now far inland. Notre Dame des Ports, also, was a harbour in 898, but is now a league from the shore. Psalmodi was an island in 815, and is now two leagues from the sea. Several old lines of towers and sea-marks occur at different distances from the present coast, all indicating the successive retreat of the sea, for each line has in its turn become useless to mariners, which may well be conceived when we state that the tower of Tignaux, erected on the shore so late as the year 1737, is already a French mile remote from it ‡.

By the confluence of the Rhone and the currents of the

* Mem. d'Astruc, cited by Von Hoff, vol. i., p. 228.

† Lib. II., c. v.

‡ Bouche, Chorographie et Hist. de Provence, vol. i., p. 23, cited by Hoff, vol. i., p. 290.

Mediterranean driven by winds from the south, sand-bars are often formed across the mouths of the river: by these means considerable spaces become divided off from the sea, and subsequently from the river also, when it shifts its channels of efflux. As some of these etangs, as they are called, are subject to the occasional ingress of the river when flooded, and of the sea during storms, they are alternately salt and fresh. Others, after being filled with salt-water, are often lowered by evaporation till they become more salt than the sea; and it has happened, occasionally, that a considerable precipitate of muriate of soda has taken place in these natural salterns. During the latter part of Napoleon's career, when the excise-laws were enforced with extreme rigour, the police was employed to prevent such salt from being used. The fluviatile and marine shells enclosed in these small lakes, often live together in brackish water; but the uncongenial nature of the fluid usually produces a dwarfish size, and sometimes gives rise to strange varieties in form and colour.

Captain Smyth, in the late survey of the coast of the Mediterranean, found the sea, opposite the mouth of the Rhone, to deepen gradually from four to forty fathoms, within a distance of six or seven miles, over which the discoloured fresh-water extends; so that the inclination of the new deposits must be too slight to be appreciable in such an extent of section as a geologist usually obtains in examining ancient formations. When the wind blew from the south-west, the ships employed in the survey were obliged to quit their moorings; and when they returned, the new sand-banks in the delta were found covered over with a great abundance of marine shells. By this means, we learn how occasional beds of drifted marine shells may become interstratified with fresh-water strata at the mouths of rivers.

That a great proportion, at least, of the new deposit in the delta of the Rhone consists of *rock*, and not of loose incoherent matter, is perfectly ascertained. In the museum at Montpellier is a cannon taken up from the sea near the mouth of the river, imbedded in a crystalline calcareous rock. Large masses, also, are continually taken up of an arenaceous rock, cemented by calcareous matter, including multitudes of broken shells of recent species. The observations recently made on this subject corroborate the former statement of Marsilli *, that the earthy deposits of the coast of Languedoc form a stony substance, for which reason he ascribed a certain bituminous, saline, and glutinous nature, to the substances brought down with sand by the Rhone. If the number of mineral springs charged with carbonate of lime which fall into the Rhone and its feeders in different parts of France be considered, we shall feel no surprise at the lapidification of the newly-deposited sediment in this delta. It should be remembered, that the fresh-water introduced by rivers, being lighter than the water of the sea, floats over the latter, and remains upon the surface for a considerable distance. Consequently, it is exposed to as much evaporation as the waters of a lake; and the area over which the river-water is spread, at the junction of great rivers and the sea, may well be compared, in point of extent, to that of considerable lakes. Now, it is well known, that so great is the quantity of water carried off by evaporation in some lakes, that it is nearly equal to the water flowing in; and in some inland seas, as the Caspian, it is quite equal. We may, therefore, well suppose that, in cases where a strong current does not interfere, the greater portion not only of the matter held mechanically in suspension, but of that also which is in chemical solution, must be precipitated within the limits of the delta. When these finer ingredients are extremely small in quantity, they may only suffice to supply crustaceous animals, corals, and marine plants, with the earthy particles necessary for their secretions; but whenever it is in excess (as generally happens if the basin of a river lie partly in a district of active or extinct volcanos), then will solid deposits be formed, and the shells will at once be included in a rocky mass.

Delta of the Po.—The Adriatic presents a great combination of circumstances favourable to the rapid formation of deltas—a gulf receding far into the land,—a sea without tides or strong currents, and the influx of two great rivers, the Po and the Adige, besides numerous minor streams draining on the one side a great crescent of the Alps, and on the other

* Hist. Phys. de la Mer.

some of the loftiest ridges of the Apennines. From the northernmost point of the Gulf of Trieste, where the Isonzo enters, down to the south of Ravenna, there is an uninterrupted series of recent accessions of land, more than one hundred miles in length, which, within the last two thousand years, have increased from *two to twenty miles in breadth.* The Isonzo, Tagliamento, Piave, Brenta, Adige, and Po, besides many other inferior rivers, contribute to the advance of the coast-line, and to the shallowing of the gulf. The Po and the Adige may now be considered as entering by one common delta, for two branches of the Adige are connected with arms of the Po. In consequence of the great concentration of the flooded waters of these streams, since the system of embankment became general, the rate of encroachment of the new land upon the Adriatic, especially at that point where the Po and Adige enter, is said to have been greatly accelerated. Adria was a seaport in the time of Augustus, and had, in ancient times, given its name to the gulf; it is now about twenty Italian miles inland. Ravenna was also a seaport, and is now about four Italian miles from the main sea. Yet even before the practice of embankment was introduced, the alluvium of the Po advanced with rapidity on the Adriatic; for Spina, a very ancient city, originally built in the district of Ravenna, at the mouth of a great arm of the Po, was, so early as the commencement of our era, eleven Italian miles distant from the sea *.

The greatest depth of the Adriatic, between Dalmatia and the mouths of the Po, is twenty-two fathoms; but a large part of the gulf of Trieste and the Adriatic, opposite Venice, is less than twelve fathoms deep. Farther to the south, where it is less affected by the influx of great rivers, the gulf deepens considerably. Donati, after dredging the bottom, discovered the new deposits to consist partly of mud and partly of rock, the latter formed of calcareous matter, encrusting shells. He also ascertained, that particular species of testacea were grouped together in certain places, and were becoming slowly incorporated with the mud, or calcareous precipitates †. Olivi, also,

* See Brocchi on the various writers on this subject. Conch. Foss. Subap., vol. i., p. 118.

† Ibid., vol. i., p. 39.

found some deposits of sand, and others of mud, extending half way across the gulf; and he states that their distribution along the bottom was evidently determined by the prevailing current *. It is probable, therefore, that the finer sediment of all the rivers at the head of the Adriatic may be intermingled by the influence of the current; and all the central parts of the gulf may be considered as slowly filling up with horizontal deposits, precisely similar to those of the Subapennine hills, and containing many of the same species of shells. The Po merely introduces at present fine sand and mud, for it carries no pebbles farther than the spot where it joins the Trebia, west of Piacenza. Near the northern borders of the basin, the Isonzo, Tagliamento, and many other streams, are forming immense beds of sand and some conglomerate, for there some high mountains of Alpine limestone approach within a few miles of the sea. In the time of the Romans, the hot baths of Monfalcone were on one of several islands of Alpine limestone, between which and the main land, on the north, was a channel of the sea, about a mile broad. This channel is now converted into a grassy plain, which surrounds the islands on all sides. Among the numerous changes on this coast, we find that the present channel of the Isonzo is several miles to the west of its ancient bed, in part of which at Ronchi, the old Roman bridge which crossed the Via Appia was lately found buried in fluviatile silt.

Notwithstanding the present shallowness of the Adriatic, it is highly probable that its original depth was very great; for if all the low alluvial tracts were taken away from its borders and replaced by sea, the high land would terminate in that abrupt manner which generally indicates, in the Mediterranean, a great depth of water near the shore, except in those spots where sediment imported by rivers and currents has diminished the depth. Many parts of the Mediterranean are now ascertained to be above two thousand feet deep, close to the shore, as between Nice and Genoa; and even sometimes six thousand feet, as near Gibraltar. When, therefore, we find near Parma, and in other districts in the interior of the peninsula, beds of horizontal tertiary marl, attaining a thickness of about two

* See Brocchi on the various writers on this subject. Conch. Foss. Subap., vol. ii., p. 94.

thousand feet, or when we discover strata of inclined conglomerate, of the same age, near Nice, measuring above a thousand feet in thickness, and extending seven or eight miles in length, we behold nothing which the analogy of the deltas in the Adriatic might not lead us to anticipate.

Delta of the Nile.—That Egypt was the gift of the Nile, was the opinion of her priests before the time of Herodotus; but we have no authentic memorials for determining, with accuracy, the additions made to the habitable surface of that country since the earliest historical period. We know that the base of the delta has been considerably modified since the days of Homer. The ancient geographers mention seven principal mouths of the Nile, of which the most eastern, the Pelusian, has been entirely silted up, and the Mendesian, or Tanitic, has disappeared. On the other hand, the Bucolic has, in modern times, been greatly enlarged, and has caused the coast to advance; so that the city of Damietta, which, in the year 1243, was on the sea, and possessed a good harbour, is now one mile inland. The Phatnitic mouth, and the Sebenitic, have been so altered, that the country immediately about them has little resemblance to that described by the ancients. The Bolbitine mouth has increased in its dimensions, so as to cause the city of Rosetta to be at some distance from the sea. But the alterations produced round the Canopic mouth are the most important. The city Foah, which, so late as the beginning of the fifteenth century, was on this embouchure, is now more than a mile inland. Canopus, which, in the time of Scylax, was a desolate insular rock, has been connected with the firm land; and Pharos, an island in the times of old, now belongs to the continent. Homer says, its distance from Egypt was one day's voyage by sea *. That this should have been the case in Homer's time, Larcher and others have, with reason, affirmed to be in the highest degree improbable; but Strabo has judiciously anticipated their objections, observing, that Homer was probably acquainted with the gradual advance of the land on this coast, and availed himself of this phenomenon to give an air of higher antiquity to the

* Ody., B. iv., 355.

remote period in which he laid the scene of his poem *. The Lake Mareotis, also, together with the canal which connected it with the Canopic arm of the Nile, has been filled with mud, and is become dry. Herodotus observes, that the country round Memphis seemed formerly to have been an arm of the sea gradually filled by the Nile, in the same manner as the Meander, Achelous, and other streams, had formed deltas. "Egypt, therefore," he says, "like the Red Sea, was once a long narrow bay, and both gulfs were separated by a small neck of land. If the Nile," he adds, "should by any means have an issue into the Arabian Gulf, it might choke it up with earth in twenty thousand, or even perhaps in ten thousand years; and why may not the Nile have filled with mud a still greater gulf, in the space of time which has passed before our age † ?"

The depth of the Mediterranean is about twelve fathoms at a small distance from the shore of the delta; it afterwards increases gradually to fifty, and then suddenly descends to three hundred and eighty fathoms, which is, perhaps, the original depth of the sea where it has not been rendered shallower by fluviatile matter. The progress of the delta, in the last two thousand years, affords, perhaps, no measure for estimating its rate of growth when it was an inland bay, and had not yet protruded itself beyond the coast-line of the Mediterranean. A powerful current now sweeps along the shores of Africa, from the Straits of Gibraltar to the prominent convexity of Egypt, the western side of which is continually the prey of the waves; so that not only are fresh accessions of land checked, but ancient parts of the delta are carried away. By this cause Canopus, and some other towns, have been overwhelmed; but to this subject we shall again refer when speaking of tides and currents.

* Lib. I., Part i., pp. 80 and 98. Consult Von Hoff, vol. i., p. 244.

† Euterpe, XI.

CHAPTER XIV.

Oceanic deltas—Delta of the Ganges and Burrampooter—Its size, rate of advance, and nature of its deposits—Formation and destruction of islands—Abundance of crocodiles—Inundations—Delta of the Mississippi—Deposits of drift wood—Gradual filling up of the Yellow Sea—Rennell's estimate of the mud carried down by the Ganges—Formation of valleys illustrated by the growth of deltas—Grouping of new strata in general—Convergence of deltas—Conglomerates—Various causes of stratification—Direction of laminæ—Remarks on the interchange of land and sea.

OCEANIC DELTAS.

THE remaining class of deltas are those in which rivers, on entering the sea, are exposed to the influence of the tides. In this case it frequently happens that an estuary is produced, or negative delta, as it has been termed by Rennell, where, instead of any encroachment of the land upon the sea, the ocean enters the river's mouth, and penetrates into the land beyond the general coast-line. Where this happens, the tides and currents are the predominating agents in the distribution of transported sediment. The phenomena, therefore, of such estuaries, will come under our examination when we treat of the movements of the ocean. But whenever the volume of fresh-water is so great as to counteract and almost neutralize the force of tides and currents, and in all cases where the latter agents have not sufficient power to remove to a distance the whole of the sediment periodically brought down by rivers, oceanic deltas are produced. Of these, we shall now select a few illustrative examples.

Delta of the Ganges.—The Ganges and the Burrampooter descend, from the highest mountains in the world, into a gulf which runs two hundred and twenty-five miles into the continent. The Burrampooter is somewhat the larger river of the two, but it first takes the name of the Megna, when joined by a smaller stream so called, and afterwards loses this again on its union with the Ganges, at the distance of about forty miles from the sea. The area of the delta of the Ganges (without including that of the Burrampooter, which has now become conterminous) is considerably more than double that of the Nile; and its head commences at a distance of two hundred and twenty miles, in a direct line from the sea. That part of the delta bordering on the sea is composed of a labyrinth of rivers and creeks, all of which are salt, except those immediately communicating with the principal arm of the Ganges. This tract, known by the name of the Woods, or Sunderbunds, a wilderness infested by tigers and alligators, is, according to Rennell, equal in extent to the whole principality of Wales *. The base of this magnificent delta is two hundred miles in length, including the space occupied by the two great arms of the Ganges which bound it on either side. On the sea-coast there are eight great openings, each of which has evidently, at some ancient periods, served in its turn as the principal channel of discharge. Although the flux and reflux of the tide extend even to the head of the delta, when the river is low, yet, when it is periodically swollen by tropical rains, the velocity of the stream counteracts the tidal current, so that, except very near the sea, the ebb and flow become insensible. During the flood-season, therefore, the Ganges almost assumes the character of a river entering a lake or inland sea; the movements of the ocean being then subordinate to the force of the river, and only slightly disturbing its operations. The great gain of the delta in height and area takes place during the inundations; and during other seasons of the year, the ocean makes reprisals, scouring out the channels, and sometimes devouring rich alluvial plains.

So great is the quantity of mud and sand poured by the Ganges into the gulf in the flood-season, that the sea only recovers its transparency at the distance of sixty miles from the coast. The general slope, therefore, of the new strata must be extremely gradual. By the charts recently published, it appears that there is a gradual deepening from four to about sixty fathoms, as we proceed from the base of the delta to the distance of about one hundred miles into the Bay of Bengal. At some few points seventy, or even one hundred fathoms are obtained at that

* Account of the Ganges and Burrampooter Rivers, by Major Rennell, Phil. Trans. 1781.

distance. One remarkable exception, however, occurs to the regularity of the shape of the bottom; for opposite the middle of the delta, at the distance of thirty or forty miles from the coast, is a nearly circular space called the "swatch of no ground," about fifteen miles in diameter, where soundings of one hundred, and even one hundred and thirty fathoms, fail to reach the bottom. This phenomenon is the more extraordinary, since the depression occurs within five miles of the line of shoals; and not only do the waters charged with Gangetic sediment pass over it continually, but, during the monsoons, the sea, loaded with mud and sand, is beaten back in that direction towards the delta. As the mud is known to extend for eighty miles farther into the gulf, we may be assured that, in the course of ages, the accumulation of strata in "the swatch" has been of enormous thickness; and we seem entitled to deduce, from the present depth at the spot, that the original inequalities of the bottom of the Bay of Bengal were on as grand a scale as are those of the main ocean. Opposite the mouth of the Hoogly river, and immediately south of Sager Island, four miles from the nearest land of the delta, a new isle was formed about thirty years ago, called Edmonston Island, where there is a lighthouse, and the surface of which is now covered with vegetation and shrubs. But while there is evidence of rapid gain at some points, the general progress of the coast is very slow, for the tides, which rise from thirteen to sixteen feet, are actively employed in removing the alluvial matter, and diffusing it over a wide area.* The new strata consist entirely of sand and fine mud; such, at least, are the only materials which are exposed to view in regular beds on the banks of the numerous creeks. No substance so coarse as gravel occurs in any part of the delta, nor nearer the sea than four hundred miles. It should be observed, however, that the superficial alluvial beds, which are thrown down rapidly from turbid

* It is stated in the chart published in the year 1825, by Captain Hosburgh, that the sands opposite the whole delta stretched between four and five miles farther south than they had done forty years previously; and this was taken as the measure of the progress of the delta itself, during the same period. But that gentleman informs me that a more careful comparison of the ancient charts, during a recent survey, has proved that they were extremely incorrect in their latitudes, so that the advance of the new sands and delta was greatly exaggerated.

waters during the floods, may be very distinct from those deposited at a greater distance from the shore, where crystalline precipitates, perhaps, are forming, on the evaporation of so great a surface, exposed to the rays of a tropical sun. The separation of sand and other matter, held in mechanical suspension, may take place where the waters are in motion; but mineral ingredients, held in chemical solution, would naturally be carried to a greater distance, where they aid in the formation of corals and shells, and, in part, perhaps, become the cementing principle of rocky masses.

Among the remarkable proofs of the immense transportation of earthy matter by the Ganges and Megna, may be mentioned the great magnitude of the islands formed in their channels during a period far short of that of a man's life. Some of these, many miles in extent, have originated in large sand-banks thrown up round the points at the angular turning of the river, and afterwards insulated by breaches of the stream. Others, formed in the main channel, are caused by some obstruction at the bottom. A large tree, or a sunken boat, is sometimes sufficient to check the current, and cause a deposit of sand, which accumulates till it usurps a considerable portion of the channel. The river then borrows on each side to supply the deficiency in its bed, and the island is afterwards raised by fresh deposits during every flood. In the great gulf below Luckipour, formed by the united waters of the Ganges and Burrampooter (or Megna), some of the islands, says Rennell, rival in size and fertility the Isle of Wight. While the river is forming new islands in one part, it is sweeping away old ones in others. Those newly formed are soon overrun with reeds, long grass, the Tamarix Indica, and other shrubs, forming impenetrable thickets, where tigers, buffaloes, deer, and other wild animals, take shelter. It is easy, therefore, to perceive, that both animal and vegetable remains must continually be precipitated into the flood, and sometimes become imbedded in the sediment which subsides in the delta.

Two species of crocodiles, of distinct genera, abound in the Ganges and its tributary and contiguous waters; and Mr. H. T. Colebrooke informs me, that he has seen both kinds in places far inland, many hundred miles from the sea. The Gangetic crocodile, or Gavial (in correct orthography, Garial), is confined to the fresh-water, but the common crocodile frequents both fresh and salt; being much larger and fiercer in salt and brackish water. These animals swarm in the brackish water along the line of sand-banks where the advance of the delta is most rapid. Hundreds of them are seen together in the creeks of the delta, or basking in the sun on the shoals without. They will attack men and cattle, destroying the natives when bathing, and tame and wild animals which come to drink. "I have not unfrequently," says Mr. Colebrooke, "been witness to the horrid spectacle of seeing a floating corpse seized by a crocodile with such avidity, that he half emerged above the water with his prey in his mouth." The geologist will not fail to observe how peculiarly the habits and distribution of these saurians expose them to become imbedded in those horizontal strata of fine mud which are annually deposited over many hundred square miles in the Bay of Bengal. The inhabitants of the land, when they happen to be submerged, are usually destroyed by these voracious reptiles; but we may suppose the remains of the saurians themselves to be continually entombed in the new formations.

It sometimes happens, at the season when the periodical flood is at its height, that a strong gale of wind, conspiring with a high spring-tide, checks the descending current of the river, and gives rise to most destructive inundations. From this cause, in the year 1763, the waters at Luckipour rose six feet above their ordinary level, and the inhabitants of a considerable district, with their houses and cattle, were totally swept away.

The population of all oceanic deltas are particularly exposed to suffer by such catastrophes, recurring at considerable intervals of time; and we may safely assume, that such tragical events have happened again and again since the Gangetic delta was inhabited by man. If human experience and forethought cannot always guard against these calamities, still less can the inferior animals avoid them; and the monuments of such disastrous inundations must be looked for in great abundance in strata of all ages, if the surface of our planet has always been governed by the same laws. When we reflect on the general order and tranquillity that reigns in the rich and populous delta of Bengal, notwithstanding the havoc occasionally committed by the depredations of the ocean, we perceive how unnecessary it is to attribute the imbedding of successive races of animals in older strata to extraordinary energy in the causes of decay and reproduction in the infancy of our planet, or to those general catastrophes and sudden revolutions resorted to by cosmogonists.

As the delta of the Ganges may be considered a type of those formed on the borders of the ocean, it will be unnecessary to accumulate examples of others on a no less magnificent scale, as at the mouths of the Orinoco and Amazon, for example. To these, indeed, it will be necessary to revert when we treat of the agency of currents. The tides in the Mexican Gulf are so feeble, that the delta of the Mississippi has somewhat of an intermediate character between an oceanic and mediterranean delta. A long narrow tongue of land is protruded, consisting simply of the banks of the river, and having precisely the same appearance as in the inland plains during the periodical inundations, when nothing appears above water but the higher part of that sloping glacis which we before described. This tongue of land has advanced many leagues since New Orleans was built. Great submarine deposits are also in progress, stretching far and wide over the bottom of the sea, which has become throughout a considerable area extremely shallow, not exceeding ten fathoms in depth. Opposite the mouth of the Mississippi large rafts of drift trees, brought down every spring, are matted together into a net-work many yards in thickness, and stretching over hundreds of square leagues*. They afterwards become covered over with a fine mud, on which other layers of trees are deposited the year following, until numerous alternations of earthy and vegetable matter are accumulated. An observation of Darby, in regard to the strata composing part of this delta, deserves attention. In the steep banks of the Atchafalaya, that arm of the Mississippi which we before alluded to when describing "the raft," the following section is observable at low water:—first, an upper stratum, consisting invariably of blueish clay, common to the banks of the Mississippi; below this a stratum of red ochreous earth peculiar to Red River, under which the blue clay of the Mississippi again appears †; and this arrange-

* Captain Hall's Travels in North America, vol. iii, p. 338.

† Darby's Louisiana, p. 103.

ment is constant, proving, as that geographer remarks, that the waters of the Mississippi and the Red River once occupied alternately considerable tracts below their present point of union. Such alternations are probably common in submarine spaces situated between two converging deltas. For before the two rivers unite, there must almost always be a certain period when an intermediate tract will be alternately occupied and abandoned by the waters of each stream; since it can rarely happen, that the season of highest flood will precisely correspond in each. In the case of the Red River, for example, and Mississippi, which carry off the waters from countries placed under widely distant latitudes, an exact coincidence in the time of greatest inundation is very improbable.

CONCLUDING REMARKS ON DELTAS.

Quantity of Sediment in River Water.—Very few satisfactory experiments have as yet been made, to enable us to determine, with any degree of accuracy, the mean quantity of earthy matter discharged annually into the sea by some one of the principal rivers of the earth. Hartsoeker computed the Rhine to contain, when most flooded, one part in a hundred of mud in suspension*. By several observations of Sir George Staunton, it appeared that the water of the Yellow River in China contained earthy matter in the proportion of one part to two hundred, and he calculated that it brought down, in a single hour, two million feet of earth, or forty-eight million daily; so that, if the Yellow Sea be taken to be one hundred and twenty feet deep, it would require seventy days for the river to convert an English square mile into firm land, and twenty-four thousand years to turn the whole sea into terra firma, assuming it to be one hundred and twenty-five thousand square miles in extent †. Manfredi, the celebrated Italian hydrographer, conceived the average proportion of sediment in all the running water on the globe, which reached the sea, to be $\frac{1}{175}$, and he imagined that it would require a thousand years for the sediment carried down to raise the general level of the sea about one foot. Some writers, on the contrary,

* Comment, Bonon., vol. ii., part i., p. 237.
† Staunton's Embassy to China. London, 1797, 4to. vol. ii., p. 408.

as De Maillet, have declared the most turbid waters to contain far less sediment than any of the above estimates would import; and there is so much contradiction and inconsistency in the facts and speculations hitherto promulgated on the subject, that we must wait for additional experiments before we can form any opinion on the question.

One of the most extraordinary statements is that of Major Rennell, in his excellent paper, before referred to, on the Delta of the Ganges. "A glass of water," he says, "taken out of this river when at its height, yields about one part in four of mud. No wonder, then, that the subsiding waters should quickly form a stratum of earth, or that the delta should encroach on the sea *!" The same hydrographer computed with much care the number of cubic feet of water discharged by the Ganges into the sea, and estimated the mean quantity through the whole year to be eighty thousand cubic feet in a second. When the river is most swollen, and its velocity much accelerated, the quantity is four hundred and five thousand cubic feet in a second. Other writers agree that the violence of the tropical rains, and the fineness of the alluvial particles in the plains of Bengal, cause the waters of the Ganges to be charged with foreign matter to an extent wholly unequalled by any large European river during the greatest floods. We have already alluded to the frequent sweeping down of large islands by the Ganges; and Major R. H. Colebrooke, in his account of the course of the Ganges, relates examples of the rapid filling up of some branches of the river, and the excavation of new channels, where the number of square miles of soil removed in a short time (the column of earth being one hundred and fourteen feet high) was truly astonishing. Forty square miles, or 25,600 acres, are mentioned as having been carried away, in one locality, in the course of a few years †. But although we can readily believe the proportion of sediment in the waters of the Ganges to exceed that of any river in northern latitudes, we are somewhat staggered by the results to which we must arrive if we compare the proportion of mud, as given by Rennell, with his computation of the quantity of water discharged, which latter is probably very correct. If it

* Phil. Trans., 1781.

† Trans. of the Asiatic Society, vol. vii., p. 14.

were true that the Ganges, in the flood-season, contained one part in four of mud, we should then be obliged to suppose that there passes down, every four days, a quantity of mud equal in volume to the water which is discharged in the course of twenty-four hours. If the mud be assumed to be equal to one-half the specific gravity of granite (it would, however, be more), the weight of matter *daily* carried down in the flood-season, would be about equal to seventy-four times the weight of the Great Pyramid of Egypt *. Even if it could be proved that the turbid waters of the Ganges contain one part *in a hundred* of mud, which is affirmed to be the case in regard to the Rhine, we should be brought to the extraordinary conclusion, that there passes down, every two days, into the Bay of Bengal, a mass about equal in weight and bulk to the Great Pyramid.

The most voluminous current of lava which has flowed from Etna within historical times, was that of 1669. Ferrara, after correcting Borrelli's estimate, calculated the quantity of cubic yards of lava in this current, at one hundred and forty millions. Now this would only equal in bulk one-seventh of the sedimentary matter which is carried down in a single year by the Ganges, assuming the average proportion of mud to water to be no more than one part in one hundred, so that, allowing seven grand eruptions in a century, it would require an hundred Etnas to transfer a mass of lava from the subterranean regions to the surface, equal in volume to the mud carried down in the same time from the Himalaya mountains into the Bay of Bengal †. As considerable labour has been bestowed

* According to Rennell, the Ganges discharges, in the flood-season, 405,000 cubic feet of water per second, which gives, in round numbers, 100,000 cubic feet of mud per second, which × 86,400, the number of seconds in twenty-four hours, = 8,641,100,000, the quantity of cubic feet of mud going down the Ganges per diem. Assuming the specific gravity of mud to be half that of granite, the matter would equal 4,320,550,000 feet of granite. Now about twelve and a half cubic feet of granite weigh one ton; and it is computed, that the Great Pyramid of Egypt, if it were a solid mass of granite, would weigh about 6,000,000 of tons.

† According to Ferrara's calculation, about 140,000,000 of cubic yards of lava were poured from the crater of Etna in 1669. This × 27, will give 3,780,000,000 of cubic feet, which would be about one-seventh of the amount of mud carried down by the Ganges in a year; for, assuming the average proportion of mud to be one part in a hundred, this would give on an average 800 cubic feet per second: 800 × 31,557,600, (the number of seconds in a Julian year,) gives 25,246,080,000.

in computing the volume of lava-streams in Sicily, Campania, and Auvergne, it is somewhat extraordinary that so few observations have been made on the quantity of matter transported by aqueous agents from one part of the earth to another. It would certainly not be difficult to approximate to the amount of sediment carried down annually by some of the largest rivers, such as the Amazon, Mississippi, Ganges, and others, because the earthy particles conveyed by them to their deltas are fine, and somewhat uniformly spread throughout the stream, and the principal efflux takes place within a limited period during the season of inundation. Arguments have been expended in vain for half a century, in controverting the opinion of those who imagine the agency of running water in the existing state of things, even if continued through an indefinite lapse of ages, to be insignificant, or at least wholly incompetent to produce considerable inequalities on the earth's surface. Some matter-of-fact data should now be accumulated, and we may confidently affirm, that when the aggregate amount of solid matter transported by rivers in a given number of centuries from a large continent, shall be reduced to arithmetical computation, the result will appear most astonishing to those who are not in the habit of reflecting how many of the mightiest operations in nature are effected insensibly, without noise or disorder. The volume of matter carried into the sea in a given time being once ascertained, every geologist will admit that the whole, with some slight exceptions, is subtracted from *valleys*, not from the tops of intervening ridges or the summits of hills; in other words, that ancient valleys have been widened and deepened, or new ones formed, to the extent of the space which the new deposits, when consolidated, would occupy.

Grouping of Strata in Deltas.—The changes which have taken place in deltas, even since the times of history, may suggest many important considerations in regard to the manner of distribution of sediment in subaqueous deposits. Notwithstanding frequent exceptions arising from the interference of a variety of causes, there are some general laws of arrangement which must evidently hold good in almost all the lakes and seas now filling up. If a lake, for example, be encircled on two sides by lofty mountains, receiving from them many rivers and torrents of different sizes, and if it be bounded on the other sides, where the surplus waters issue, by a comparatively low country, it is not difficult to define some of the leading geological features which will characterize the lacustrine formation when this basin shall have been gradually converted into dry land by influx of fluviatile sediment. The strata would be divisible into two principal groups; the *older* comprising those deposits which originated on the side adjoining the mountains, where numerous deltas first began to form; and the *newer* group consisting of beds deposited in the more central parts of the basin, and towards the side farthest from the mountains. The following characters would form the principal marks of distinction between the strata in each series. The more ancient system would be composed, for the most part, of coarser materials, containing many beds of pebbles and sand often of great thickness, and sometimes dipping at a considerable angle. These, with associated beds of finer ingredients, would, if traced round the borders of the basin, be seen to vary greatly in colour and mineral composition, and would also be very irregular in thickness. The beds, on the contrary, in the newer group, would consist of finer particles, and would be horizontal, or very slightly inclined. Their colour and mineral composition would be very homogeneous throughout large areas, and would differ from almost all the separate beds in the older series.

The following are the causes of the diversity here alluded to between the two great members of the lacustrine formation. When the rivers and torrents first reach the edge of the lake, the detritus washed down by them from the adjoining heights sinks at once into deep water, all the heavier pebbles and sand subsiding near the shore. The finer mud is carried somewhat farther out, but not to the distance of many miles, for the greater part may be seen, where the Rhone enters the Lake of Geneva, to fall down in clouds to the bottom not far from the river's mouth. Certain alluvial tracts are soon formed at the mouths of every torrent and river, and many of these, in the course of ages, become several miles in length. Pebbles and sand are then transported farther from the mountains, but in their passage they decrease in size by attrition, and are in part converted into mud and sand. At length some of the numerous deltas, which are all directed towards a common centre, approach near to each other—those of adjoining torrents become united, and are merged, in their turn, in the delta of the largest river, which advances most rapidly into the lake, and renders all the minor streams, one after the other, its tributaries. The various mineral ingredients of each are thus blended together into one homogeneous mixture, and the sediment is poured out from a common channel into the lake. As the average size of the transported particles decreases continually, so also the force and volume of the current augments, and thus the newer deposits are diffused over a wider area, and are consequently more horizontal than the older. When there were many independent deltas near the borders of the basin, their separate deposits differed entirely from each other. We may suppose that one was charged, like the Arve where it joins the Rhone, with white sand and sediment, chiefly derived from decomposed granite—that another was black, like many streams in the Tyrol, flowing from incoherent rocks of dark slate—that a third was coloured by ochreous sediment, like the Red River in Louisiana—and that a fourth, like the Elsa in Tuscany, held much carbonate of lime in solution. At first, they would each form distinct deposits of sand, gravel, limestone, marl, or other materials; but after their junction, new chemical combinations and distinct colours would be the result, and the particles, having been conveyed ten, twenty, or a greater number of miles over alluvial plains, would become finer.

In deltas where the causes are more complicated, and where tides and currents partially interfere, the above description would only be applicable, with certain modifications; but if a series of earthquakes accompany the growth of a delta, and change the levels of the land from time to time, as in the region where the Indus now enters the sea, and others hereafter to be mentioned, the phenomena will then depart widely from the ordinary type. If we possessed an accurate series of maps of the Adriatic for many thousand years, our retrospect would, without doubt, carry us gradually back to the time when the number of rivers descending from the mountains into that gulf by independent deltas, was far greater in number. The deltas of the Po and the Adige, for instance, would separate themselves within the *human* era, as, in all probability, would those of the Isonzo and the Torre. If, on the other hand, we speculate on future changes, we may anticipate the period when the number of deltas will greatly diminish; for the Po cannot continue to encroach at the rate of a mile in a century, and other rivers to gain as much in six or seven centuries upon the shallow gulf, without new junctions occurring from time to time, so that Eridanus, "the king of rivers," will continually boast a greater number of tributaries. The Ganges and Burrampooter have probably become confluent within the historical era; and the date of the junction of the Red River and the Mississippi would, in all likelihood, have been known, if America had not been so recently discovered. The union of the Tigris and the Euphrates must undoubtedly have been one of the modern geographical changes on our earth, and similar remarks might be extended to many other regions.

Along the base of the Maritime Alps, between Toulon and Genoa, the rivers, with few exceptions, are now forming strata of conglomerate and sand. Their channels are often several miles in breadth, some of them being dry, and the rest easily forded for nearly eight months in the year; whereas during the melting of the snow they are swollen, and a great transportation of mud and pebbles takes place. In order to keep open the main road from France to Italy, now carried along the sea-coast, it is necessary to remove annually great masses of shingle, brought down during the flood-season. A portion of the pebbles are seen in some localities, as near Nice, to form beds of shingle along the shore, but the greater part are swept into a deep sea. The small progress made by the deltas of minor rivers on this coast need not surprise us, when we recollect that there is sometimes a depth of two thousand feet at a few hundred yards from the beach, as near Nice. Similar observations might be made respecting a large proportion of the rivers in Sicily, and, among others, respecting that which, immediately north of the port of Messina, hurries annually vast masses of granitic pebbles into the sea.

When the deltas of rivers having many mouths converge, a partial union at first takes place by the confluence of some one or more of their arms; but it is not until the main trunks are connected above the head of the common delta, that a complete intermixture of their joint waters and sediment takes place.

The union, therefore, of the Po and Adige, and of the Ganges and Burrampooter, is still incomplete. If we reflect on the geographical extent of surface drained by rivers such as now enter the Bay of Bengal, and then consider how complete the blending together of the greater part of their transported matter has already become, and throughout how vast a delta it is spread by numerous arms, we no longer feel so much surprise at the area occupied by some ancient formations of homogeneous mineral composition. But our surprise will be still farther lessened when we afterwards inquire into the action of tides and currents, in disseminating the matter accumulated in various deltas.

Stratification of Deposits in Deltas.—That the matter carried by rivers into seas and lakes is not thrown in confused and promiscuous heaps, but is spread out far and wide along the bottom, is well ascertained; and that it must for the most part be divided into distinct strata, may in part be inferred where it cannot be proved by observation. The horizontal arrangement of the strata, when laid open to the depth of twenty or thirty feet in the delta of the Ganges and in that of the Mississippi, is alluded to by many writers; and the same disposition is well known to obtain in all modern deposits of lakes and estuaries. Natural divisions are often occasioned by the interval of time which separates annually the deposition of matter during the periodical rains, or melting of the snow upon the mountains. The deposit of each year acquires some degree of consistency before that of the succeeding year is superimposed. A variety of circumstances also give rise annually to slight variations in colour, fineness of the particles, and other characters. Alternations of strata distinct in texture, mineral ingredients, or organic contents, are produced by numerous causes. Thus, for example, at one period of the year, drift wood may be carried down, and at another mud, as was before stated to be the case in the delta of the Mississippi; or at one time when the volume and velocity of the stream are greatest, pebbles and sand may be spread over a certain area, over which, when the waters are low, fine matter or chemical precipitates are formed. During inundations the current of fresh-water often repels the sea for many miles; but when the river is low, salt-water again occupies the same space. When two deltas are converging, the intermediate space is often, for reasons before explained, alternately the receptacle of different sediment derived from the converging streams. The one is, perhaps, charged with calcareous, the other with argillaceous matter; or one may sweep down sand and pebbles, the other impalpable mud. These differences may be repeated with considerable regularity, until a thickness of hundreds of feet of alternating beds is accumulated.

An examination of the strata of shell-marl now forming in the Scotch lakes, or of the sediment termed "warp," which subsides from the muddy water of the Humber, and other rivers, shows that recent deposits are often composed of a great number of extremely thin layers, either even or slightly undulating, and parallel to the planes of stratification. Sometimes, however, the laminæ in modern strata are disposed diagonally at a considerable angle, which appears to take place where there are conflicting movements in the waters. In January, 1829, I visited, in company with Professor L. A. Necker, of Geneva, the confluence of the Rhone and Arve, when those rivers were very low, and were cutting channels through the vast heaps of debris thrown down from the waters of the Arve, in the preceding spring. One of the sand-banks which had formed, in the spring of 1828, where the opposing currents of the two rivers neutralized each other, and caused a retardation in the motion, had been undermined; and the following is an exact representation of the arrangement of laminæ exposed in a vertical section. The length of the portion here seen is about twelve feet, and

No. 6.

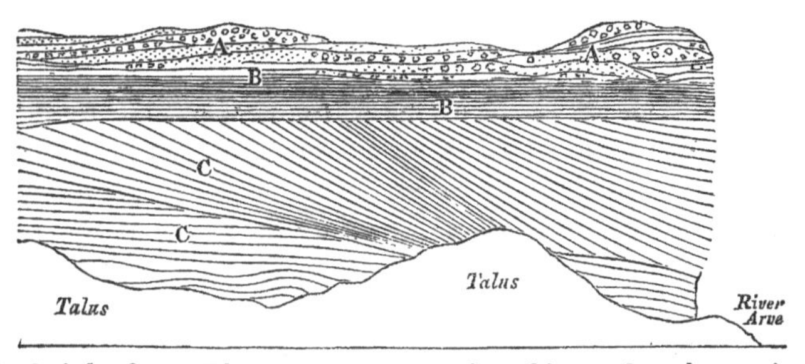

the height five. The strata A A consist of irregular alternations of pebbles and sand in undulating beds: below these are seams of very fine sand, B B, some as thin as paper, others about a quarter of an inch thick. The strata c c are composed of layers of fine greenish-grey sand, as thin as paper. Some of the inclined beds will be seen to be thicker at their upper, others at their lower extremity, the inclination of some being very considerable. These layers must have accumulated one on the other by lateral apposition, probably when one of the rivers was very gradually increasing or diminishing in velocity, so that the point of greatest retardation caused by their conflicting currents shifted slowly, allowing the sediment to be thrown down in successive layers on a sloping bank. The same phenomenon is exhibited in older strata of all ages; and when we treat of them, we shall endeavour more fully to illustrate the origin of such a structure.

We may now conclude our remarks on deltas, observing that, imperfect as is our information of the changes which they have undergone within the last three thousand years, they are sufficient to show how constant an interchange of sea and land is taking place on the face of our globe. In the Mediterranean alone, many flourishing inland towns, and a still greater number of ports, now stand where the sea rolled its waves since the era when civilized nations first grew up in Europe. If we could compare with equal accuracy the ancient and actual state of all the islands and continents, we should probably discover that millions of our race are now supported by lands situated where deep seas prevailed in earlier ages. In many districts not yet occupied by man, land animals and forests now abound where the anchor once sank into the oozy bottom. We shall find, on inquiry, that inroads of the ocean have been no less considerable; and when to these revolutions produced by aqueous causes, we add analogous changes wrought by igneous agency, we shall, perhaps, acknowledge the justice of the conclusion of a great philosopher of antiquity, when he declared that the whole land and sea on our globe periodically changed places *.

* See an account of the Aristotelian system, p. 16, *ante*.

CHAPTER XV.

Destroying and transporting effects of tides and currents—Shifting of their position—Differences in the rise of the tides—Causes of currents—Action of the sea on the British coast—Shetland Islands—Large blocks removed—Effects of lightning—Breach caused in a mass of porphyry—Isles reduced to clusters of rocks—Orkney Isles—East coast of Scotland—Stones thrown up on the Bell Rock—East coast of England—Waste of the cliffs of Holderness, Norfolk, and Suffolk—Silting up of estuaries—Origin of submarine forests—Yarmouth estuary—Submarine forests—Suffolk coast—Dunwich—Essex coast—Estuary of the Thames—Goodwin Sands—Coast of Kent—Formation of Straits of Dover—Coast of Hants—Coast of Dorset—Portland—Origin of the Chesel Bank—Cornwall—Lionnesse tradition—Coast of Brittany.

Although the movements of great bodies of water, termed tides and currents, are in general due to very distinct causes, we cannot consider their effects separately, for they produce, by their joint action, those changes which are subjects of geological inquiry. We may view these forces as we before considered rivers, first, as employed in destroying portions of the solid crust of the earth, and removing them to other localities; secondly, as reproductive of new strata. Some of the principal currents which traverse large regions of the globe depend on permanent winds, and these on the rotation of the earth on its axis, and its position in regard to the sun:—they are causes, therefore, as constant as the tides themselves, and, like them, depend on no temporary or accidental circumstances, but on the laws which preside over the motions of the heavenly bodies. But, although the sum of their influence in altering the surface of the earth may be very constant throughout successive epochs, yet the points where these operations are displayed in the fullest energy shift perpetually. The height to which the tides rise, and the violence and velocity of currents, depend in a great measure on the actual configuration of the land, the contour of a long line of continental or insular coast, the depth and breadth of channels, the peculiar form of the bottom of seas—in a word, on a combination of circumstances which are made to vary continually by many igneous and aqueous causes, and, among the rest, by tides and currents. Although these agents, therefore, of decay and reproduction, are local in reference to periods of short duration, such as those which history in general embraces, they are nevertheless universal, if we extend our views to a sufficient lapse of ages.

The tides, as is well known, rise in certain channels, bays, and estuaries, to an elevation far above the average height of the same tides in more open parts of the coast, or on islands in the main ocean. In all lakes, and in most inland seas, the tides are not perceptible. In the Mediterranean, even, deep and extensive as is that sea, they are only sensible in certain localities, and they then rarely rise more than six inches above the mean level. In the Straits of Messina, however, there is an ebb and flow every six hours, to the amount of two feet, but this elevation is partly due to the peculiar set of the currents. In islands remote from the shore, the rise of the tides is slight, as at St. Helena, for example, where it rarely exceeds three feet *. In the estuary of the Severn, the rise at King's Road near Bristol is forty-two feet; and at Chepstow on the Wye, a small river which opens into the same estuary, about fifty feet. All the intermediate elevations may be found at different places on our coast. Thus, at Milford Haven, the rise is thirty-six feet; at London, and the promontory of Beachy Head, eighteen feet; at the Needles, in the Isle of Wight, nine feet; at Weymouth, seven; at Lowestoff about five; at Great Yarmouth, still less.

That movements of no inconsiderable magnitude should be impressed on an expansive ocean, by winds blowing for many months in one direction, may easily be conceived, when we observe the effects produced in our own seas by the temporary action of the same cause. It is well known that a strong south-west or north-west wind, invariably raises the tides to an unusual height along the east coast of England, and in the Channel; and that a north-west wind of any continuance causes the Baltic to rise two feet and upwards above its ordinary level. Smeaton ascertained, by experiment, that in a canal four miles in length, the water was kept up four inches higher at one end than at the other, merely by the action of the wind along the canal; and Rennel informs us, that a large piece

* Romme, Vents et Courans, vol. ii., p. 2. Rev. F. Fallows, Quart. Journ. of Sci., March, 1829.

of water, ten miles broad, and generally only three feet deep, has, by a strong wind, had its waters driven to one side, and sustained so as to become six feet deep, while the windward side was laid dry *. As water, therefore, he observes, when pent up so that it cannot escape, acquires a higher level, so, in a place *where it can escape*, the same operation produces a current; and this current will extend to a greater or less distance, according to the force by which it is produced. The most extensive and best determined of all currents, is the gulf stream, which sets westward in tropical regions; and, after doubling the Cape of Good Hope, where it runs nearly at the rate of two miles an hour, inclines considerably to the northward, along the western coast of Africa, then crosses the Atlantic, and, having accumulated in the Gulf of Mexico, passes out at the Straits of Bahama, with a velocity of four miles an hour, which is not reduced to two miles until the stream has proceeded to the distance of eighteen hundred miles in the direction of Newfoundland: near that island it meets with a current setting southward from Baffin's Bay, on the coast of Greenland, and is thereby deflected towards the east. One branch extends in that direction, while another runs towards the north; so that fruits, plants, and wood, the produce of America and the West Indies, are drifted to the shores of Ireland, the Hebrides, and even to Spitzbergen.

In describing the destroying effects of tides and currents, it will be necessary to enter into some detail, because we have not the advantage here, as in the case of the deltas of many rivers, of viewing the aggregate mass which has resulted from the continual transportation of matter, for many centuries, at certain points. We must infer the great amount of accumulation as a corollary from the proofs adduced of the removing force; and this it will not be difficult to show is, on the whole, greater than that of running water on the land.

If we follow the eastern and southern shores of the British islands, from our Ultima Thule in Shetland, to the Land's End in Cornwall, we shall find evidence of a series of changes since the historical era, very illustrative of the kind and degree of force exerted by the agents now under consideration. In this survey we shall have an opportunity of tracing the power of the sea on islands, promontories, bays, and estuaries; on bold,

* Rennel on the Channel-current.

lofty cliffs as well as on low shores; and on every description of rock and soil, from granite to blown sand. We shall afterwards explain by reference to other regions, some phenomena of which our own coast furnishes no examples.

Shetland Islands.—The northernmost group of the British islands, the Shetland, are composed of a great variety of primary and trap rocks, including granite, gneiss, mica-slate, serpentine, greenstone, and many others, with some secondary rocks, chiefly sandstone and conglomerate. These isles are exposed continually to the uncontrolled violence of the Atlantic, for no land intervenes between their western shores and America. The prevalence, therefore, of strong westerly gales, causes the waves to be sometimes driven with irresistible force upon the coast, while there is also a current setting from the north. The spray of the sea aids the decomposition of the rocks, and prepares them to be breached by the mechanical force of the waves. Steep cliffs are hollowed out into deep caves and lofty arches; and almost every promontory ends in a cluster of rocks, imitating the forms of columns, pinnacles, and obelisks. Modern observations show that the reduction of continuous tracts to such insular masses is a process in which Nature is still actively engaged. "The Isle of Stenness," says Dr. Hibbert, "presents a scene of unequalled desolation. In stormy winters, huge blocks of stones are overturned, or are removed from their native beds, and hurried up a slight acclivity to a distance almost incredible. In the winter of 1802, a tabular-shaped mass, eight feet two inches by seven feet, and five feet one inch thick, was dislodged from its bed, and removed to a distance of from eighty to ninety feet. I measured the recent bed from which a block had been carried away the preceding winter (A. D. 1818), and found it to be seventeen feet and a half by seven feet, and the depth two feet eight inches. The removed mass had been borne to a distance of thirty feet, when it was shivered into thirteen or more lesser fragments, some of which were carried still farther, from thirty to one hundred and twenty feet. A block, nine feet two inches by six feet and a half, and four feet thick, was hurried up the acclivity to a distance of one hundred and fifty feet *."

* Dr. Hibbert, Description of the Shetland Islands, p. 527. Edin., 1822.

At Northmavine, also, angular blocks of stone have been removed in a similar manner to considerable distances, by the waves of the sea*, some of which are represented in the annexed figure, No. 7.

No. 7.

Stony fragments drifted by the sea. Northmavine, Shetland.

In addition to numerous examples of masses detached and driven by the tides and currents from their place, some remarkable effects of lightning are recorded in these isles. At Funzie, in Fetlar, about the middle of the last century, a rock of mica schist, one hundred and five feet long, ten feet broad, and in some places four feet thick, was in an instant torn by a flash of lightning from its bed, and broken into three large, and several lesser fragments. One of these, twenty-six feet long, ten feet broad, and four feet thick, was simply turned over. The second, which was twenty-eight feet long, seventeen broad, and five feet in thickness, was hurled across a high point to the distance of fifty yards. Another broken mass, about forty feet long, was thrown still farther, but in the same direction, quite into the sea. There were also many lesser fragments scattered up and down †.

When we thus see electricity co-operating with the violent movements of the ocean in heaping up piles of shattered rocks on dry land, and beneath the waters, we cannot but admit that a region which shall be the theatre, for myriads of ages, of the action of such disturbing causes, will present, at some future period, a scene of havoc and ruin that may compare with any

* For this and the three following representations of rocks in the Shetland Isles, I am indebted to Dr. Hibbert's work before cited, which is rich in antiquarian and geological research.

† Dr. Hibbert, from MSS. of Rev. George Low, of Fetlar.

now found by the geologist on the surface of our continents; raised as they all have been in former ages from the bosom of the deep. We have scarcely begun, as yet, to study the effects of a single class of the mighty instruments of change and disorder now operating on our globe; and yet geologists have presumed to resort to a nascent order of things, or to revolutions in the economy of Nature, to explain every obscure phenomenon!

In some of the Shetland Isles, as on the west of Meikle Roe, dikes, or veins of soft granite have mouldered away; while the matrix in which they were inclosed, being of the same substance, but of a firmer texture, has remained unaltered. Thus, long narrow ravines, sometimes twenty feet wide, are laid open, and often give access to the waves. After describing some huge cavernous apertures into which the sea flows for two hundred and fifty feet in Roeness, Dr. Hibbert enumerates other ravages of the ocean. "A mass of rock, the average dimensions of which may perhaps be rated at twelve or thirteen feet square, and four and a half or five in thickness, was first moved from its bed, about fifty years ago, to a distance of thirty feet, and has since been twice turned over. But the most sublime scene is where a mural pile of porphyry, escaping the process of disintegration that is devastating the coast, appears to have been left as a sort of rampart against the inroads of the ocean;—the Atlantic, when provoked by wintry gales, batters against it with all the force of real artillery—the

No. 8.

Grind of the Navir—Passage forced by the sea through rocks of hard porphyry.

waves having, in their repeated assaults, forced themselves an entrance. This breach, named the Grind of the Navir (No. 8), is widened every winter by the overwhelming surge that, finding a passage through it, separates large stones from its sides, and forces them to a distance of no less than one hundred and eighty feet. In two or three spots, the fragments which have been detached are brought together in immense heaps, that appear as an accumulation of cubical masses, the product of some quarry *."

It is evident from this example, that although the greater indestructibility of some rocks may enable them to withstand, for a longer time, the action of the elements, yet they cannot permanently resist. There are localities in Shetland, in which rocks of almost every variety of mineral composition are suffering disintegration: thus the sea makes great inroads on the clay slate of Fitfel Head, on the serpentine of the Vord Hill in Fetlar, and on the mica-schist of the Bay of Triesta, on the east coast of the same isle, which decomposes in angular blocks. The quartz rock on the east of Walls, and the gneiss and mica-schist of Garthness, suffer the same fate.

Such devastation cannot be incessantly committed for thousands of years without dividing islands, until they become at last mere clusters of rocks, the last shreds of masses once continuous. To this state many appear to have been reduced, and innumerable fantastic forms are assumed by rocks adjoining these isles, to which the name of Drongs is applied, as it is to those of similar shape in Feroe.

No. 9.

Granitic rocks named the Drongs, between Papa Stour and Hillswick Ness.

* Hibbert, p. 528.

The granitic rocks (No. 9) between Papa Stour and Hillswick Ness afford an example. A still more singular cluster of rocks is seen to the south of Hillswick Ness (No. 10), which presents a variety of forms as viewed from different points, and has often been likened to a small fleet of vessels with spread sails *.

No. 10.

Granitic Rocks to the south of Hillswick Ness, Shetland.

We may imagine that in the course of time Hillswick Ness itself may present a similar wreck, from the unequal decomposition of the rocks whereof it is composed, consisting of gneiss and mica schist, traversed in all directions by veins of felspar porphyry.

Midway between the groups of Shetland and Orkney is Fair Island, said to be composed of sandstone with high perpendicular cliffs. The current runs with such velocity, that during a calm, and when there is no swell, the rocks on its shores are white with the foam of the sea driven against them. The Orkneys, if carefully examined, would probably afford as much illustration of our present topic as the Shetland Islands. The north-east promontory of Sanda, one of these isles, has been cut off in modern times by the sea, so that it became what is now called Start Island, where a lighthouse was erected in 1807, since which time the new strait has grown broader.

East Coast of Scotland.—To pass over to the main land of Scotland, we find that, in Inverness-shire, there have been inroads of the sea at Fort George, and others in Murrayshire,

* Hibbert, p. 519.

which have swept away the old town of Findhorn. On the coast of Kincardineshire, an illustration was afforded, at the close of the last century, of the effect of promontories in protecting a line of low shore. The village of Mathers, two miles south of Johnshaven, was built on an ancient shingle beach, protected by a projecting ledge of limestone rock. This was quarried for lime to such an extent, that the sea broke through, and in 1795 carried away the whole village in one night, and penetrated one hundred and fifty yards inland, where it has maintained its ground ever since, the new village having been built farther inland on the new shore. In the Bay of Montrose, we find the North Esk and the South Esk rivers pouring annually into the sea large quantities of sand and pebbles, yet they have formed no deltas; for the tides scour out the channels, and the current, setting across their mouths, sweeps away all the materials. Considerable beds of shingle, brought down by the North Esk, are seen along the beach. Proceeding southwards, we find that at Arbroath, in Forfarshire, which stands on a rock of red sandstone, gardens and houses have been carried away within the last thirty years by encroachments of the sea. It has become necessary to remove the lighthouses at the mouth of the estuary of the Tay, in the same county, at Button Ness, which were built on a tract of blown sand, the sea having encroached for three-quarters of a mile.

A good illustration was afforded, during the building of the Bell Rock Lighthouse, at the mouth of the Frith of Tay, of the power which currents in estuaries can exert at considerable depths, in scouring out the channel. The Bell Rock is a sunken reef, consisting of red sandstone, being from twelve to sixteen feet under the surface at high water, and about twelve miles from the main land. At the distance of one hundred yards, there is a depth, in all directions, of two or three fathoms at low water. The perpendicular rise and fall of the spring-tides is fifteen feet, and at neap-tides, eight feet; their velocity varying from one to three miles per hour. In 1807, during the erection of the lighthouse, six large blocks of granite, which had been landed on the reef, were removed by the force of the sea, and thrown over a rising ledge to the distance of twelve or fifteen paces; and an anchor, weighing about

22 cwt., was thrown up upon the rock *. Mr. Stevenson informs us moreover, that drift-stones, measuring upwards of thirty cubic feet, or more than two tons weight, have, during storms, been often thrown upon the rock from the deep water †.

Among the proofs that the sea has encroached both in the estuaries of the Tay and Forth, may be mentioned the submarine forests which have been traced for several miles by Dr. Fleming, along the margins of those estuaries on the north and south shores of the county of Fife ‡. The alluvial tracts, however, on which such forests grow, generally occupy spaces which may be said to be in dispute between the river and the sea, and to be alternately lost and won. *Estuaries* (a term which we confine to inlets entered both by rivers and tides of the sea) have a tendency to become silted up *in parts*; but the same tracts, after remaining dry, perhaps, for thousands of years, are again liable to be overflowed, for they are always low, and, if inhabited, must generally be secured by artificial embankments. Meanwhile the sea devours, as it advances, the high as well as the low parts of the coast, breaking down, one after another, the rocky bulwarks which protect the mouths of estuaries. The changes of territory, therefore, within the general line of coast are all of a subordinate nature, in no way tending to arrest the march of the great ocean, nor to avert the destiny eventually awaiting the whole region; they are like the petty wars and conquests of the independent states and republics of Greece, while the power of Macedon was steadily pressing on, and preparing to swallow up the whole.

On the coast of Fife, at St. Andrew's, a tract of land which intervened between the castle of Cardinal Beaton and the sea has been entirely swept away, as were the last remains of the Priory of Crail, in the same county, in 1803. On both sides of the Frith of Forth, land has been consumed; at North Berwick in particular, and at Newhaven, where an arsenal and dock, built in the reign of James IV., in the fifteenth century has been overflowed.

* Account of the Erection of the Bell Rock Lighthouse, p. 163.

† Ed. Phil. Journ., vol. iii., p. 54, 1820.

‡ Quarterly Journal of Science, &c., No. XIII. new series, March, 1830.

East Coast of England.—If we now proceed to the English coast, we find records of numerous lands having been destroyed in Northumberland, as those near Banborough and Holy Island, and at Tynemouth Castle, which now overhangs the sea, although formerly separated from it by a strip of land. At Hartlepool, and several other parts of the coast of Durham composed of magnesian limestone, the sea has made considerable inroads. Almost the whole coast of Yorkshire, from the mouth of the Tees to that of the Humber, is in a state of gradual dilapidation. That part of the cliffs which consists of lias, the oolite series, and chalk, decays slowly. They present abrupt and naked precipices, often three hundred feet in height; and it is only at a few points that the grassy covering of the sloping talus marks a temporary relaxation of the erosive action of the sea. The chalk cliffs are washed into caves in the projecting headland of Flamborough, where they are decomposed by the salt vapours, and slowly crumble away. But the waste is most rapid between that promontory and Spurn Point, or the coast of Holderness, as it is called. This tract consists chiefly of beds of clay, gravel, sand, and chalk rubble. The irregular intermixture of the argillaceous beds causes many springs to be thrown out, and this facilitates the undermining process, the waves beating against them, and a strong current setting chiefly from the north. The wasteful action is very conspicuous at Dimlington Height, the loftiest point in Holderness, where the beacon stands on a cliff one hundred and forty-six feet above high water, the whole being composed of clay, with pebbles scattered through it *.

In the old maps of Yorkshire, we find spots, now sandbanks in the sea, marked as the ancient sites of the towns and villages of Auburn, Hartburn, and Hyde. "Of Hyde," says Pennant, "only the tradition is left; and near the village of Hornsea, a street called Hornsea Beck has long since been swallowed †." Owthorne and its church have also been in great part destroyed, and the village of Kilnsea; but these places are now removed farther inland. The rate of enroachment at Owthorne, at present, is about *four yards a year* ‡.

* Phillips's Geology of Yorkshire, p. 61.
† Arctic Zoology, vol. i., p. 10, Introduction.
‡ For this information I am indebted to Mr. Phillips, of York.

Not unreasonable fears are entertained that at some future time the Spurn Point will become an island, and that the ocean, entering into the estuary of the Humber, will cause great devastation *. Pennant, after speaking of the silting up of some ancient ports in that estuary, observes, "But, in return, the sea has made most ample reprisals; the site, and even the very names of several places, once towns of note upon the Humber, are now only recorded in history; and Ravensper was at one time a rival to Hull (Madox, Ant. Exch. i., 422), and a port so very considerable in 1332, that Edward Balliol and the confederated English barons sailed from hence to invade Scotland; and Henry IV., in 1399, made choice of this port to land at, to effect the deposal of Richard II., yet the whole of it has long since been devoured by the merciless ocean: extensive sands, dry at low water, are to be seen in their stead †."

Pennant describes Spurn Head as a promontory in the form of a sickle, and says the land, for some miles to the north, was "perpetually preyed on by the fury of the German Sea, which devours whole acres at a time, and exposes on the shores considerable quantities of beautiful amber ‡."

According to Bergmann, a strip of land, with several villages, was carried away near the mouth of the Humber in 1475.

The maritime district of Lincolnshire consists chiefly of lands which lie below the level of the sea, being protected by embankments. Great parts of this fenny tract were, at some unknown period, a woody country, but were afterwards inundated, and are now again recovered from the sea. Some of the fens were embanked and drained by the Romans; but after their departure the sea returned, and large tracts were covered with beds of silt containing marine shells, now again converted into productive lands. Many dreadful catastrophes are recorded by incursions of the sea, whereby several parishes have been at different times overwhelmed.

We come next to the cliffs of Norfolk and Suffolk, where the decay is in general incessant and rapid. At Hunstanton, on the north, the undermining of the lower arenaceous beds at the foot of the cliff causes masses of red and white chalk to be precipitated from above. Between Hunstanton and Wey-

* Phillips's Geology of Yorkshire, p. 60.

† Arct. Zool., vol. i., p. 10, Introduction. ‡ Ibid.

bourne, low hills, or dunes, of blown sand, are formed along the shore, from fifty to sixty feet high. They are composed of dry sand, bound in a compact mass by the long creeping roots of the plant called Marram (Arundo arenaria). Such is the present set of the tides, that the harbours of Clay, Wells, and other places, are securely defended by these barriers; affording a clear proof that it is not the strength of the material at particular points that determines whether the sea shall be progressive or stationary, but the general contour of the coast. The waves constantly undermine the low chalk cliffs, covered with sand and clay, between Weybourne and Sheringham, a certain portion of them being annually removed. At the latter town I ascertained, in 1829, some facts which throw light on the rate at which the sea gains upon the land. It was computed, when the present inn was built, in 1805, that it would require seventy years for the sea to reach the spot; the mean loss of land being calculated, from previous observations, to be somewhat less than one yard annually. The distance between the house and the sea was fifty yards; but no allowance was made for the slope of the ground being *from* the sea, in consequence of which, the waste was naturally accelerated every year, as the cliff grew lower, there being at each succeeding period less matter to remove when portions of equal area fell down. Between the years 1824 and 1829, no less than seventeen yards were swept away, and only a small garden was then left between the building and the sea. There is now a depth of twenty feet (sufficient to float a frigate) at one point in the harbour of that port, where, only forty-eight years ago, there stood a cliff fifty feet high, with houses upon it! If once in half a century an equal amount of change were produced at once by the momentary shock of an earthquake, history would be filled with records of such wonderful revolutions of the earth's surface, but, if the conversion of high land into deep sea be gradual, it excites only local attention. The flag-staff of the Preventive Service station, on the south side of this harbour, has, within the last fifteen years, been thrice removed inland, in consequence of the advance of the sea.

Farther to the south we find cliffs, composed, like those of Holderness before mentioned, of alternating strata of blue clay, gravel, loam, and fine sand. Although they sometimes exceed two hundred feet in height, the havoc made on the coast is most formidable. The whole site of ancient Cromer now forms part of the German Ocean, the inhabitants having gradually retreated inland to their present situation, from whence the sea still threatens to dislodge them. In the winter of 1825, a fallen mass was precipitated from near the lighthouse, which covered twelve acres, extending far into the sea, the cliffs being two hundred and fifty feet in height *. The undermining by springs has sometimes caused large portions of the upper part of the cliffs, with houses still standing upon them, to give way, so that it is impossible, by erecting breakwaters at the base of the cliffs, permanently to ward off the danger. On the same coast, the ancient villages of Shipden, Wimpwell, and Eccles, have disappeared; several manors and large portions of neighbouring parishes having, piece after piece, been swallowed up; nor has there been any intermission, from time immemorial, in the ravages of the sea along a line of coast twenty miles in length, in which these places stood †. Hills of blown sand, between Eccles and Winterton, have barred up and excluded the tide for many hundred years from the mouths of several small estuaries; but there are records of nine breaches, from twenty to one hundred and twenty yards wide, having been made through these, whereby immense damage was done to the low grounds in the interior. A few miles south of Happisburgh, also, are hills of blown sand, which extend to Yarmouth; and these are supposed to protect the coast, but in fact their formation proves that a temporary respite of the incursions of the sea on this part is permitted by the present set of the tides and currents. Were it otherwise, the land, as we have seen, would give way, though made of solid rock.

At Yarmouth, the sea has not advanced upon the sands in the slightest degree since the reign of Elizabeth. In the time of the Saxons, a great estuary extended as far as Norwich, which city is represented, even in the thirteenth and fourteenth centuries, as "situated on the banks of an arm of the sea." The sands whereon Yarmouth is built first became firm and habitable ground about the year 1008, from which time a line of dunes has gradually increased in height and breadth, stretching

* Taylor's Geology of East Norfolk, p. 32.

† Ibid.

across the whole entrance of the ancient estuary, and obstructing the ingress of the tides so completely, that they are only admitted by the narrow passage which the river keeps open, and which has gradually shifted several miles to the south. The tides at the river's mouth only rise, at present, to the height of three or four feet. By the exclusion of the sea, thousands of acres in the interior have become cultivated lands; and, exclusive of smaller pools, upwards of sixty fresh water lakes have been formed, varying in depth from fifteen to thirty feet, and in extent from one acre to twelve hundred *. The Yare, and other rivers, frequently communicate with these sheets of water; and thus they are liable to be filled up gradually with lacustrine and fluviatile deposits, and to be converted into land covered with forests. When the sea at length returns (for as the whole coast gives way, this must inevitably happen sooner or later), these tracts will be again submerged, and submarine forests may then be found, as along the margins of many estuaries. We may easily conceive that such natural embankments as those thrown by the waves, and subsequently raised by winds, across the entrance of this river, may so shut out the tide, that inland places may become dry which, on the breaching of the barrier, might again be permanently overflowed even at low water; for the tides are now so depressed, even outside the barrier, that the river is almost in the condition of one which enters an inland sea. Were high tides to recur, the fresh-water would be ponded back during the flow, and would perhaps not entirely escape during the ebb. It has been observed, by Dr. Fleming, that the roots of the trees in several submarine forests in Scotland are in lacustrine silt. The stumps of the trees evidently occupy the position in which they formerly grew, and are sometimes from eight to ten feet below high water mark. The horizontality of the strata and other circumstances, preclude the supposition of a slide, and the countries in question have been from time immemorial free from violent earthquakes, which might have produced subsidences. He has, therefore, attributed the depression, with much probability, to the drainage of peaty soil on the removal of a seaward barrier. Suppose a lake (like one of those in the

* Taylor's Geology of East Norfolk, p. 10.

valley of the Yare,) to become a marsh, and a stratum of vegetable matter to be formed on the surface, of sufficient density to support trees. Let the outlet of the marsh be elevated a few feet only above the rise of the tide. All the strata below the level of the outlet would be kept constantly wet or in a semifluid state, but if the tides rise in the estuary, and the sea encroaches, portions of the gained lands are swept away, and the extremities of the alluvial and peaty strata, whereon the forest grew, are exposed to the sea, and at every ebb tide left dry to a depth equal to the increased fall of the tide. Much water, formerly prevented from escaping by the altitude of the outlet, now oozes out from the moist beds,—the strata collapse and the surface of the morass instead of remaining at its original height, sinks below the level of the sea *.

Yarmouth does not project beyond the general line of coast which has been rounded off by the predominating current from the north-west. It must not be imagined, therefore, that the acquisition of new land fit for cultivation in Norfolk and Suffolk indicates any permanent growth of the eastern limits of our island, to compensate its reiterated losses. No *delta* can form on such a shore.

The cliffs of Suffolk, to which we next proceed, are somewhat less elevated than those of Norfolk, but composed of similar accumulations of alternating clay, sand, and gravel. From Gorleston in Suffolk, to within a few miles north of Lowestoff, the cliffs are slowly undermined. Near the last-mentioned town, there is an inland cliff about sixty feet high, the talus being covered with turf and heath, between which and the sea is a low, flat tract of sand, called the Ness, which gains slowly on the sea. It does not seem difficult to account for the retreat of the sea at this point from its ancient limits, the base of the inland cliff. About a mile off Lowestoff lies the Holm Sand, the highest part of which is dry at low water. The current in the intervening passage, called Lowestoff Roads, is a back-water, wherein the tide, instead of obeying the general rule along this coast, runs nine hours towards the north, and only three towards the south. Here, therefore, we have an

* See two papers on Submarine Forests by the Rev. Dr. Fleming, in the Trans. Roy. Soc. Edin., vol. ix., p. 419, and Quarterly Journ. of Sci., No. 13, new series, March, 1830.

eddy, and the Holm Sand is a bank caused by the meeting of currents, where, as usual, sediment subsides. The channel called Lowestoff Roads is about a mile broad, and the depth varies from twenty to fifty-nine feet at low water. On one side, the current has hollowed out of the Holm Sand a deep curve, called the Hook, and on the other side precisely opposite, is the projecting point of the Ness *. As the *points* and *bends* of a river correspond to each other, sand-bars being thrown up at each point, and the greatest depth being where the river is wearing into the bend, so we find here a shoal increasing at the Ness, and deep water preserved in the Hook. We cannot doubt that, at a modern period in the history of this coast, the high cliffs on which Lowestoff stands, were once continuous across the space where the roadstead now is, and where we have stated the present depth to be fifty-nine feet at low water.

By the mean of thirty-eight observations, it has been found that the difference of high and low tide at Lowestoff is only five feet eight inches †—a remarkably slight oscillation for our eastern coast, and which naturally suggests the inquiry whether, at other points where there are inland cliffs, the rise of the tides is below their average level.

The sea undermines the high cliffs a few miles north of Lowestoff, near Corton; as also two miles south of the same town, at Pakefield, a village which has been in part swept away during the present century. From thence to Dunwich the destruction is constant. At the distance of two hundred and fifty yards from the wasting cliff at Pakefield, the sea is sixteen feet deep at low water, and in the roadstead beyond, twenty-four feet. Of the gradual destruction of Dunwich, once the most considerable sea-port on this coast, we have many authentic records. Gardner, in his History of that borough, published in 1754, shows, by reference to documents beginning with Doomsday Book, that the cliffs at Dunwich, Southwold, Eastern, and Pakefield, have been always subject to wear away. At Dunwich, in particular, two tracts of land which had been taxed in the eleventh century, in the time of King

* See Plan of proposed Canal at Lowestoff, by Cubitt and Taylor, 1826.

† These observations were made by Mr. R. C. Taylor.

Edward the Confessor, are mentioned, in the Conqueror's survey, made but a few years afterwards, as having been devoured by the sea. The losses, at a subsequent period, of a monastery—at another of several churches—afterwards of the old port—then of four hundred houses at once—of the church of St. Leonard, the high road, town-hall, gaol, and many other buildings, are mentioned, with the dates when they perished. It is stated that, in the sixteenth century, not one quarter of the town was left standing; yet the inhabitants retreating inland, the name was preserved, as has been the case with many other ports when their ancient site has been blotted out. There is, however, a church, of considerable antiquity, still standing, the last of twelve mentioned in some records. In 1740, the laying open of the churchyard of St. Nicholas and St. Francis, in the sea cliffs, is well described by Gardner, with the coffins and skeletons exposed to view,—some lying on the beach, and rocked

In cradle of the rude imperious surge.

Of these cemeteries, no remains can now be seen. Ray also says, "that ancient writings make mention of a wood a mile and a half to the east of Dunwich, the site of which must at present be so far within the sea *." This city, once so flourishing and populous, is now a small village, with about twenty houses and one hundred inhabitants.

There is an old tradition, "that the tailors sat in their shops at Dunwich, and saw the ships in Yarmouth Bay;" but when we consider how far the coast at Lowestoff Ness projects between these places, we cannot give credit to the tale, which, nevertheless, proves how much the inroads of the sea in times of old prompted men of lively imagination to indulge a taste for the marvellous.

Gardner's description of the cemeteries laid open by the waves reminds us of the scene which has been so well depicted by Bewick †, and of which numerous points on our coast might have suggested the idea. On the verge of a cliff, which the sea has undermined, are represented the unshaken tower and western end of an abbey. The eastern aisle is gone,

* Consequences of the Deluge, Phys. Theol. Discourses.

† History of British Birds, vol. ii., p. 220, Ed. 1821.

and the pillars of the cloister are soon to follow. The waves have almost isolated the promontory, and invaded the cemetery, where they have made sport with the mortal relics, and thrown up a skull upon the beach. In the foreground is seen a broken tombstone, erected, as its legend tells, "to *perpetuate* the memory" of one whose name is obliterated, as is that of the county for which he was "Custos Rotulorum." A cormorant is perched on the monument, defiling it, as if to remind some moralizer like Hamlet of "the base uses" to which things sacred may be turned. Had this excellent artist desired to satirize certain popular theories of geology, he might have inscribed the stone to the memory of some philosopher who taught "the permanency of existing continents"—"the era of repose"—"the impotence of modern causes."

South of Dunwich are two cliffs, called Great and Little Cat Cliff. That which bears the name of Great, has become the smallest of the two, and is only fifteen feet high, the more elevated portion of the hill having been carried away; on the other hand, the Lesser Cat Cliff has gained in importance, for the sea has here been cutting deeper into a hill which slopes towards it. But at no distant period, the ancient names will again become appropriate, for at Great Cat Cliff, the base of another hill will soon be reached, and at Little Cat Cliff, the sea will, at about the same time, arrive at a valley.

The incursions of the sea at Aldborough were formerly very destructive, and this borough is known to have been once situated a quarter of a mile east of the present shore. The inhabitants continued to build farther inland, till they arrived at the extremity of their property, and then the town decayed greatly, but two sand-banks, thrown up at a short distance, now afford a temporary safeguard to the coast. Between these banks and the present shore, where the current now flows, the sea is twenty-four feet deep on the spot where the town formerly stood. Continuing our survey of the Suffolk coast to the southward, we find that the cliffs of Bawdsey and Felixtow are foundering slowly, and that the point on which Landguard Fort is built suffers gradual decay. It appears that, within the memory of persons now living, the Orwell river continued its course in a more direct line to the sea, and entered to the north instead of the south of the low bank on which the fort last mentioned is built.

Harwich, in Essex, stands on an isthmus, which will probably become an island in little more than half a century; for the sea will then have made a breach near Lower Dover Court, should it continue to advance as rapidly as it has done during the last fifty years. Within ten years, there was a considerable space between the battery at Harwich, built twenty-three years ago, and the sea; part of the fortification has already been swept away, and the rest overhangs the water.

At Walton Naze, in the same county, the cliffs, composed of London clay, capped by the shelly sands of the crag, reach the height of about one hundred feet, and are annually undermined by the waves. The old churchyard of Walton has been washed away, and the cliffs to the south are continually disappearing.

On the coast bounding the estuary of the Thames, there are numerous examples both of the gain and loss of land. The Isle of Sheppey, which is now about six miles long by four in breadth, is composed of London clay. The cliffs on the north, which are from sixty to eighty feet high, decay rapidly, fifty acres having been lost within the last twenty years. The church at Minster, now near the coast, is said to have been in the middle of the island fifty years ago; and it is computed that, at the present rate of destruction, the whole isle will be annihilated in about another half century *. On the coast to the east of Sheppey stands the Church of Reculver, upon a sandy cliff about twenty feet high. In the reign of Henry VIII. it is said to have been nearly a mile distant from the sea. In the "Gentleman's Magazine," there is a view of it about the middle of the last century, which still represents a considerable space as intervening between the north wall of the churchyard and the cliff. About twenty years ago, the waves came within one hundred and fifty feet of the boundary of the churchyard, half of which has since been washed away. The church is now dismantled (1829), and is in great danger; several houses in a field immediately adjoining having been washed away.

* For this information I am indebted to W. Gunnell, Esq.

In the Isle of Thanet, Bedlam Farm, belonging to the hospital of that name, has lost eight acres in the last twenty years, the land being chalk from forty to fifty feet above the level of the sea. It has been computed, that the average waste of the cliff between the North Foreland and the Reculvers, a distance of about eleven miles, is not less than two feet per annum. The chalk cliffs on the south of Thanet, between Ramsgate and Pegwell Bay, have, on an average, lost three feet per annum for the ten last years. The Goodwin Sands lie opposite this part of the Kentish coast. They are about ten miles in length, and are in some parts three, and in others, seven miles distant from the shore, and, for a certain space, are laid bare at low water. When the erection of a lighthouse on these sands was in contemplation by the Trinity Board, twelve years since, it was found, by borings, that the bank consisted of fifteen feet of sand, resting on blue clay. An obscure tradition has come down to us, that the estates of Earl Goodwin were situated here, and some have conjectured that they were overwhelmed by the flood mentioned in the Saxon Chronicle, *sub anno* 1099. The last remains of an island, consisting, like Sheppey, of clay, may, perhaps, have been carried away about that time.

In the county of Kent, there are other records of waste, at Deal; and at Dover, Shakspeare's Cliff, composed entirely of chalk, has suffered greatly, and continually diminishes in height, the slope of the hill being towards the land. About twenty years ago, there was an immense land-slip from this cliff, by which Dover was shaken as if by an earthquake. In proceeding from the northern parts of the German Ocean towards the Straits of Dover, the water becomes gradually more shallow, so that in the distance of about two hundred leagues, we pass from a depth of one hundred and twenty, to that of fifty-eight, thirty-eight, twenty-four, and eighteen fathoms. In the same manner, the English Channel deepens progressively from Dover to its entrance, formed by the Land's End of England, and the Isle of Ushant on the coast of France; so that the strait between Dover and Calais may be said to form a point of partition between two great inclined planes, forming the bottom of these seas*.

* Stevenson on the Bed of the German Ocean.—Ed. Phil. Journ., No. V., p. 45.

Whether England was formerly united with France has often been a favourite subject of speculation; and in 1753 a society at Amiens proposed this as the subject of a prize essay, which was gained by the celebrated Desmarest, then a young man. He founded his principal arguments on the identity of composition of the cliffs on the opposite sides of the Channel, on a submarine chain extending from Boulogne to Folkestone, only fourteen feet under low water, and on the identity of the noxious animals in England and France, which could not have swam across the straits, and would never have been introduced by man. He also attributed the rupture of the isthmus to the preponderating violence of the current from the north*. It will hardly be disputed that the ocean might have effected a breach through the land which, in all probability, once united our country to the continent, in the same manner as it now gradually forces a passage through rocks of the same mineral composition, and often many hundred feet high, upon our coast. Although the time required for such an operation was probably very great, yet we cannot estimate it by reference to the present rate of waste on both sides of the Channel. For when, in the thirteenth century, the sea burst through the isthmus of Staveren, which formerly united Friesland with North Holland, it opened in about one hundred years a strait more than half as wide as that which divides England from France, after which the dimensions of the new channel remained almost stationary. The greatest depth of the straits between Dover and Calais is twenty-nine fathoms, which only exceeds, by one fathom, the greatest depth of the Mississippi at New Orleans. If the moving column of water in the great American river, which, as we before stated, does not flow rapidly, can maintain an open passage to that depth in its alluvial accumulations, still more might a channel of the same magnitude be excavated by the resistless force of the tides and currents of "the ocean stream,"

ποταμοιο μεγα σθενος Ωκεανοιο.

At Folkestone, the sea eats away the chalk and subjacent strata. About the year 1716, there was a remarkable sinking

* Cuvier, Éloge de Desmarest.

of a tract of land near the sea, so that houses became visible at points near the shore, from whence they could not be seen previously. In the description of this subsidence in the Philosophical Transactions, it is said, "that the land consisted of a solid stony mass (chalk), resting on wet clay (galt), so that it slid forwards towards the sea, just as a ship is launched on tallowed planks." It is also stated that, within the memory of persons then living, the cliff there had been washed away to the extent of ten rods *. Encroachments of the sea at Hythe are also on record; but between this point and Rye there has been a gain of land, within the times of history; the rich level tract called Romney Marsh, about ten miles in width and five in breadth, consisting of silt, having received great accession. It has been necessary, however, to protect it, from the earliest periods, by a wall from the sea. These additions of land are exactly opposite that part of the English Channel where the conflicting tides meet; for as those from the north are the most powerful, they do not neutralize each other's force till they arrive at this distance from the straits. Rye, on the south of this tract, was once destroyed by the sea, but it is now two miles distant from it. The neighbouring town of Winchelsea was destroyed in the reign of Edward I., the mouth of the Rother stopped up, and the river diverted into another channel. In its old bed an ancient vessel, apparently a Dutch merchantman, was recently found. It was built entirely of oak, and much blackened †.

South Coast of England.—To pass over some points near Hastings, where the cliffs have wasted at several periods, we arrive at the promontory of Beachy Head. Here a mass of chalk, three hundred feet in length, and from seventy to eighty in breadth, fell, in the year 1813, with a tremendous crash; and similar slips have since been frequent ‡.

About a mile to the west of the town of Newhaven the remains of an ancient entrenchment are seen, on the brow of Castle Hill. This earth-work was evidently once of considerable extent, but the greater part has been cut away. The cliffs, which are undermined here, are high; more than one hundred feet of chalk being covered by tertiary clay and sand,

* Phil. Trans., 1716. † Edin. Journ. of Sci., No. xix., p. 56.
‡ Webster, Geol. Trans., vol. iii., p. 192.

from sixty to seventy feet in thickness. In a few centuries the last vestiges of the plastic clay formation on the southern borders of the chalk of the South Downs on this coast will be annihilated, and future geologists will learn, from historical documents, the ancient geographical boundaries of groups of strata then no more. On the opposite side of the estuary of the Ouse, on the east of Newhaven harbour, a bed of shingle, composed of chalk flints, derived from the waste of the adjoining cliffs, had accumulated at Seaford for several centuries. In the great storm of November, 1824, this bank was entirely swept away, and the town of Seaford inundated. Another great beach of shingle is now forming from fresh materials.

The whole coast of Sussex has been incessantly encroached upon by the sea from time immemorial; and, although sudden inundations only, which overwhelmed fertile or inhabited tracts are noticed in history, the records attest an extraordinary amount of loss. During a period of no more than eighty years, there are notices of about *twenty* inroads in which tracts of land of from twenty to *four hundred acres* in extent were overwhelmed at once; the value of the tithes being mentioned by Nicholas, in his Taxatio Ecclesiastica *. In the reign of Elizabeth, the town of Brighton was situated on that tract where the chain-pier now extends into the sea. In the year 1665, twenty-two tenements had been destroyed under the cliff. At that period there still remained under the cliff one hundred and thirteen tenements, the whole of which were overwhelmed in 1703 and 1705. No traces of the ancient town are now perceptible, yet there is evidence that the sea has merely resumed its ancient position at the base of the cliffs, the site of the old town having been merely a beach abandoned by the ocean for ages. It would be endless to allude to all the localities on the Sussex and Hampshire coasts, where the land has been destroyed; but we may point to the relation of the present shape and geological structure of the Isle of Wight, as attesting that it owes its present outline to the continued action of the sea. Through the middle of the island a high ridge of chalk strata, in a vertical position, runs in a direction east and west. This chalk forms the projecting promontory of Culver Cliff on the east,

* Mantell, Geology of Sussex, p. 293.

and of the Needles on the west; while Sandown Bay on the one side, and Compton Bay on the other, have been hollowed out of the softer sands and argillaceous strata, which are inferior to the chalk. The same phenomena are repeated in the Isle of Purbeck, where the line of vertical chalk forms the projecting promontory of Handfast Point; and Swanage Bay marks the deep excavation made by the waves in the softer strata, corresponding to those of Sandown Bay.

The entrance of the Channel called the Solent is becoming broader by the waste of the cliffs in Colwell Bay; it is crossed for more than two-thirds of its width by the shingle bank of Hurst Castle, which is about seventy yards broad, and twelve feet high, presenting an inclined plane to the west. This singular bar consists of a bed of rounded chalk flints, resting on an argillaceous base, always covered by the sea. The flints and a few other pebbles, intermixed, are exclusively derived from the waste of Hordwell, and other cliffs to the westward, where fresh-water marls, capped with a covering of chalk flints from five to fifty feet thick, are rapidly undermined.

In the great storm of November, 1824, this bank of shingle was moved bodily forwards for forty yards towards the north-east; and certain piles which served to mark the boundaries of two manors were found, after the storm, on the opposite side of the bar. At the same time many acres of pasture land were covered by shingle, on the farm of Westover, near Lymington. This bar probably marks the line where the opposing tides meet, for there is a second, or half-tide, of eighteen inches, three hours after the regular tide in this channel.

The cliffs between Hurst Shingle Bar and the mouth of the Stour and Avon are undermined continually. Within the memory of persons now living, it has been necessary thrice to remove the coast-road farther inland. The tradition, therefore, is probably true, that the church of Hordwell was once in the middle of that parish, although now very near the sea. The promontory of Christ Church Head gives way slowly. It is the only point between Lymington and Poole Harbour where any hard stony masses occur in the cliffs. Five layers of large ferruginous concretions, somewhat like the septaria of the London clay, have occasioned a resistance at this point, to which we may ascribe the existence of this headland. In the mean time, the waves have cut deeply into the soft sands and loam of Poole Bay; and, after severe frosts, great land-slips take place, which by degrees become enlarged into narrow ravines, or chines, as they are called, with vertical sides. One of these chines, near Boscomb, has been deepened twenty feet within a few years. At the head of each there is a spring, the waters of which have been chiefly instrumental in producing these narrow excavations, which are sometimes from one hundred to one hundred and fifty feet deep.

The peninsulas of Purbeck and Portland are continually wasting away. In the latter, the soft argillaceous substratum (Kimmeridge clay) hastens the dilapidation of the superincumbent mass of limestone.

In 1665, the cliffs adjoining the principal quarries gave way to the extent of one hundred yards, and fell into the sea; and in December, 1734, a slide to the extent of one hundred and fifty yards occurred on the east side of the isle, by which several skeletons, buried between slabs of stone, were discovered. But a much more memorable occurrence of this nature, in 1792, is thus described in Hutchins's History of Dorsetshire. "Early in the morning the road was observed to crack: this continued increasing, and before two o'clock the ground had sunk several feet, and was in one continued motion, but attended with no other noise than what was occasioned by the separation of the roots and brambles, and now and then a falling rock. At night it seemed to stop a little, but soon moved again; and before morning, the ground, from the top of the cliff to the water-side, had sunk in some places fifty feet perpendicular. The extent of ground that moved was about *a mile and a quarter* from north to south, and six hundred yards from east to west."

Portland is connected with the main land by the Chesil Bank, a ridge of shingle about seventeen miles in length, and, in most places, nearly a quarter of a mile in breadth. The pebbles forming this immense barrier are chiefly of limestone; but there are many of quartz, jasper, chert, and other substances, all loosely thrown together. What is singular, they gradually diminish in size, from the Portland end of the bank to that which attaches to the main land. The formation of this bar may probably be ascribed, like that of Hurst Castle, to a meeting of tides, or to a great eddy, between the peninsula and the land. We have seen that slight obstructions in the course of the Ganges will cause, in the course of a man's life, islands many times larger than the whole of Portland, and which, in some cases, consist of a column of earth more than one hundred feet deep. In like manner, we may expect the slightest impediment in the course of that tidal wave, which is sweeping away annually large tracts of our coast, to give rise to banks of sand and shingle many miles in length, if the transported materials be intercepted in their passage to those submarine receptacles whither they are borne by the current. The gradual diminution in the size of the gravel, as we proceed eastward, might probably admit of explanation, if the velocity of the tide or eddy at different points was ascertained; the rolled masses thrown up being largest where the motion of the water is most violent, or where they are deposited at the least distance from the rocks from which they were detached. The storm of 1824 burst over this bar with great fury, and the village of Chesilton, built upon the southern extremity of the bank, was overwhelmed, with many of the inhabitants*. The fundamental rocks whereon the shingle rests are found at the depth of a few yards only below the level of the sea.

At Lyme Regis, in Dorsetshire, the "Church Cliffs," as they are called, consisting of lias, about one hundred feet in height, have gradually fallen away, at the rate of one yard a year, since 1800†. The cliffs of Devonshire and Cornwall, which are chiefly composed of hard rocks, decay less rapidly. Near Penzance, in Cornwall, there is a projecting tongue of land, called the "Green," formed of granitic sand,

* This same storm carried away part of the Breakwater at Plymouth, and huge masses of rock were lifted from the bottom of the weather side, and rolled fairly to the top of the pile. It was in the same month, and also during a spring tide, that a great flood is mentioned on the coasts of England, in the year 1099. Florence of Worcester says, "On the third day of the nones of Nov., 1099, the sea came out upon the shore, and buried towns and men very many, and oxen and sheep innumerable." Also the Saxon Chronicle, already cited, for the year 1099, "This year eke on St. Martin's-mass day, the 11th of Novembre, sprung up so much of the sea flood, and so myckle harm did, as no man minded that it ever afore did, and there was the ylk day a new moon."

† This ground was measured by Dr. Carpenter, of Lyme, in 1800, and again in 1829.

from which more than thirty acres of pasture land have been gradually swept away in the course of the last two or three centuries *. It is also said that St. Michael's Mount, now an insular rock, was formerly situated in a wood several miles from the sea; and its old Cornish name, according to Carew, signifies the Hoare Rock in the Wood. Between the Mount and Newlyn, there is seen, under the sand, black vegetable mould, full of hazel nuts, and the branches, leaves, roots, and trunks of forest trees, all of indigenous species. This vegetable stratum has been traced seaward as far as the ebb permits, and seems to indicate some ancient estuary on that shore.

The oldest historians mention a celebrated tradition in Cornwall of the submersion of the Lionnesse, a country which formerly stretched from the Land's End to the Scilly Islands. The tract, if it existed, must have been thirty miles in length, and perhaps ten in breadth. The land now remaining on either side is from two hundred to three hundred feet high; the intervening sea about three hundred feet deep. Although there is no evidence for this romantic tale, it probably originated in some catastrophe occasioned by former inroads of the Atlantic upon this exposed coast †.

Having now laid before the reader an ample body of proofs of the destructive operations of tides and currents on our eastern and southern shores, it will be unnecessary to enter into details of changes on the western coast, for they present merely a repetition of the same phenomena, and in general on an inferior scale. On the borders of the estuary of the Severn, the flats of Somersetshire and Gloucestershire have received enormous accessions, while, on the other hand, submarine forests on the coast of Lancashire indicate the overflowing of alluvial tracts. There are traditions in Pembrokeshire ‡ and Cardiganshire § of far greater losses of territory than that which the Lionnesse tale of Cornwall pretends to commemorate. They are all important, as demonstrating that the earliest inhabitants were familiar with the phenomenon of incursions of the sea.

The French coast, particularly that of Brittany, where the

* Boase, Trans. Royal Geol. Soc. of Cornwall, vol. ii., p. 129. † Ibid., p. 130.

‡ Camden, who cites Gyraldus, also Ray, "On the Deluge." Phys. Theol. p. 228.

§ Meyrick's Cardigan.

tides rise to an extraordinary height, is the constant prey of the waves. In the ninth century many villages and woods are reported to have been carried away, the coast undergoing great change, whereby the hill of St. Michael was detached from the main land. The parish of Bourgneuf, and several others in that neighbourhood, were overflowed in the year 1500. In 1735, during a great storm, the ruins of Palnel were seen uncovered in the sea *. A romantic tradition, moreover, has descended from the fabulous ages, of the destruction of the south-western part of Brittany, whence we may probably infer some great inroad of the sea at a remote period †.

* Hoff, Geschichte, &c., vol. i., p. 49.

† Ibid., p. 48.

CHAPTER XVI.

Action of Tides and Currents, *continued*—Inroads of the sea upon the delta of the Rhine in Holland—Changes in the arms of the Rhine—Estuary of the Bies Bosch, formed in 1421—Formation of the Zuyder Zee, in the 13th century—Islands destroyed—Delta of the Ems converted into a bay—Estuary of the Dollart formed—Encroachment of the sea on the coast of Sleswick—Inroads on the eastern shores of North America—Tidal wave, called the Bore—Influence of tides and currents on the mean level of seas—Action of currents on inland lakes and seas—Baltic—Cimbrian deluge—Straits of Gibraltar—Under-currents—Shores of Mediterranean—Rocks transported on floating icebergs—Dunes of blown sand—Sands of the Libyan Desert—De Luc's natural chronometers.

THE line of British coast considered in the preceding chapter offered no example of the conflict of two antagonist forces; the entrance, on the one hand, of a river draining a large continent, and on the other, the flux and reflux of the tide, aided by a strong current setting across a river's mouth. But when we pass over by the Straits of Dover to the continent, and proceed northwards, we find an admirable illustration of such a contest, where the Rhine and the ocean are opposed to each other, each disputing the ground now occupied by Holland; the one striving to shape out an estuary, the other to form a delta. There was evidently a period when the river obtained the ascendency, but for the last two thousand years, during which man has witnessed and actively participated in the struggle, the result has been in favour of the ocean, the area of the whole territory having become more and more circumscribed; natural and artificial barriers having given way, one after another, and many hundred thousand human beings having perished in the waves.

The Rhine, after flowing from the Grison Alps, copiously charged with sediment, first purges itself in the Lake of Constance, where a large delta is formed: then, swelled by the Aar and numerous other tributaries, it flows for more than six hundred miles towards the north; when, entering a low tract, it divides into two arms, not far north of Cleves—a point which must therefore be considered the head of its delta. The left arm takes the name of the Waal, and the right, retaining that of the Rhine, sends off another branch to the left, called the Leck, and still lower down, another named the Yssel. After this last division, the smallest stream, still called the Rhine, passes by Utrecht, and loses itself in the sands before reaching the German Sea, a few miles below Leyden. It is common, in all great deltas, that the principal channels of discharge should shift from time to time; but in Holland so many magnificent canals have been constructed, and have diverted, from time to time, the course of the waters, that the geographical changes in this delta are endless; and their history, since the Roman era, forms a complicated topic of antiquarian research. The present head of the delta is about forty geographical miles from the nearest part of the gulf called the Zuyder Zee, and more than twice that distance from the general coast-line. The present head of the Nilotic delta is about eighty or ninety geographical miles from the sea; that of the Ganges, as we before stated, two hundred and twenty; and that of the Mississippi about one hundred and eighty, reckoning from the point where the Atchafalaya branches off, to the extremity of the new tongue of land in the Gulf of Mexico. But the comparative distance between the heads of deltas and the sea affords scarcely any data for estimating the relative magnitude of the alluvial tracts formed by their respective rivers. For the ramifications depend on many varying and temporary circumstances, and the area over which they extend does not hold any constant proportion to the volume of water in the river.

We may consider the Rhine, at present, as having three mouths; the southernmost or left arm being the Waal; the Leck, the largest of the three, being in the centre; and the Yssel forming the right or northern arm. As the whole coast to the south, as far as Calais, and on the north, to the entrance of the Baltic, has, from time immemorial, yielded to the force of the waves, it is evident that the delta of the Rhine, if it had advanced, would have become extremely prominent, and even if it had remained stationary, would long ere this have projected, like that strip of land already described, at the mouth of the Mississippi, beyond the rounded outline of the coast. But we find, on the contrary, that a line of islands which skirts the coast have not only lessened in size, but in number also, while great bays in the interior have been formed by incursions of the sea. We shall confine ourselves to the enumeration of some of the leading facts, in confirmation of these views, and begin with the southernmost part of the delta where the Waal enters, which is at present united with the Meuse, in the same manner as an arm of the Po, before mentioned, has become confluent with the Adige. The Meuse itself had once a common embouchure with the Schelde, by Sluys and Ostburg, but this channel was afterwards sanded up, as were many others between Walcheren, Beveland, and other isles at the mouths of these rivers. The new accessions were almost all within the coast-line, and were far more than counterbalanced by inroads of the sea, whereby large tracts of land, and dunes of blown sand, together with towns and villages, were swept away between the fourteenth and eighteenth centuries. Besides the destruction of parts of Walcheren, Beveland, and populous districts in Kadzand, the island Orisant was in the year 1658 annihilated.

One of the most memorable eruptions occurred in 1421, where the tide, pouring into the mouth of the united Meuse and Waal, burst through a dam in the district named Bergse-Veld, and overflowed twenty-two villages, forming that large sheet of water called the Bies Bosch. No vestige even of the ruins of these places could ever afterwards be seen, but a small portion of the new bay became afterwards silted up, and formed an island. The Leck, or central arm of the Rhine, which enters the sea a little to the north of this new estuary, has, at present, a communication with it. The island Grunewert, which in the year 1228 existed not far from Houten, has been entirely destroyed. Farther to the north is a long line of shore, covered with sand dunes, where great depredations have been made from time to time. The church of Scheveningen, not far from the Hague, was once in the middle of the village, and now stands on the shore; half the place having been overwhelmed by the waves in 1570. Catwyck, once far from the sea, is now upon the shore; two of its streets having been overflowed, and land torn away to the extent of two hundred yards in 1719. It is only by aid of embankments, that Petten, and several other places farther north, have been defended against the sea.

We may next examine the still more important changes which have taken place on the coast opposite the right arm of the Rhine or the Yssel, where the ocean has burst through a large isthmus, and entered the inland lake Flevo, which, in ancient times, was, according to Pomponius Mela, formed by the overflowing of the Rhine over certain low lands. It appears, that in the time of Tacitus, there were several lakes in the present site of the Zuyder Zee, between Friesland and Holland. The successive inroads by which these, and a great part of the adjoining territory, were transformed into a great gulf, began about the beginning of the thirteenth century, and were completed about the end of the same. Alting gives the following relation of the occurrence, drawn from manuscript documents of contemporary inhabitants of the neighbouring provinces. In the year 1205, the island now called Wieringen, to the south of the Texel, was still a part of the main land, but during several high floods, of which the dates are given, ending in December, 1251, it was separated from the continent. By subsequent incursions, the sea consumed great parts of the rich and populous isthmus, a low tract which stretched on the north of Lake Flevo, between Staveren in Friesland, and Medemblick in Holland, till at length a breach was completed about the year 1282, and afterwards widened. Great destruction of land took place when the sea first broke in, and many towns were destroyed; but there was afterwards a reaction to a certain extent, large tracts at first submerged having been gradually redeemed. The new straits south of Staveren are more than half the width of those of Dover, but are very shallow, the greatest depth not exceeding two or three fathoms. The new bay is of a somewhat circular form, and between *thirty* and *forty* miles in diameter. How much of this space may formerly have been occupied by Lake Flevo, is unknown.

A series of isles, stretching from the Texel to the mouths of the Weser and Elbe, are evidently the last relics of a tract once continuous. They have greatly diminished in size, and have lost about a third of their number, since the time of Pliny*. While the delta of the Rhine has suffered so mate-

* Some few of them have extended their bounds, or become connected with others, by the sanding up of channels; but even these, like Juist, have generally

rially from the action of tides and currents, it cannot be supposed that minor rivers should have been permitted to extend their deltas. It appears, that in the time of the Romans there was an alluvial plain of great fertility, where the Ems entered the sea by three arms. This low country stretched between Groningen and East Friesland, and sent out a peninsula to the north-east towards Emden. A flood, in 1277, first destroyed part of the peninsula. Other inundations followed at different periods throughout the fifteenth century. In 1507, a part only of Torum, a considerable town, remained standing; and in spite of the erection of dams, the remainder of that place, together with market-towns, villages, and monasteries, to the number of fifty, were finally overwhelmed. The new gulf, called the Dollart, although small in comparison to the Zuyder Zee, occupied no less than six square miles at first; but part of this space was, in the course of the two following centuries, again redeemed from the sea. The small bay of Leybucht, farther north, was formed in a similar manner in the thirteenth century, and the bay of Harlbucht in the middle of the sixteenth. Both of these have since been partially reconverted into dry land. Another new estuary, called the Gulf of Jahde, near the mouth of the Weser, scarcely inferior in size to the Dollart, has been gradually hollowed out since the year 1016, between which era and 1651 a space of about four square miles has been added to the sea. The rivulet which now enters this inlet is very small; but Arens conjectures, that an arm of the Weser had once an outlet in that direction.

Farther north we find so many records of waste on the western coast of Sleswick, as to lead us to anticipate, that, at no distant period in the history of the physical geography of Europe,

given way as much on the north towards the sea, as they have gained on the south, or land side. Osterdun, Borkun, and several others, have been continually wasting away. Buissen is reduced to a sand-bank. Langeroog has been divided into three parts, and Wangeroog cut in two, many buildings having been carried away. Pliny counted twenty-three islands between the Texel and Eider, whereas there are now only sixteen, including Heligoland and Neuwerk.—Hoff, vol. i., p. 364. Heligoland at the mouth of the Elbe began in the year 800 to be much consumed by the waves. In the years 1300, 1500, and 1649, other parts were swept away, till at last only a rock and some low ground remained. Since 1770, a current has cut a passage sufficiently deep to admit large ships through this remaining portion, and has formed two islands.—Hoff, vol. i., p. 57.

Jütland will become an island, and the ocean will obtain a more direct entrance into the Baltic.

Northstrand, up to the year 1240, was, with the islands Sylt and Föhr, so nearly connected with the main land as to appear a peninsula, and was called North Friesland, a highly cultivated and populous district. It measured from nine to eleven geographical miles from north to south, and six to eight from east to west. In the above-mentioned year it was torn asunder from the continent, and in part overwhelmed. The Isle of Northstrand, thus formed, was, towards the end of the sixteenth century, only four geographical miles in circumference, and was still celebrated for its cultivation and numerous population. After many losses, it still contained nine thousand inhabitants. At last, in the year 1634, on the evening of the 11th of October, a flood passed over the whole island, whereby one thousand three hundred houses, with many churches, were lost; fifty thousand head of cattle perished, and above six thousand men. Three small isles, one of them still called Northstrand, alone remained, which are now continually wasting..

A review of the ravages committed during the last two thousand years on the French, Dutch, and Danish coasts, naturally leads us to inquire how it happened that the Rhine was enabled, at some former period, to accumulate so large a delta. We might, perhaps, in reply to this question, repeat our former observation, that the set of tides and currents necessarily varies from time to time; and that different coasts become, each in their turn, exposed to their fury, and then again restored to a state of quiescence. Islands and promontories, moreover, may have disappeared, which once protected the present site of Holland; and that region may afterwards have been laid open, as the Baltic would be, if the ocean, by renewing its attacks, should finally breach the isthmus by Sleswick. It may also be suggested that if, in former times, the Straits of Dover were closed, the Rhine must have entered at the bottom of a deep bay, on the one side of which was Great Britain, and on the other the coasts of Norway, Denmark, the Netherlands, and France. The transporting power of the current might then have been much inferior to that afterwards exerted, when the tide ran freely through the channel. Pliny expressed his wonder that the new lands at the mouths of the Tigris and Euphrates grew so rapidly, and "that the fluviatile matter was not swept away by the tide, which penetrated far above the tracts where great accessions were made *." The remark proves that he had considered the different condition of rivers in inland seas, and those discharging their waters into the ocean; but he did not reflect, that at the bottom of a deep bay where there is no current setting across the river's mouth, the ebbing and flowing of the waters cannot remove the sedimentary matter to a great distance.

After so many authentic details respecting the destruction of the coast in parts of Europe best known, it will be unnecessary to multiply examples of analogous changes in more distant regions of the world. It must not, however, be imagined that our own seas form any exception to the general rule. Thus, for example, if we pass over to the Eastern coast of North America, where the tides rise to a great elevation, we find many facts attesting the incessant demolition of land. At Cape May, for example, on the north side of Delaware Bay, in the United States, the encroachment of the sea was shown by observations made consecutively for sixteen years, from 1804 to 1820, to average above nine feet a year†; and at Sullivan's Island, which lies on the north side of the entrance of the harbour of Charlestown, in South Carolina, the sea carried away a quarter of a mile of land in three years, ending in 1786‡.

Of oceanic deltas in general, it may be said that, even where they advance, a large portion of the sediment is carried away by the movements of the sea. In the case of the great river of Amazons, the effects of the tides are still sensible at the Straits of Pauxis, five hundred miles from the sea, after an interval of several days spent in their passage up. The ponding back, therefore, of this great body of fresh-water, and the resistance opposed by the spring-tides to its descent, cause a rapid acceleration during the ebb, whereby the sediment is

* "Nec ullâ in parte plus aut celerius profecere terræ fluminibus invectæ. Magis id mirum est, æstu longè ultra id accedente non repercussas."—Hist. Nat., lib. vi., c. 27.

† New Monthly Mag., vol. vi., p. 69.

‡ Hoff, vol. i., p. 96.

carried far from the mouth of the river, and then borne by a current towards the north. Captain Sabine * found that the sea was discoloured by the waters of the Amazon, at the distance of not less than three hundred miles from its mouth, where they were still running, with considerable rapidity, in a direction inclined to that of the equatorial current of the ocean. The deposits derived from this source appear to have formed a large portion of the maritime districts of Guiana, and are said to extend even to the mouths of the Orinoco, ten degrees of latitude farther north, where that river also is pouring an annual tribute of earthy matter into the sea.

Before we conclude our remarks on the action of the tides, we must not omit to mention the wave called "the Bore," which is a sudden and abrupt influx of the tide into a river or narrow strait. Those rivers are most subject to this wave which have the greatest embouchures in proportion to the size of their channels; because, in that case, a larger proportion of tide is forced through a passage comparatively smaller. For this reason, the Bristol Channel is very subject to the Bore, where it is of almost daily occurrence, and at spring-tides rushes up the estuary with extraordinary rapidity. The same phenomena is frequently witnessed in the principal branches of the Ganges, and in the Megna. "In the Hoogly, or Calcutta river," says Rennell, "the Bore commences at Hoogly Point, the place where the river first contracts itself, and is perceptible above Hoogly Town; and so quick is its motion, that it hardly employs four hours in travelling from one to the other, though the distance is near seventy miles. At Calcutta it sometimes occasions an instantaneous rise of five feet; and both here, and in every other part of its track, the boats, on its approach, immediately quit the shore, and make for safety to the middle of the river. In the channels, between the islands in the mouth of the Megna, the height of the Bore is said to exceed twelve feet; and is so terrific in its appearance, and dangerous in its consequences, that no boat will venture to pass at spring-tide †." These waves may sometimes cause inundations, undermine cliffs, and still more frequently sweep away trees and

* Account of Experiments to determine the Figure of the Earth, &c., p. 446.

† Rennell, Phil. Trans., 1781.

land animals from low shores, whereby they may be carried down, and ultimately imbedded in submarine deposits.

There is another question, in regard to the effects of tides and currents, not yet fully determined—how far they may cause the mean level of the ocean to vary at particular parts of the coast. It has been supposed, that the waters of the Red Sea maintain a constant elevation of between four and five fathoms above the neighbouring waters of the Mediterranean, at all times of the tide; and that there is an equal, if not greater diversity, in the relative levels of the Atlantic and Pacific, on the opposite sides of the isthmus of Panama. But the levellings recently carried across that isthmus by Mr. Lloyd, to ascertain the relative height of the Pacific Ocean at Panama, and of the Atlantic at the mouth of the river Chagres, have shown, that the difference of mean level between those oceans is not considerable. According to the result of this survey, on which great dependence may be placed *, the mean height of the Pacific is three feet and a half, or 3.52 above the Atlantic, if we assume the mean level of the sea to coincide with the mean between the extremes of the elevation and depression of the tides. For between the extremes of elevation and depression of the greatest tides in the Pacific, at Panama, there is a difference of 27.44 feet; but the mean difference at the usual spring-tides is 21.22 feet: whereas at Chagres this difference is only 1.16 feet, and is the same at all seasons of the year. The tides, in short, in the Caribbean Sea are scarcely perceptible, not exceeding those in some parts of the Mediterranean; whereas the rise is very high in the bay of Panama. But astronomers are agreed, that, on mathematical principles, the rise of the tidal wave above the mean level of a particular sea must be greater than the fall below it; and although the difference has been hitherto supposed insufficient to cause an appreciable error, it is, nevertheless, worthy of observation, that the error, such as it may be, would tend to reduce the difference now inferred, from the observations of Mr. Lloyd,

* Mr. Lloyd received from General Bolivar a special commission to survey the isthmus of Panama, with the view of ascertaining the most eligible line of communication between the two seas. He was assisted by Capt. Falmarck, a Swedish officer of engineers; and the result of their labours will appear in the Philosophical Transactions.

to exist between the levels of the two oceans. It is scarcely necessary to remark how much all points relating to the permanence of the mean level of the sea must affect our reasoning on the phenomena of estuary deposits; and it is to be hoped, that further experiments will be made to ascertain the amount of irregularity, if any exist.

ACTION OF CURRENTS IN INLAND LAKES AND SEAS.

Coast of the Baltic.—In such large bodies of water as the North American lakes, the continuance of a strong wind in one direction often causes the elevation of the water and its accumulation on the leeward side; and while the equilibrium is being restored, powerful currents are occasioned. By this means the finer sedimentary particles, as we before mentioned, are borne far out from the deltas, and argillaceous and calcareous marls are formed far from the shores. In the Euxine, also, although free from tides, we learn from Pallas, that there is a sufficiently strong current to undermine the cliffs in many parts, and particularly in the Crimea. But the force of currents is exerted in a much more powerful degree in seas like the Mediterranean and the Baltic, where strong currents set in from the ocean, whether driven in during tempests or from other more constant causes. The current which runs through the Cattegat, or channel of communication between the German Ocean and the Baltic, not only commits dreadful devastations on the isles of the Danish Archipelago, but acts, though with less energy, on the coasts far in the interior, as, for example, in the vicinity of Dantzic *. The continuance of north-westerly gales and storms in the Atlantic, during the height of the spring-tides, has often been attended with the most fatal disasters on the Danish coast, where, during the last ten centuries, we find authentic accounts of the wearing down of promontories, the deepening of gulfs, the conversion of peninsulas into

* Thus, in the year 1800, near the village of Jershoft a great mass was projected by a landslip into the sea. Hela, a point of land running out before Dantzic, was formerly much broader than at present; and farther north, in Samland, woods and territories have been torn away by the sea.—Hoff, vol. i., p. 73, who cites Pisansky.

islands, and the waste of isles; while in several cases marsh land, defended for centuries by dikes, has at last been overflowed, and thousands of the inhabitants whelmed in the waves *.

We have before enumerated the ravages of the ocean on the eastern shores of Sleswick; and as we find memorials of a series of like catastrophes on the western coast of that peninsula, we can scarcely doubt that a large opening will, at some future period, connect the Baltic with the North Sea. Jutland was the Cimbrica Chersonesus of the ancients, and was then evidently the theatre of similar calamities; for Florus says, "Cimbri, Theutoni, atque Tigurini, ab extremis Galliæ profugi, cum terras eorum inundasset Oceanus, novas sedes toto orbe quærebant†." Some have wished to connect this "Cimbrian deluge" with the bursting of the isthmus between England and France, and with other supposed convulsions; but when we consider the annihilation of Heligoland and Northstrand, and the other terrific inundations in Jutland and Holstein since the Christian era, wherein thousands have perished, we need not resort to hypothetical agents to account for the historical relation. The wave which, in 1634, devastated the whole coast of Jutland, committed such havoc, that we must be cautious how we reject hastily the traditions of like catastrophes on the coasts of Kent, Cornwall, Pembrokeshire, and Cardigan; for, however sceptical we may be as to the amount of territory destroyed, it is very possible that former inroads of the sea may have been greater on those shores than any witnessed in modern times.

* Thus the island Barsoe, on the coast of Sleswick, has lost year after year an acre at a time. The island Alsen suffers in like manner. The peninsula Zingst was converted into an island in 1625. There is a tradition, that the isle of Rugen (which is composed of tertiary limestone) was originally torn by a storm from the main land of Pomerania; and it is known, in later times, to have lost ground, as in the year 1625, when a tract of land was carried away. Some of the islands which have wasted away consist of ancient alluvial accumulations, containing blocks of granite, which are also spread over the neighbouring main land. The Marsh Islands are mere banks, like the lands formed of the "warp" in the Humber, protected by dikes. Some of them, after having been inhabited with security for more than ten centuries, have been suddenly overwhelmed. In this manner, in 1216, no less than ten thousand of the inhabitants of Eyderstede and Ditmarsch perished; and on the 11th of October, 1634, the islands and the whole coast as far as Jutland suffered by a dreadful deluge.

† Lib. iii., cap. 3.

Straits of Gibraltar.—It is well known that a powerful current sets constantly from the Atlantic into the Mediterranean, and its influence extends along the whole southern borders of that sea, and even to the shores of Asia Minor. Captain Smyth found, during his survey, that the central current ran constantly at the rate of from three to six miles an hour eastward into the Mediterranean, the body of water being three miles and a half wide. But there are also two lateral currents—one on the European, and one on the African side; each of them about two miles and a half broad, and flowing at about the same rate as the central stream. These lateral currents ebb and flow with the tide, setting alternately into the Mediterranean and into the Atlantic. The escape of the great body of water, which is constantly flowing in, has usually been accounted for by evaporation, which must be very rapid and copious in the Mediterranean; for the winds blowing from the shores of Africa are hot and dry, and hygrometrical experiments recently made in Malta and other places show that the mean quantity of moisture in the air investing the Mediterranean is equal only to one half of that in the atmosphere of England. It is, however, objected that evaporation carries away only fresh-water, and that the current is continually bringing in salt-water: why, then, do not the component parts of the waters of the Mediterranean vary? or, why do they remain apparently the same as those of the ocean? Some have imagined that the excess of salt might be carried away by an under-current running in a contrary direction to the superior; and this hypothesis appeared to receive confirmation from a late discovery that the water taken up about fifty miles within the Straits, from a depth of six hundred and seventy fathoms, contained a quantity of salt *four times greater* than the water of the surface. Dr. Wollaston*, who analysed this water obtained by Captain Smyth, truly inferred that an under-current of such denser water, flowing outward, if of equal breadth and depth with the current near the surface, would carry out as much salt below as is brought in above, although it moved with less than one-fourth part of the velocity, and would thus prevent a perpetual increase of saltness in the Mediterranean beyond that

* On the Water of the Mediterranean, by W. H. Wollaston, M.D., F.R.S., Phil. Trans. 1829, part I., p. 29.

existing in the Atlantic. It was also remarked by others, that the result would be the same if, the swiftness being equal, the inferior current had only a fourth of the volume of the superior. At the same time there appeared reason to conclude that this great specific gravity was only acquired by water at immense depths; for two specimens of the water taken at the distance of some hundred miles from the Straits, and at depths of four hundred, and even four hundred and fifty fathoms, were found by Dr. Wollaston not to exceed in density that of many ordinary samples of sea-water. Such being the case, we can now prove that the vast amount of salt brought into the Mediterranean *does not* pass out again by the Straits. For it appears by Captain Smyth's soundings, which Dr. Wollaston had not seen, that between the Capes of Trafalgar and Spartel, which are twenty-two miles apart, and where the Straits are shallowest, the deepest part, which is on the side of Cape Spartel, is only *two hundred and twenty fathoms.* It is therefore evident that if water sinks in certain parts of the Mediterranean, in consequence of the increase of its specific gravity, to greater depths than two hundred and twenty fathoms, it can never flow out again into the Atlantic, since it must be stopped by the submarine barrier which crosses the shallowest part of the Straits of Gibraltar.

What, then, becomes of the excess of salt?—for this is an inquiry of the highest geological interest. The Rhone, the Po, and many hundred minor streams and springs, pour annually into the Mediterranean large quantities of carbonate of lime, together with iron, magnesia, silica, alumina, sulphur, and other mineral ingredients, in a state of chemical solution. To explain why the influx of this matter does not alter the composition of this sea has never been thought to present a great difficulty; for it is known that calcareous rocks are forming in the delta of the Rhone, in the Adriatic, on the coast of Asia Minor, and in other localities. Precipitation is acknowledged to be the means whereby the surplus mineral matter is disposed of, after the consumption of a certain portion in the secretions of testacea and zoophytes. But some have imagined that, before muriate of soda can, in like manner, be precipitated, the whole Mediterranean ought to become as much saturated with salt as the brine-springs of Cheshire, or Lake Aral, or the

Dead Sea. There is, however, an essential difference between these cases; for the Mediterranean is not only incomparably greater in extent than the two last-mentioned basins, but its depth is enormous. In the narrowest part of the Straits of Gibraltar, where they are about nine miles broad, between the Isle of Tariffa and Alcanzar Point, the depth varies from one hundred and sixty to five hundred fathoms; but between Gibraltar and Ceuta, Captain Smyth sounded to the extraordinary depth of *nine hundred and fifty fathoms!* where he found a gravelly bottom, with fragments of broken shells. Saussure sounded to the depth of two thousand feet, within a few yards of the shore, at Nice. What profundity, then, may we not expect some of the central abysses of this sea to reach! The evaporation being, as we before stated, very rapid, the surface water becomes impregnated with a slight excess of salt; and its specific gravity being thus increased, it instantly falls to the bottom, while lighter water rises to the top, or that introduced by rivers, and by the current from the Atlantic, flows over it. But the heavier fluid does not merely fall to the bottom, but flows on till it reaches the lowest part of one of those submarine basins into which we must suppose the bottom of this inland sea to be divided. By the continuance of this process, additional supplies of brine are annually carried to deep repositories, until the lower strata of water are fully saturated, and precipitation takes place—not in thin films, such as are said to cover the alluvial marshes along the western shores of the Euxine, nor in minute layers, like those of the salt "étangs" of the Rhone, but on the grandest scale—continuous masses of pure rock-salt, extending, perhaps, for hundreds of miles in length, like those in the mountains of Poland, Hungary, Transylvania, and Spain*.

* As to the existence of an inferior current flowing westward, none of the experiments made in the late survey give any countenance whatever to this popular notion; and it seems most unnecessary to resort to it, not only because the expenditure of the Mediterranean, by evaporation, must be immense, but because it is not yet proved that the two lateral currents, which conjointly exceed in breadth that of the centre, do not restore the equilibrium, if occasionally disturbed. They ebb and flow with the tide, but they may carry more water to the west than to the east. The opinion, that in the middle of the Straits the water returned into the

The Straits of Gibraltar are said to become gradually wider by the wearing down of the cliffs on each side at many points; and the current sets along the coast of Africa so as to cause considerable inroads in various parts, particularly near Carthage. Near the Canopic mouth of the Nile, at Aboukir, the coast was greatly devastated in the year 1784, when a small island was nearly consumed. By a series of similar operations, the old site of the cities of Nicopolis, Taposiris, Parva, and Canopus, have become a sandbank*.

Floating Icebergs.—Marine currents are sometimes instrumental in the transportation of rock and soil, by floating large masses of ice to great distances from the shore. When glaciers in northern latitudes descend the valleys burdened with alluvial debris, and arrive at the shore, they are frequently detached, and float off. Scoresby† counted five hundred icebergs in latitude 69° and 70° north, rising above the surface from the height of one to two hundred feet, and measuring from a few hundred yards to a mile in circumference. Many of these contained strata of earth and stones, or were loaded with beds of rock of great thickness, of which the weight was conjectured to be from fifty thousand to one hundred thousand tons. As the mass of ice below the level of the water is between seven and eight times greater than that above, these masses may sometimes take the ground in great numbers, in particular parts of the sea, and may, as they dissolve, deposit such masses of matter on particular parts of the bottom of the deep, or on the shores of some isles, as may offer perplexing problems to future geologists. Some ice islands have been known to drift from Baffin's

Atlantic by a submarine counter-current, first originated in the following circumstance. M. Du l'Aigle, commander of a privateer called the Phœnix, of Marseilles, gave chase to a Dutch merchant ship, near Ceuta Point, and came up with her in the middle of the gut, between Tariffa and Tangier, and there gave her one broadside, which directly sunk her. A few days after, the sunk ship, with her cargo of brandy and oil, arose on the shore near Tangier, which is at least four leagues to the westward of the place where she sunk, and directly against the strength of the *central* current.—Phil. Trans., 1724. It seems obvious that the ship in this case was brought back by one of the *lateral* currents, not by an *under* current.

* Clarke's Travels in Europe, Asia, and Africa, vol. iii., pp. 340 and 363, 4th edition.

† Voyage in 1822, p. 233.

Bay to the Azores, as we before stated, and from the South Pole to the immediate neighbourhood of the Cape of Good Hope.

Sand Hills.—It frequently happens, where the sea is encroaching on a coast, that perpendicular cliffs of considerable height, composed of loose sand, supply, as they crumble away, large quantities of fine sand, which, being in mid-air when detached, are carried by the winds to great distances, covering the land or barring up the mouths of estuaries. This is exemplified in Poole Bay, in Hampshire, and in many points of the coast of Norfolk and Suffolk. But a violent wind will sometimes drift the sand of a sea-beach, and carry it up with fragments of shells to great heights, as in the case of the sands of Barry, at the northern side of the estuary of the Tay, where hills of this origin attain the extraordinary height of from two hundred and fifty to three hundred feet. On the coast of France and Holland long chains of these dunes have been formed in many parts, and often give rise to very important geological changes, by barring up the mouths of estuaries, and preventing the free ingress of the tides, or free efflux of river water. The Bay of Findhorn, in Morayshire, has been blocked up in this manner since the beginning of the seventeenth century, so that large vessels can no longer enter; and we have already mentioned changes of a similar kind at Great Yarmouth, in Norfolk. Chains of sand-hills have also accumulated on the shores of the delta of the Nile, especially opposite the Lakes of Brulos and Menzala, forming mounds whereby the waters of these lakes are retained*. By the alternate formation and destruction of such barriers, fresh-water and marine deposits may sometimes be formed in succession on the same spots, and afterwards be laid dry by the exclusion of the tides, and be again submerged when high tides break into the estuary again. Many of the phenomena of submarine forests may, perhaps, admit of explanation, when the effects of such barriers of sand have been more carefully studied. The loose sand often forms a firm mass when bound together by the roots of plants fitted for such a soil, particularly the Arundo arenaria, and Elymus arenarius.

* Rennell's Herodotus.

A considerable tract of cultivated land on the north coast of Cornwall has been inundated by drift-sand, forming hills several hundred feet above the level of the sea, and composed of comminuted marine shells. By the shifting of the sands, the ruins of ancient buildings have been discovered; and, in some cases where they have been bored to a great depth, distinct strata, separated by a vegetable crust, are visible. In some localities, as at New Quay, large masses have become sufficiently indurated to be used for architectural purposes. The lapidification, which is still in progress, appears to be due to oxide of iron held in solution by the water which percolates the sand *. Terrestrial shells are found enclosed entire in this rock.

The moving sands of the African deserts have been driven by the west winds over all the lands capable of tillage on the western banks of the Nile, except such as are sheltered by mountains †. The ruins of ancient cities are buried under these sands between the Temple of Jupiter Ammon and Nubia. De Luc attempted to infer the recent origin of our continents, from the fact that the sands of the desert have only arrived in modern times at the fertile plains of the Nile. This scourge, he said, would have afflicted Egypt for ages anterior to the times of history, if the continents had risen above the level of the sea several hundred centuries before our era ‡. But the author proceeded in this, as in all his other chronological computations, on a multitude of gratuitous assumptions, not one of which he had the candour to state explicitly. He ought, in the first place, to have demonstrated that the whole continent of Africa was raised above the level of the sea at one period; for unless this point was established, the region from whence the sands began to move might have been the last addition made to Africa, and the commencement of the sand-flood might have been long posterior to the formation of the greater portion of that continent. That the different parts of Europe were not all elevated at one time, is now generally admitted. De Luc should also have pointed out the depth of

* Boase on the Submersion of part of the Mount's Bay, &c. Trans. Roy. Geol. Society of Cornwall, vol. ii., p. 140.

† De Luc, Mercure de France, Sept. 1809. ‡ Ibid.

drift sand in various parts of the great Libyan deserts, and have shown whether any valleys of large dimensions had been filled up,—how long these arrested the progress of the sands, and how far the flood had upon the whole advanced since the times of history. If, in the absence of all these necessary elements of the computation, the doctrines of this author, respecting "natural chronometers," were extremely popular, and that, too, in an age when close reasoning and rigorous investigation were applied to other branches of physical science, it only proves how strong were the prepossessions in regard to time which impeded the progress of geology.

There is not one great question relating to the former changes of the earth and its inhabitants into which considerations of time do not enter; and so long as the public mind was violently prejudiced in regard to this important topic, men of superior talent alone, who thought for themselves, and were not blinded by authority, could deduce any just conclusions from geological evidence. It ought not, therefore, to be matter either of surprise or discouragement to us, that at the commencement of the present century, when for three hundred years much labour had been devoted to these investigations, so few sound and enlightened views had met with general reception.

CHAPTER XVII.

Reproductive effects of Tides and Currents—Silting up of Estuaries does not compensate the loss of land on the borders of the ocean—Bed of the German Ocean—Composition and extent of its sand-banks—Strata formed by currents on the southern and eastern shores of the Mediterranean—Transportation by currents of the sediment of the Amazon, Orinoco, and Mississippi—Stratification—Concluding remarks.

From the facts enumerated in the last chapter, it will appear that, on the borders of the ocean, currents co-operating with tides are most powerful instruments in the destruction and transportation of rocks; and as numerous tributaries discharge their alluvial burden into the channel of one great river, so we find that many great rivers often deliver their earthy contents to one marine current, to be borne by it to a distance, and deposited in some deep receptacles of the ocean. The current not only receives this tribute of sedimentary matter from streams draining the land, but acts also itself on the coast, as does a river on the cliffs which bound a valley. The course of currents on the British shores is ascertained to be as tortuous as that of ordinary rivers. Sometimes they run between sand-banks which consist of matter thrown down at certain points where the velocity of the stream had been retarded; but it very frequently happens, that as in a river one bank is made of alluvial gravel, while the other is composed of some hard rock constantly undermined, so the current, in its bends, strikes here and there upon a coast which then forms one bank, while a shoal under water forms the other. If the coast be composed of solid materials, it yields slowly, or if of great height, it does not lose ground rapidly, since a large quantity of matter must then be removed before the sea can penetrate to any distance. But the openings where rivers enter are generally the points of least resistance, and it is here, therefore, that the ocean makes the widest and longest breaches.

But a current alone cannot shape out and keep open an

estuary, because it holds in suspension, like the river, during certain seasons of the year, a large quantity of sediment; and when their waters, flowing in opposite directions, meet, this matter subsides. For this reason, in inland seas, and even on the borders of the ocean, where the rise of the tide happens to be slight, it is scarcely possible to prevent a harbour from silting up; and it is often expedient to carry out a jetty to beyond the point where the marine current and the river neutralise each other's force, for beyond this point a free channel is maintained by the superior force of the current. The formation and keeping open of large estuaries are due to the *combined influence* of tides and currents; for when the tide rises, a large body of water suddenly enters the mouth of the river, where, becoming confined within narrower bounds, while its momentum is not destroyed, it is urged on, and, having to pass through a contracted channel, rises and runs with increased velocity, just as a swollen river, when it reaches the arch of a bridge scarcely large enough to give passage to its waters, is precipitated in a cataract, while rushing through the arch. During the ascent of the tide, a stream of fresh-water is flowing down from the higher country, and is arrested for several hours; and thus a large lake of brackish water is accumulated, which, as soon as the ebb causes the sea to fall, is let loose, as on the removal of an artificial sluice or dam. By the force of this retiring body of water, the alluvial sediment, both of the river and of the sea, is swept away, and transported to such a distance from the mouth of the estuary, that a small part only can return with the next tide. In many estuaries, as in the Thames, for example, the tide requires about five hours to flow up, and about seven to flow down; so that the preponderating force is always in the direction which tends to keep open a deep and broad passage. But as it is evident that both the river and the tidal current are ready to part with their sediment whenever their velocity is checked, there is naturally a tendency in all estuaries to silt up partially, since the causes of retardation are very numerous, and constantly change their position.

The new lands acquired within the mouth of an estuary are only a few feet above the mean level of the sea, whereas cliffs of great height are consumed every year. If, therefore, the area of land annually abandoned by the sea were equal to that invaded by it, there would still be no compensation *in kind.*

Many writers have declared that the gain on our eastern coast, since the earliest periods of history, has more than counterbalanced the loss; but they have been at no pains to calculate the amount of the latter, and have often forgotten that, while the new acquisitions are manifest, there are rarely any natural monuments to attest the former existence of what is now no more. They have also taken into their account those tracts, artificially recovered, which are often of great agricultural importance, and may remain secure, perhaps, for thousands of years, but which are nevertheless exposed to be overflowed again by a small proportion of the force required to remove the high lands of our shores. It will seem, at first sight, somewhat paradoxical, but it is nevertheless true, that the greater number of estuaries, although peculiarly exposed to the invasion of the sea, are usually contracting in size, even where the whole line of coast is giving way. But the fact is, that the inroads made by the ocean upon estuaries, although extremely great, are completed during periods of comparatively short duration; and in the intervals between these visitations, the mouths of rivers, like other parts of the coast, usually enjoy a more or less perfect respite. All the estuaries, taken together, constitute but a small part of a great line of coast; it is, therefore, most probable, that if our observations extend to a few centuries only, we shall not see any, and very rarely all, of this small part exposed to the fury of the ocean. The coast of Holland and Friesland, if studied for several consecutive centuries since the Roman era, would generally have led to the conclusion that the land was encroaching fast upon the sea, and that the aggrandizement within the estuaries far more than compensated the losses on the open coast. But when our retrospect embraces the whole period, an opposite inference is drawn; and we find that the Zuyder Zee, the Bies Bosch, Dollart, and Yahde, are modern gulfs and bays, and that these points have been the principal theatres of the retreat, instead of the advance, of the land. If we possessed records of the changes on our coast for several thousand years, they would probably present us with similar results; and although we have hitherto seen our estuaries, for the most part, become partially converted into dry land, and portions of bold cliffs intervening between the mouths of rivers consumed by the sea, this has merely arisen from the accidental set of the currents and tides during a brief period.

The current which flows from the north-west and bears against our eastern coast, transports, as we have seen, materials of various kinds. It undermines and sweeps away the granite, gneiss, trap rocks, and sandstone of Shetland, and removes the gravel and loam of the cliffs of Holderness, Norfolk, and Suffolk, which are between fifty and two hundred feet in height, and which waste at the rate of from one to six yards annually. It bears away the strata of London-clay on the coast of Essex and Sheppey—consumes the chalk with its flints for many miles continuously on the shores of Kent and Sussex—commits annual ravages on the fresh-water beds, capped by a thick covering of chalk flints, in Hampshire, and continually saps the foundations of the Portland limestone. It receives, besides, during the rainy months, large supplies of pebbles, sand and mud, which numerous streams from the Grampians, Cheviots and other chains, send down to the sea. To what regions, then, is all this matter consigned? It is not retained in mechanical suspension by the waters of the sea, nor does it mix with them in a state of chemical solution,—it is deposited *somewhere*, yet certainly not in the immediate neighbourhood of our shores; for, in that case, there would soon be a cessation of the encroachment of the sea, and large tracts of low land, like Romney Marsh, would everywhere encircle our island. As there is now a depth of water, exceeding thirty feet, in some spots where cities flourished but a few centuries ago, it is clear that the current not only carries far away the materials of the wasted cliffs, but tears up besides many of the regular strata at the bottom of the sea.

The German Ocean is deepest on the Norwegian side, where the soundings give one hundred and ninety fathoms; but the mean depth of the whole basin may be stated at only about thirty-one fathoms*. The bed of this sea is encumbered in an extraordinary degree with accumulations of debris, especially in the middle or central parts. One of the great central banks trends from the Frith of Forth, in a north-

* Stevenson, on the Bed of the German Ocean, or North Sea.—Ed. Phil. Journ., No. V., p. 44; 1820.

easterly direction, to a distance of one hundred and ten miles; others run from Denmark and Jutland upwards of one hundred and five miles to the north-west; while the greatest of all, the Dogger Bank, extends for upwards of three hundred and fifty-four miles from north to south. The whole superficies of these enormous shoals is equal to about one-fifth of the whole area of the German Ocean, or to about one-third of the whole extent of England and Scotland *. The average height of the banks measures, according to Mr. Stevenson, about seventy-eight feet; and, assuming that the mass is uniformly composed to this depth of the same drift matter, the debris would cover the whole of Great Britain to the depth of twenty-eight feet, supposing the surface of the island to be a level plain. A great portion of these banks consists of fine and coarse siliceous sand, mixed with fragments of corals and shells ground down, the proportion of these calcareous matters being extremely great†. As we know not to what distance our continents formerly extended, we cannot conjecture, from any data at present obtained, how much of the space occupied by these sands was formerly covered with strata, subsequently removed by the encroachments of the sea, or whether certain tracts were originally of great depth, and have since been converted into shoals by matter drifted by currents. But as the sea is moved to and fro with every tide, portions of these loose sands must, from time to time, be carried into those deep parts of the North Sea where they are beyond the reach of waves or currents.

So great is the quantity of matter held in suspension by the tidal current on our shores, that the waters are in some places artificially introduced into certain lands below the level of the sea; and by repeating this operation, which is called "warping," for two or three years, considerable tracts have been raised, in the estuary of the Humber, to the height of about six feet. Large quantities of coarse sand and pebbles are also drifted along at the bottom: and when such a current meets with any deep depression in the bed of the ocean, it must necessarily fill it up; just as a river, when it meets with a lake in its course, fills it gradually with sediment. But in the one

* Stevenson, on the Bed of the German Ocean, or North Sea.—Ed. Phil. Journ., No. V., p. 47; 1820. † Ibid.

case, the sheet of water is converted into land, whereas, in the other, a shoal only will be raised, overflowed at high water, or at least by spring-tides. The only records which we at present possess of the gradual shallowing of seas are confined, as might be expected, to estuaries, havens, and certain channels of no great depth; and to some inland seas, as the Baltic, Adriatic, and Arabian Gulf. It is only of late years that accurate surveys and soundings have afforded data of comparison in very deep seas, of which future geologists will avail themselves.

It appears extraordinary that in some tracts of the sea, adjoining our coast, where we know that currents are not only sweeping along rocky masses, thrown down, from time to time, from the high cliffs, but scouring out also deep channels in the regular strata, there should exist fragile shells and tender zoophytes in abundance, which live uninjured by these violent movements. The ocean, however, is in this respect a counterpart of the land; and as, on our continents, rivers may undermine their banks, uproot trees, and roll along sand and gravel, while their waters are inhabited by testacea and fish, and their alluvial plains are adorned with rich vegetation and forests, so the sea may be traversed by rapid currents, and its bed may suffer great local derangement, without any interruption of the general order and tranquillity.

One important character in the formations produced by currents, is the immense extent over which they are the means of diffusing homogeneous mixtures; for these are often coextensive with a great line of coast, and, by comparison with their deposits, the deltas of rivers must shrink into insignificance. In the Mediterranean the same current which is rapidly destroying many parts of the African coast, between the Straits of Gibraltar and the Nile, preys also upon the Nilotic delta, and drifts the sediment of that great river to the eastward. To this source the rapid accretions of land on parts of the Syrian shores where rivers do not enter, may be attributed. The ruins of ancient Tyre are now far inland, and those of ancient Sidon are two miles distant from the coast, the modern town having been removed towards the sea*. But the south coast of Asia Minor affords far more striking examples of advances of the land upon the

* Hoff, vol. i., p. 253.

sea, where small streams co-operate with the current before mentioned. Captain Beaufort, in his Survey of that coast, has pointed out the great alterations effected on these shores since the time of Strabo, where havens are filled up, islands joined to the main land, and where the whole continent has increased many miles in extent. Strabo himself, on comparing the outline of the coast in his time with its ancient state, was convinced, like our countryman, that it had gained very considerably upon the sea. The new-formed strata of Asia Minor consist *of stone*, not of loose, incoherent materials. Almost all the streamlets and rivers, like many of those in Tuscany and the south of Italy, hold abundance of carbonate of lime in solution, and precipitate travertin, or sometimes bind together the sand and gravel into solid sandstones and conglomerates: every delta and sand-bar thus acquires solidity, which often prevents streams from forcing their way through them, so that their mouths are constantly changing their position *.

Among the greatest deposits now in progress, and of which the distribution is chiefly determined by currents, we may class those between the mouths of the Amazon and the southern coast of North America. It is well known that a great current is formed along the coast of Africa, by the water impelled by the Trade Winds blowing from the south. When this current reaches the head of the Gulf of Guinea, it is opposed by the waters brought to the same spot by the Guinea current, and it then streams off in a westerly direction, and pursues its rapid course quite across the Atlantic to the continent of South America. Here one portion proceeds along the northern coast of Brazil to the Caribbean Sea and the Gulf of Mexico. Captain Sabine found that this current was running with the astonishing rapidity of four miles an hour where it crosses the stream of the Amazon, which river preserves part of its original impulse, and its waters not wholly mingled with those of the ocean at the distance of three hundred miles from its mouth †. The sediment of the Amazon is thus constantly carried to the north-west as far as to the mouths of the Ori-

* Karamania, or a brief Description of the Coast of Asia Minor, &c. London, 1817.

† Experiments to determine the Figure of the Earth, &c., p. 445.

noco, and an immense tract of swamp is formed along the coast of Guiana, with a long range of muddy shoals bordering the marshes and becoming converted into land*. The sediment of the Orinoco is partly detained, and settles near its mouth, causing the shores of Trinidad to extend rapidly, and is partly swept away into the Caribbean Sea by the equatorial current. According to Humboldt, much sediment is carried again out of the Caribbean Sea into the Gulf of Mexico. The rivers, also, which descend from the high plateau of Mexico, between the mouths of the Norte and Tampico, when they arrive at the edge of the plateau, swollen by tropical rains, bear down an enormous quantity of rock and mud to the sea; but the current, setting across their mouths, prevents the growth of deltas, and preserves an almost uniform curve in that line of coast†. It must, therefore, exert a great transporting power, and it cannot fail to sweep away part of the matter which is discharged from the mouths of the Norte and the Mississippi. It follows from these observations, that, in certain parts of the globe, continuous formations are now accumulated over immense spaces along the bottom of the ocean. The materials undoubtedly must vary in different regions, yet for thousands of miles they may often retain some common characters, and be simultaneously in progress throughout a space stretching 30° of latitude from south-east to north-west, from the mouths of the Amazon for example, to those of the Mississippi—as far as from the Straits of Gibraltar to Iceland. At the same time, great coral reefs are growing around the West Indian islands; and in some parts, streams of lava are occasionally flowing into the sea, which become covered again, in the intervals between eruptions, with other beds of corals. The various rocks, therefore, stratified and unstratified, now forming in this part of the globe, may occupy, perhaps, far greater areas than any group of our ancient secondary series which has yet been traced through Europe.

In regard to the internal arrangement of "pelagian" formations deposited by currents far from the land, we may infer that in them, as in deltas, there is usually a division into strata;

* Lochead's Observations on the Nat. Hist. of Guiana. Edin. Trans., vol. iv.

† This coast has been recently examined by Captain Vetch.—See also Bauza's new chart of the Gulf of Mexico.

for, in both cases, the accumulations are successive, and, for the most part, interrupted. The waste of cliffs on the British coast is almost entirely confined to the winter months; so that running waters in the sea, like those on the land, are periodically charged with sediment, and again become pure. It will happen, in many cases, that the melting of snow will yield an annual tribute of fluviatile sediment in spring or summer, while violent gales of wind will cause the principal dilapidations on the shores to occur in autumn and winter; so that distinct materials may be arranged in alternate strata in deep depressions of the bed of the ocean.

Those geologists who are not averse to presume that the course of Nature has been uniform from the earliest ages, and that causes now in action have produced the former changes of the earth's surface, will consult the ancient strata for instruction in regard to the reproductive effects of tides and currents. It will be enough for them to perceive clearly that great effects now annually result from the operations of these agents, in the inaccessible depths of lakes, seas, and the ocean; and they will then search the ancient lacustrine and marine strata for manifestations of analogous effects in times past. Nor will it be necessary for them to resort to very ancient monuments; for in certain regions where there are active volcanos, and where violent earthquakes prevail, we may examine submarine formations many thousand feet in thickness, belonging to our own era, or, at least, to the era of contemporary races of organic beings.

CHAPTER XVIII.

Division of igneous agents into the volcano and the earthquake—Distinct regions of subterranean disturbance—Region of the Andes—System of volcanos extending from the Aleutian isles to the Moluccas—Polynesian archipelago—Volcanic region extending from the Caspian Sea to the Azores—Former connexion of the Caspian with Lake Aral and the Sea of Azof—Low steppes skirting these seas—Tradition of Deluges on the shores of the Bosphorus, Hellespont, and the Grecian archipelago—Periodical alternation of earthquakes in Syria and Southern Italy—Western limits of the European region—Earthquakes rarer and more feeble in proportion as we recede from the centres of volcanic action—Extinct volcanos not to be included in lines of active vents.

We have hitherto considered the changes wrought, since the times of history and tradition, by the continued action of aqueous causes on the earth's surface; and we have next to examine those resulting from igneous agency. As the rivers and springs on the land, and the tides and currents in the sea, have, with some slight modifications, been fixed and constant to certain localities from the earliest periods of which we have any records, so the volcano and the earthquake have, with few exceptions, continued, during the same lapse of time, to disturb the same regions. But as there are signs, on almost every part of our continent, of great power having been exerted by running water on the surface of the land, and by tides and currents on cliffs bordering the sea, where, in modern times, no rivers have excavated, and no tidal currents undermined—so we find signs of volcanic vents and violent subterranean movements in places where the action of fire has long been dormant. We can explain why the intensity of the force of aqueous causes should be developed in succession in different districts. Currents, for example, and tides, cannot destroy our coasts, shape out or silt up estuaries, break through isthmuses, and annihilate islands, form shoals in one place and remove them from another, without the direction and position of their destroying and transporting power becoming transferred to new localities. Neither can the relative levels of the earth's crust, above and beneath the waters, vary from time to time, as they are admitted to have varied at former periods, and as we shall demonstrate that they still do, without the continents being, in the course of ages, modified, and even entirely altered, in their external configuration. Such events must clearly be accompanied by a complete change in the volume, velocity, and direction of the streams and land floods to which certain regions give passage. That we should find, therefore, cliffs where the sea once committed ravages, and from which it has now retired—estuaries where high tides once rose, but which are now dried up—valleys hollowed out by water, where no streams now flow;—all these and similar phenomena are the necessary consequences of physical causes now in operation; and we may affirm that, if there be no instability in the laws of Nature, similar fluctuations must recur again and again in time to come.

But however natural it may be that the force of running water in numerous valleys, and of tides and currents in many tracts of the sea, should now be *spent*, it is by no means so easy to explain why the violence of the earthquake and the fire of the volcano should also have become locally extinct, at successive periods. We can look back to the time when the marine strata, whereon the great mass of Etna rests, had no existence; and that time is extremely modern in the earth's history. This alone affords ground for anticipating that the eruptions of Etna will one day cease.

> Nec quæ sulfureis ardet fornacibus Ætna
> Ignea semper erit, *neque enim fuit ignea semper*,

are the memorable words which are put into the mouth of Pythagoras by the Roman poet, and they are followed by speculations as to the causes of volcanic vents shifting their position. Whatever doubts the philosopher expresses as to the nature of these causes, it is assumed, as incontrovertible, that the points of eruption will hereafter vary, *because they have formerly done so.*

We have endeavoured to show, by former chapters, how utterly this principle of reasoning is set at nought by the modern schools of geology, which not only refuse to conclude that great revolutions in the earth's surface are now in progress, or that they will take place *because* they have often been repeated in former ages, but assume the improbability of such a conclusion and throw the whole weight of proof on those by whom that doctrine is embraced.

In our view of igneous causes we shall consider, first, the volcano, and afterwards the earthquake; for although both are probably the effects of the same subterranean process, they give rise to very different phenomena on the surface of the globe. Both are confined to certain regions, but the subterranean movements are least violent in the immediate proximity of volcanic vents, especially where the discharge of aëriform fluids and melted rock is made constantly from the same crater. We say that there are certain regions to which both the points of eruption, and the movements of great earthquakes are confined; and we shall begin by tracing out the geographical boundaries of some of these, that the reader may be aware of the magnificent scale on which the agency of subterranean fire is now simultaneously developed. Over the whole of the vast tracts alluded to, active volcanic vents are distributed at intervals, and most commonly arranged in a linear direction. Throughout the intermediate spaces there is abundant evidence that the subterranean fire is at work continuously, for the ground is convulsed from time to time by earthquakes; gaseous vapours, especially carbonic acid gas, are disengaged plentifully from the soil; springs often issue at a very high temperature, and their waters are very commonly impregnated with the same mineral matters which are discharged by volcanos during eruptions.

Of these great regions, that of the Andes is one of the best defined. Respecting its southern extremity, we are still in need of more accurate information, some conceiving it to extend into Terra del Fuego and Patagonia*. But if we begin with Chili, in the forty-sixth degree of south latitude, we find that, in proceeding from this point towards the north to the twenty-seventh degree, there is a line of volcanos so uninterrupted, that it is rare to find any intervening degree of latitude in which there is not an active vent. About twenty of these are now enumerated, but we may expect the number to augment greatly when the country

* Hoff, vol. ii., p. 476.

has been more carefully examined, and throughout a longer period. How long an interval of rest entitles us to consider a volcano extinct, cannot yet be determined; but we know that, in Ischia, there intervened, between two consecutive eruptions, a pause of *seventeen centuries*; and a much longer period, perhaps, elapsed between the eruptions of Vesuvius before the earliest Greek colonies settled in Campania, and the renewal of its activity in the reign of Titus. It will be necessary, therefore, to wait for at least six times as many centuries as have elapsed since the discovery of America, before any one of the dormant craters of the Andes can be presumed to be entirely spent, unless there are some *geological* proofs of the last eruptions having belonged to a remote era. The Chilian volcanos rise up through granitic mountains. Villarica, one of the principal, continues burning without intermission, and is so high that it may be distinguished at the distance of one hundred and fifty miles. A year never passes in this province without some slight shocks of earthquakes; and about once in a century, or oftener, tremendous convulsions occur, by which, as we shall afterwards see, the land has been shaken from one extremity to the other, and continuous tracts, together with the bed of the Pacific, have been raised permanently from one to twenty feet and upwards above their former level. Hot springs are numerous in this district, as well as springs of naphtha and petroleum, and mineral waters of various kinds. If we pursue our course northwards, we find in Peru only one active volcano as yet known; but the province is so subject to earthquakes, that scarcely a week happens without a shock, and many of these have been so considerable as to create great changes of the surface. Proceeding farther north, we find in the middle of Quito, where the Andes attain their highest elevation, from the second degree of south, to the third degree of north latitude, Tunguragua, Cotopaxi, Antisana, and Pichinca, the three former of which throw out flames not unfrequently. From fissures on the side of Tunguragua, a deluge of mud (moya) descended in 1797, and filled valleys a thousand feet wide to the depth of six hundred feet, forming barriers whereby rivers were dammed up, and lakes occasioned. Earthquakes have, in the same province, caused great revolutions in the physical features of the surface. Farther north, there are three volcanos in the province of Pasto, and three others in that of Popayan. In the provinces of Guatimala and Nicaragua, which lie between the Isthmus of Panama and Mexico, there are no less than twenty-one active volcanos, all of them contained between the tenth and fifteenth degrees of north latitude. The great volcanic chain, after having pursued its course for several thousand miles from south to north, turns off in a side direction in Mexico, and is prolonged in a great plateau, between the eighteenth and twenty-second degrees of north latitude. This high table-land owes its present form to the circumstance of an ancient system of valleys, in a chain of primary mountains, having been filled up, to the depth of many thousand feet, with various volcanic products. Five active volcanos traverse Mexico from west to east—Tuxtla, Orizaba, Popocatepetl, Jorullo, and Colima. Jorullo, which is in the centre of the great plateau, is no less than forty leagues from the nearest ocean—an important circumstance, as showing that the proximity of the sea is not a necessary condition, although certainly a very general characteristic, of the position of active volcanos. The extraordinary eruption of this mountain, in 1759, will be described in the sequel. If the same parallel line which connects these five vents be prolonged, in a westerly direction, it cuts the volcanic group of islands, called the Isles of Revillagigedo. To the north of Mexico there are three, or according to some, five volcanos, in the peninsula of California, but of these we have at present no detailed account. We have before mentioned the violent earthquakes which, in 1812, convulsed the valley of the Mississippi at New Madrid, for the space of three hundred miles in length. As this happened exactly at the same time as the great earthquake of Caraccas, it is probable that these two points are parts of one continuous volcanic region; for the whole circumference of the intervening Caribbean Sea must be considered as a theatre of earthquakes and volcanos. On the north lies the island of Jamaica, which, with a tract of the contiguous sea, has often experienced tremendous shocks; and these are frequent along a line extending from Jamaica to St. Domingo, and Porto Rico. On the south of the same basin the shores and mountains of Colombia are perpetually convulsed. On the west, is the volcanic chain of Guatimala and Mexico, before traced out; and on the east the West Indian isles, where, in St. Vincent's and Guadaloupe, are active vents.

Thus it will be seen that volcanos and earthquakes occur uninterruptedly, from Chili to the north of Mexico; and it seems probable, that they will hereafter be found to extend from Cape Horn to California, or even perhaps to New Madrid, in the United States—a distance as great as from the pole to the equator. In regard to the eastern limits of the region, they lie deep beneath the waves of the Pacific, and must continue unknown to us. On the west they do not appear, except where they include the West Indian islands, to be prolonged to a great distance, for there seem to be no indications of volcanic disturbances in Guiana, Brazil, and Buenos Ayres.

On an equal, if not a still grander scale, is another continuous line of volcanic action, which commences, on the north, with the Aleutian Isles in Russian America, and extends, first in an easterly direction for nearly two hundred geographical miles, and then southwards, without interruption, throughout a space of between sixty and seventy degrees of latitude to the Moluccas, and there branches off in different directions both towards the east and north-west. The northern extremity of this volcanic region is the Peninsula of Alaska, in about the fifty-fifth degree of latitude. From thence the line is continued through the Aleutian or Fox Islands, to Kamtschatka. In that archipelago eruptions are frequent; and a new isle rose in 1814, which, according to some reports, is three thousand feet high and four miles round*. Earthquakes of the most terrific description agitate and alter the bed of the sea and surface of the land throughout this tract. The line is continued in the southern extremity of the peninsula of Kamtschatka, where there are seven active volcanos, which, in some eruptions, have scattered ashes to immense distances. The Kurile chain of isles constitutes the prolongation of the range, where a train of volcanic mountains, nine of which are known to have been in eruption, trends in a southerly direction. In these, and in the bed of the adjoining sea, alterations of level have resulted from earthquakes since the middle of the last century. The line is then continued to the south-west in the great Island of Jesso, where

* Hoff, vol. ii., p. 414.

there are active volcanic vents, as also in Nipon, the principal of the Japanese group, where the number of burning mountains is very great; slight shocks of earthquakes being almost incessant, and violent ones experienced at distant intervals. Between the Japanese and Philippine Islands, the communication is preserved by several small insular vents. Sulphur Island, in the Loo Choo archipelago, emits sulphureous vapour; and Formosa suffers greatly from earthquakes. In Luzon, the most northern and largest of the Philippines, are three active volcanos; Mindinao also was in eruption in 1764. The line is then prolonged through Sanguir and the north-eastern extremity of Celebes, by Ternate and Tidore, to the Moluccas, and, amongst the rest, Sumbawa. Here a great transverse line may be said to run from east to west. On the west it passes through the whole of Java, where there are thirty-eight large volcanic mountains, many of which continually discharge smoke and sulphureous vapours. In the volcanos of Sumatra, the same linear arrangement is preserved; but the line inclines gradually to the north-west in such a manner as to point to the active volcano in Barren Island in the Bay of Bengal, in about the twelfth degree of north latitude. In another direction the volcanic range is prolonged through Borneo, Celebes, Banda, and New Guinea; and farther eastward in New Britain, New Ireland, and various parts of the Polynesian archipelago. The Pacific Ocean, indeed, seems, in equatorial latitudes, to be one vast theatre of igneous action; and its innumerable archipelagos, such as the New Hebrides, Friendly Islands, and Georgian Isles, are all composed either of coralline limestones, or volcanic rocks with active vents here and there interspersed. The abundant production of carbonate of lime, in solution, would alone raise a strong presumption of the volcanic constitution of these tracts, even if there were not more positive proofs of igneous agency.

If we now turn our attention to the principal region in the Old World, which, from time immemorial, has been agitated by earthquakes, and has given vent at certain points to subterranean fires, we find that it possesses the same general characters. This region extends from east to west for the distance of about one thousand geographical miles, from the Caspian Sea to the Azores; including within its limits the greater part

100
110
20
10
0
PEGU
CAMBODIA
Hainan
Barren I.
Domel
MALAY
Salanga
Tantalam
Nicobar I^s.
SUMATRA
Delli
Vol. of Allas
Hog I.
Nias I.
G^t. Natuna
Mid. Anambas
Volcano
Sambas
BOR
Singapore
Bintang
Lingen
P^o. Battooa
Berapi
Padang
P^o. Sebeeroo
Gunong Penkalan
Banka
Good Fortune
N. Poggy
S. Poggy
Gunong Dempo
Billiton
Bencoolen
Cracatoa
Princes I.
Karang
Batavia
Salak
Tankubanprahu
Cheremai
Gagak
Tiluga-bodas
JAVA
Merapi
Tegal
Lawu
Willis
Kawi
Arjuno
Semiru
Lamang
Volcanic Band of the Greek Islands.
Athens
Andros
Negropont
Tine
Egina
Poros
Zea
Syra
Rhenia
Mycone
Hydra
Thermia
Serpho
Paros
Siphanto
Naxia
Anti Paro
Argentiera
Milo
Policandro
Therasia
Santorin
Longitude E

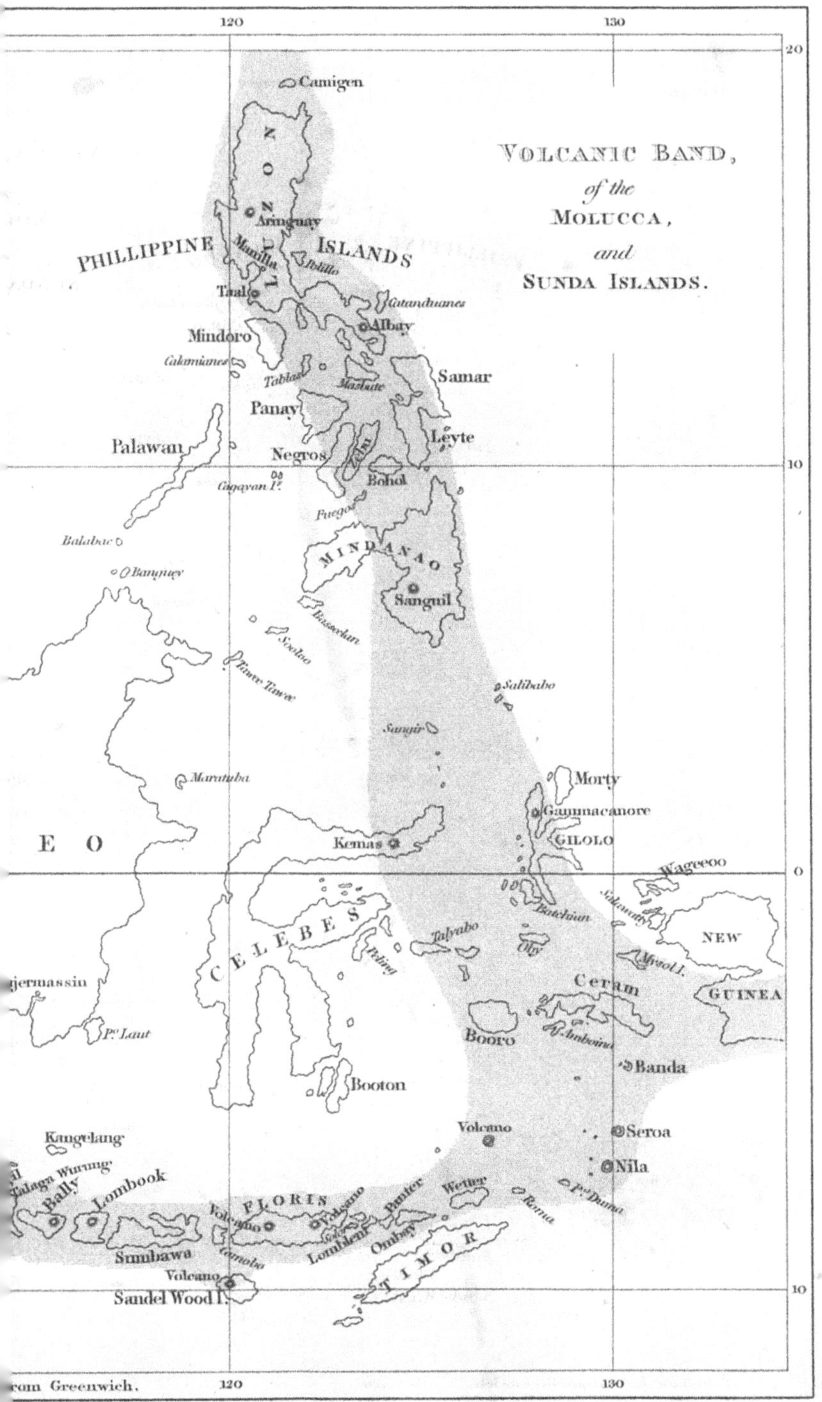
VOLCANIC BAND,
of the
MOLUCCA,
and
SUNDA ISLANDS.
120
130
20
10
0
10
Camigen
LUZON
Aringuay
PHILLIPPINE
ISLANDS
Manilla
Polillo
Taal
Catanduanes
Albay
Mindoro
Calamianes
Tablas
Masbate
Samar
Panay
Leyte
Palawan
Negros
Zebu
Bohol
Cagayan I.
Fuego
Balabac
Banguey
MINDANAO
Sanguil
Basselan
Sooloo
Tawe Tawe
Salibabo
Sangir
Maratuba
Morty
Gammacanore
GILOLO
Kemas
E O
Wageeoo
Salawaty
Batchian
CELEBES
Talyabo
Oby
NEW
Peling
Mysol I.
Ceram
GUINEA
jermassin
Po Laut
Booro
Amboina
Banda
Booton
Volcano
Seroa
Kangelang
Nila
Talaga Wurung
Bally
Lombook
Po Dama
Roma
Wetter
Panter
FLORIS
Volcano
Volcano
Lomblem
Ombay
TIMOR
Sumbawa
Comoba
Volcano
Sandel Wood I.
rom Greenwich.

of the Mediterranean, and its most prominent peninsulas. From south to north, it reaches from about the thirty-fifth to the forty-fifth degree of latitude. Its northern boundaries are Caucasus, the Black Sea, the mountains of Thrace, Transylvania, and Hungary,—the Austrian, Tyrolian, and Swiss Alps,—the Cevennes and Pyrenees, with the mountains which branch off from the Pyrenees westward, to the north side of the Tagus. Its western limits are the ocean, but it is impossible to determine how far it may be prolonged in that direction; neither can we assign with precision its extreme eastern limit, since the country beyond the Caspian and Sea of Aral is scarcely known. The great steppe of Tartary, in particular, is unexplored; and we are almost equally ignorant of the physical constitution of China, in which country, however, many violent earthquakes have been felt.

The southern boundaries of the region include the most northern parts of Africa, and part of the Desert of Arabia*. We may trace, through the whole area comprehended within these extensive limits, numerous points of volcanic eruptions, hot springs, gaseous emanations, and other signs of igneous agency; while few tracts, of any extent, have been entirely exempt from earthquakes throughout the last three thousand years.

To begin on the Asiatic side, we find that, on the western shores of the Caspian, in the country round Baku, there is a tract called the Field of Fire, which continually emits inflammable gas, and springs of naphtha and petroleum occur in the same vicinity, as also mud volcanos. In the chain of Elburs, to the south of this sea, is a lofty mountain, which, according to Morier, sometimes emits smoke, and at the base of which are several small craters, where sulphur and saltpetre are procured in sufficient abundance to be used in commerce. Violent subterranean commotions have been experienced along the borders of the Caspian; and it is reported that, since 1556, the waters of that sea have encroached on the Russian territory to the north; but the fact, as Malte-Brun observes, requires confirmation. According to Engelhard and Parrot, the depth of the water has increased in places, while the general surface has been lowered; and they say that the bottom of the sea has, in modern

* Hoff, vol. ii., p. 99.

times, varied in form; and that, near the south coast, the Isle of Idak, north from Astrabat, formerly high land, has now become very low*. Any indications of a change in the relative levels of the land in this part of Asia are of more than ordinary interest, because a succession of similar variations would account for many prominent features in the physical geography of the district between the salt lake Aral, and the western shores of the Euxine—a district well known to have been always subject to great earthquakes. The level of the Caspian is lower than that of the Black Sea, by more than fifty feet. A low and level tract, called the Steppe, abounding in saline plants, and said to contain shells of species now common in the adjoining sea, skirts the shores of the Caspian, on the north-west. This plain often terminates abruptly by a line of inland cliffs, at the base of which runs a kind of beach, consisting of fragments of limestone and sand, cemented together into a conglomerate. Pallas has endeavoured to show that there is an old line of sandy country, which indicates the ancient bed of a strait, by which the Caspian Sea was once united to that of Azof. On similar grounds, it is inferred that the salt lake Aral was formerly connected with the Caspian. However modern in the earth's history the convulsions may be which have produced the phenomena of the steppes, it is consistent with analogy to suppose that a very minute portion of the whole change has happened in the last twenty or thirty centuries. Yet, if we possessed more authentic records of physical events, we should probably discover that some small portion of those great revolutions have fallen within such recent periods. Remote traditions have come down to us of inundations, in which the waters of the Black Sea were forced through the Thracian Bosphorus, and through the Hellespont, into the Ægean. In the deluge of Samothrace, it appears that that small island, and the adjoining coast of Asia, were inundated; and in the Ogygian, which happened at a different time, Bœotia and Attica were overflowed. Notwithstanding the mixture of fable, and the love of the marvellous, in those rude ages, and the subsequent inventions of Greek poets and historians,

* Travels in the Crimea and Caucasus, in 1815, vol. i., pp. 257 and 264.—Hoff, vol. i., p. 137.

it may be distinctly perceived that the floods alluded to were local and transient, and that they happened in succession near the borders of that chain of inland seas. They seem, therefore, to have been nothing more than great waves, which, about fifteen centuries before our era, devastated the borders of the Black Sea, the Sea of Marmora, the Archipelago and neighbouring coasts, in the same manner as the western shores of Portugal, Spain, and Northern Africa were inundated, during the great earthquake at Lisbon, by a wave which rose, in some places, to the height of fifty or sixty feet; or as happened in Peru, in 1746, where two hundred violent shocks followed each other in the space of twenty-four hours, and the ocean broke with impetuous force upon the land, destroying the town of Callao, and four other seaports, and converting a considerable tract of inhabited country into a bay.

In the country between the Caspian and the Black Seas, and in the chain of Caucasus, numerous earthquakes have, in modern times, caused fissures and subsidences of the soil, especially at Tiflis*. The Caucasian territories abound in hot-springs and mineral waters. So late as 1814, a new island was raised by volcanic explosions, in the Sea of Azof; and Pallas mentions that, in the same locality, opposite old Temruk, a submarine eruption took place in 1799, accompanied with dreadful thundering, emission of fire and smoke, and the throwing up of mire and stones. Violent earthquakes were felt at the same time at great distances from Temruk. The country around Erzerum exhibits similar phenomena, as does that around Tauris and the lake of Urmia, in which latter we have already remarked the rapid formation of travertin. The lake of Urmia, which is about two hundred and eighty English miles in circumference, resembles the Dead Sea, in having no outlet, and in being more salt than the ocean. Between the Tigris and Euphrates, also, there are numerous springs of naphtha, and frequent earthquakes agitate the country.

Syria and Palestine abound in volcanic appearances, and very extensive areas have been shaken, at different periods, with great destruction of cities and loss of lives.

It has been remarked, by Von Hoff, that from the commencement of the thirteenth to the latter half of the seventeenth century, there was an almost entire cessation of earthquakes

* Hoff, vol. ii., p. 210.

in Syria and Judea; and, during this interval of quiescence, the Archipelago, together with part of the adjacent coast of Lesser Asia, as also Southern Italy and Sicily, suffered extraordinary convulsions; while volcanic eruptions in those parts were unusually frequent. A more extended comparison, also, of the history of the subterranean convulsions of these tracts seems to confirm the opinion, that a violent crisis of commotion never visits both at the same time. It is impossible for us to declare, as yet, whether this phenomenon is constant in this, or general in other regions, because we can rarely trace back a connected series of events farther than a few centuries; but it is well known that, where numerous vents are clustered together within a small area, as in many archipelagos for instance, two of them are never in violent eruption at once. If the action of one becomes very great for a century or more, the others assume the appearance of spent volcanos. It is, therefore, not improbable that separate provinces of the same range of volcanic fires may hold a relation to one deep-seated focus, analagous to that which the apertures of a small group bear to some one rent or cavity. Thus, for example, we may conjecture that, at a comparatively small distance from the surface, Ischia and Vesuvius mutually communicate with certain fissures, and that each afford relief alternately to elastic fluids and lava there generated. So we may suppose Southern Italy and Syria to be connected, at a much greater depth, with a lower part of the very same system of fissures; in which case any obstruction occurring in one duct may have the effect of causing almost all the vapour and melted matter to be forced up the other, and if they cannot get vent, they may be the cause of violent earthquakes.

Continual mention is made in history of the ravages committed by earthquakes in Sidon, Tyre, Berytus, Laodicea, and Antioch, as also in the island of Cyprus. The country around the Dead Sea appears evidently, from the accounts of modern travellers, to be volcanic; and there are similar appearances, according to Burckhardt, in Arabia Petrea. A district near Smyrna, in Asia Minor, was termed by the Greeks Catacecaumene, or the burnt, where there is a large arid territory, without trees, and with a cindery soil *.

Proceeding westwards, we reach the Grecian archipelago,

* Strabo, Ed. Fal., p. 900.

where Santorin, afterwards to be described, is the grand centre of volcanic action. To the north-west of Santorin is another volcano, in the island of Milo, of recent aspect, having a very active solfatara in its central crater, and many sources of boiling water and steam. Continuing precisely the same line, we arrive at that part of the Morea, where we learn, from ancient writers, that Helice and Bura were, in the year 373 B. C., submerged beneath the sea by an earthquake; and the walls, according to Ovid, were to be seen beneath the waters. Near the same spot, in our times (1817), Vostizza was laid in ruins by a subterranean convulsion *. At Methone, also (now Modon), in Messenia, about three centuries before our era, an eruption threw up a great volcanic mountain, which is represented by Strabo as being nearly four thousand feet in height; but the magnitude of the hill requires confirmation. Some suppose that the accounts of the formation of a hill near Træzene, of which the date is unknown, may refer to the same event. Macedonia, Thrace, and Epirus, have always been subject to earthquakes, and the Ionian Isles are continually convulsed. Respecting Southern Italy, Sicily, and the Lipari Isles, we need not enlarge here, as the existence of volcanos in that region is known to all, and we shall have occasion again to allude to them.

The north-eastern portion of Africa, including Egypt, which lies six or seven degrees south of the volcanic line already traced, has been almost always exempt from earthquakes; but the north-western portion, especially Fez and Morocco, which fall within the line, suffer greatly from time to time. The southern part of Spain, also, and Portugal, have generally been exposed to the same scourge simultaneously with Northern Africa. The provinces of Malaga, Murcia, and Grenada, and in Portugal, the country round Lisbon, are recorded at several periods to have been devastated by great earthquakes. It will be seen, from Michell's account of the great Lisbon shock in 1755, that the first movement proceeded from the bed of the ocean ten or fifteen leagues from the coast. So late as February 2, 1816, when Lisbon was vehemently shaken, two ships felt a shock in the ocean west from Lisbon; one of them at the distance of one hundred and twenty, and the other two hundred and sixty-two French leagues from the

* Hoff, vol. ii., p. 172.

coast *—a fact which is the more interesting, because a line drawn through the Grecian archipelago, the volcanic region of Southern Italy, Sicily, Southern Spain, and Portugal, will, if prolonged westward through the ocean, strike the volcanic group of the Azores, which has, therefore, in all probability, a submarine connexion with the European line. How far the isles of Madeira and the Canaries, in the former of which violent earthquakes, and in the latter great eruptions, frequently happen, may communicate beneath the waters with the same region, must for the present be mere matter of conjecture.

Besides the continuous spaces of subterranean disturbance of which we have merely sketched the outline, there are other disconnected volcanic groups, of which the geographical extent is as yet very imperfectly known. Among these may be mentioned Iceland, which belongs, perhaps, to the same region as the volcano in Jan Mayen's Island, situated five degrees to the northeast. With these, also, part of the nearest coast of Greenland, which is sometimes shaken by earthquakes, may be connected. The island of Bourbon belongs to another theatre of volcanic action, of which Madagascar probably forms a part, if the alleged existence of burning volcanos in that island shall, on further examination, be substantiated. In following round the borders of the ocean to the north, we find the volcano of Gabel Tor, within the entrance of the Arabian Gulf. In the province of Cutch, in Bombay, and the adjoining districts of Hindostan, violent earthquakes repeatedly devastate an extensive territory.

Respecting the volcanic system of Southern Europe, it may be observed, that there is a central tract where the greatest earthquakes prevail, in which rocks are shattered, mountains rent, the surface elevated or depressed, and cities laid in ruins. On each side of this line of greatest commotion, there are parallel bands of country, where the shocks are less violent. At a still greater distance (as in Northern Italy, for example, extending to the foot of the Alps), there are spaces where the shocks are much rarer and more feeble, yet possibly of sufficient force to cause, by continued repetition, some appreciable alteration in the external form of the earth's crust. Beyond these limits, again, all countries are liable to slight tremors at distant intervals of time, when some great crisis of subterranean movement agitates an adjoining volcanic region; but these may be consi-

* Verneur, Journal des Voyages, vol. iv., p. 111. Hoff, vol. ii., p. 275.

dered as mere vibrations, propagated mechanically through the external crust of the globe, as sounds travel almost to indefinite distances through the air. Shocks of this kind have been felt in England, Scotland, Northern France, and Germany—particularly during the Lisbon earthquake. But these countries cannot, on this account, be supposed to constitute parts of the southern volcanic region, any more than the Shetland and Orkney Isles can be considered as belonging to the Icelandic circle, because the sands ejected from Hecla have been wafted thither by the winds.

We must also be careful to distinguish between lines of extinct and active volcanos, even where they appear to run in the same direction; for ancient and modern systems may cross and interfere with each other. Already, indeed, we have proof that this is the case; so that it is not by geographical position, but by reference to the species of organic beings alone, whether aquatic or terrestrial, whose remains occur in beds interstratified with lavas, that we can clearly distinguish the relative age of volcanos of which no eruptions are recorded. Had Southern Italy been known to civilized nations for as short a period as America, we should have had no record of eruptions in Ischia; yet we might have assured ourselves that the lavas of that isle had flowed since the Mediterranean was inhabited by the species of testacea now living in the Neapolitan seas. With this assurance it would not have been rash to include the numerous vents of that isle in the modern volcanic group of Campania. On similar grounds we may class, without much hesitation, the submarine lavas of the Val di Noto in Sicily, in the modern circle of subterranean commotion, of which Etna and Calabria form a part. But the lavas of the Euganean hills and the Vicentin, although not wholly beyond the range of earthquakes in Northern Italy, must not be confounded with any existing volcanic system; for when they flowed, the seas were inhabited with animals entirely distinct from those now known to live, whether in the Mediterranean or other parts of the globe. But we cannot enter into a full development of our views on these subjects in the present volume, as they would carry us into the consideration of changes in the earth's surface far anterior to the times of history, to which our present examination is exclusively confined.

CHAPTER XIX.

History of the volcanic eruptions in the district round Naples—Early convulsions in the island of Ischia—Numerous cones thrown up there—Epomeo not an habitual volcano—Lake Avernus—The Solfatara—Renewal of the eruptions of Vesuvius A.D. 79—Pliny's description of the phenomena—Remarks on his silence respecting the destruction of Herculaneum and Pompeii—Subsequent history of Vesuvius—Lava discharged in Ischia in 1302—Pause in the eruptions of Vesuvius—Monte Nuovo thrown up—Uniformity of the volcanic operations of Vesuvius and the Phlegræan Fields in ancient and modern times.

WE shall next present the reader with a sketch of the history of some of the volcanic vents dispersed throughout the great regions before described, and consider attentively the composition and arrangement of their lavas and ejected matter. The only volcanic region known to the ancients, was that of which the Mediterranean forms a part; and they have transmitted to us very imperfect records of the eruptions in three principal provinces of that region, namely, the district round Naples; that of Sicily and its isles; and that of the Grecian Archipelago. By far the most connected series of records throughout a long period relates to the first of these districts; and these cannot be too attentively considered, as much historical information is indispensable in order to enable us to obtain a clear view of the connexion and alternate mode of action of the different vents in a single volcanic group. The Neapolitan volcanos extend from Vesuvius, through the Phlegræan Fields, to Procida and Ischia, in a somewhat linear arrangement, ranging from the north-east to the south-west, as will be seen in the annexed map. (Pl. 3.) Within the space above limited, the volcanic force is sometimes developed in single eruptions from a considerable number of irregularly scattered points; but a great part of its action has been confined to one principal and habitual vent, Vesuvius or Somma. Before the Christian era, from the remotest periods of which we have any tradition, this principal

Fig. 1. Pl. 1.

Fig. 2.

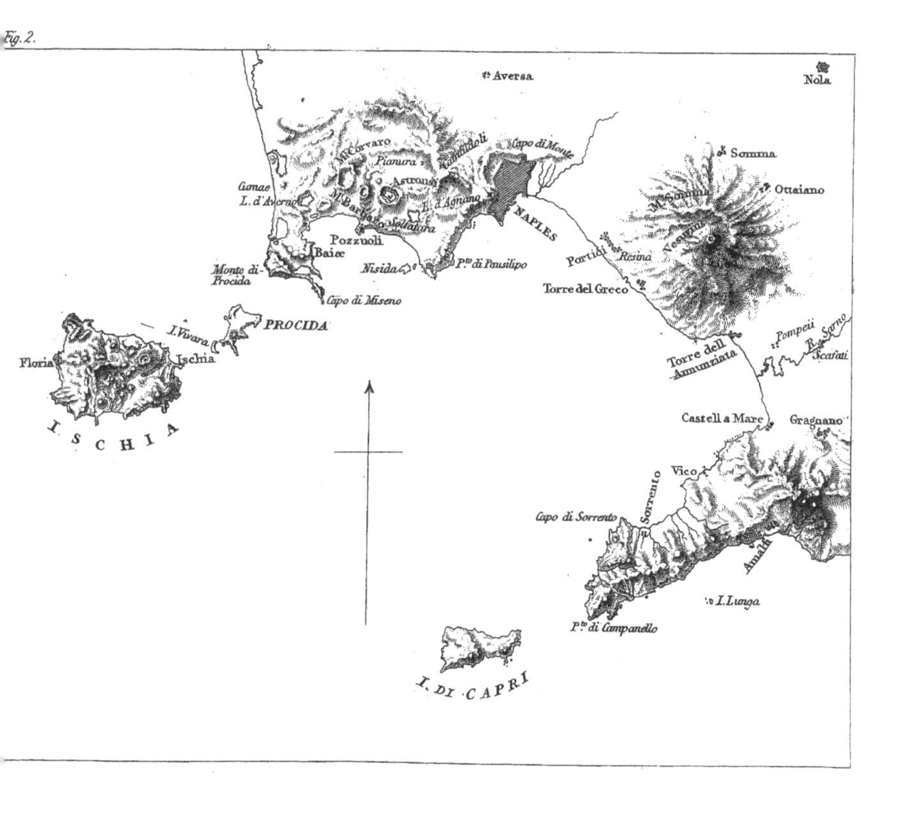

vent was in a state of inactivity. Terrific convulsions then took place from time to time in Ischia (Pithecusa), and seem to have extended to the neighbouring isle of Procida (Prochyta), for Strabo* mentions a story of the latter having been torn asunder from Ischia; and Pliny † derives its name from its having been poured forth by an eruption from Ischia. So violent were the earthquakes and volcanic explosions to which Ischia was subject, that Typhon the giant, "from whose eyes and mouth fire proceeded, and who hurled stones to heaven with a loud and hollow noise," was said to lie buried beneath it. The present circumference of the island along the water's edge is eighteen miles, its length from west to east about five, and its breadth from north to south three miles. Several Greek colonies which settled there before the Christian era were compelled to abandon it in consequence of the violence of the eruptions. First the Erythræans, and afterwards the Chalcidians, are mentioned as having been driven out by earthquakes and igneous exhalations. A colony was afterwards established by Hiero, king of Syracuse, about three hundred and eighty years before the Christian era; but when they had built a fortress, they were compelled by an eruption to fly, and never again returned. Strabo tells us that Timæus recorded a tradition that a little before his time Epomeus, the principal mountain in the centre of the island, vomited fire during great earthquakes; that the land between it and the coast had ejected much fiery matter which flowed into the sea, and that the sea receded for the distance of three stadia, and then returning, overflowed the island. This eruption is supposed by some to have been that which formed the crater of Monte Corvo on one of the higher flanks of Epomeo, above Foria, the lava-current of which may still be traced, by aid of the scoriæ on its surface, from the crater to the sea. To one of the subsequent eruptions in the lower parts of the isle, which caused the expulsion of the first Greek colony, Monte Rotaro has been attributed, and it bears every mark of recent origin. The cone is remarkably perfect, and has a crater on its summit precisely resembling that of Monte Nuovo; but the hill is larger, and resembles some of the more considerable cones of single eruption near Clermont in

* Lib. v. † Nat. Hist., lib. iii., c. 6.

Auvergne, and, like some of them, it has given vent to a lava-stream at its base, instead of its summit. A small ravine swept out by a torrent exposes the structure of the cone, which is composed of innumerable inclined and slightly undulating layers of pumice, scoriæ, white lapilli, and enormous angular blocks of trachyte. These last have evidently been thrown out by violent explosions, like those which, in 1822, launched from Vesuvius a mass of pyroxenic lava, of many tons weight, to the distance of three miles, which fell in the garden of Prince Ottajano. The cone of Rotaro is covered with the arbutus, and other beautiful evergreens. Such is the strength of the virgin soil, that the shrubs have become almost arborescent; and the growth of some of the smaller wild plants has been so vigorous, that botanists have scarcely been able to recognise the species. The eruption whereby the Syracusan colony was dislodged, is supposed to have given rise to that mighty current which forms the promontory of Zaro and Caruso. The surface of these lavas is still very arid and bristling, and is covered with black scoriæ ; so that it is not without great labour that human industry has redeemed some small spots, and converted them into vineyards. From the date of the great eruption last alluded to, down to our own time, Ischia has enjoyed tranquillity, with the exception of one emission of lava hereafter to be described, which, although it occasioned much local damage, does not appear to have devastated the whole country, in the manner of more ancient explosions.

The population of the isle amounts at present to about twenty-five thousand, and is on the increase. They are supported almost entirely on the production of their vineyards. The lofty central hill, Epomeo or S. Nichola, on this island, is composed of greenish indurated tuff, of a prodigious thickness, interstratified in some parts with argillaceous marl, and, here and there, with great streams of indurated lava. Visconti ascertained, by trigonometrical measurement, that this mountain was 2605 feet above the level of the sea. In mineral composition and in form, as seen from many points of view, it resembles the hill to the north of Naples, on the summit of which stands the convent of Camaldoli, which is 1643 feet in height. Both these mountains, like the greater part of those in the Terra di Lavoro, are of subaqueous origin; although it has frequently happened to them, as to Epomeo, that, after being elevated above the level of the sea, fresh eruptions have broken through at different points. I found more than one argillaceous stratum containing marine shells, within eight hundred feet of the summit of Epomeo; and from this circumstance, and from the general structure of the mountain, I am compelled to dissent from the opinion expressed by Mr. Scrope, who supposed it to have been once a great habitual volcano, like Vesuvius*. At least it is certain, that if any one of the cones on the present mountain gave vent to several streams of lava in succession, this happened when the whole mass was still beneath the level of the sea. Brocchi long ago announced, that the igneous rocks of this island rest on a plastic clay containing shells. Of these a considerable number have now been obtained, and identified with species still living in the Mediterranean. There are, upon the whole, on different parts of Epomeo, or scattered through the lower tracts of the island, twelve considerable volcanic cones, which have been thrown up since the island was raised above the surface of the deep; and many streams of lava may have flowed, like that of "Arso" in 1302, without cones having been produced; so that this isle may, for ages before the period of the remotest traditions, have served as a safety-valve to the whole Terra di Lavoro, while the fires of Vesuvius were dormant. It seems also clear, that Avernus, a circular lake near Puzzuoli, about half a mile in diameter, which is now a salubrious and cheerful spot, once exhaled mephitic vapours, such as are often emitted by craters after eruptions. There is no reason for discrediting the account of Lucretius†, that birds could not fly over it without being stifled, although they may now frequent it uninjured. There must have been a time when this crater was in action; and for many centuries afterwards it may have deserved the appellation of "atri janua Ditis," emitting, perhaps, gases as destructive of animal life as those suffocating vapours which were given out by Lake Quilotoa, in Quito, in 1797, by which whole herds of cattle on its shores were killed ‡, or as those deleterious

* Geol. Trans., vol. ii., part 3, p. 388; second series.

† De Rerum Nat., VI., 740.—Mr. Forbes on the Bay of Naples, Edin. Journ. of Science, No. 3, *new series*, p. 87, Jan. 1830.

‡ Humboldt, Voy., p. 317.

emanations which annihilated all the cattle in the island of Lancerote, one of the Canaries, in 1730*. Bory St. Vincent mentions, that in the same isle birds fell lifeless to the ground; and Sir William Hamilton informs us that he picked up dead birds on Vesuvius during an eruption. The Solfatara, also, near Puzzuoli, which may be still considered as a half-extinguished crater, appears, by the accounts of Strabo and others, to have been before the Christian era in very much the same state as at present, giving vent continually to aqueous vapour, together with sulphureous and muriatic acid gases, similar to those evolved by Vesuvius.

Such, then, were the points where the subterranean fires obtained vent, from the earliest period to which tradition reaches back, down to the first century of the Christian era; but we then arrive at a crisis in the volcanic action of this district —one of the most interesting events witnessed by man during the brief period throughout which he has observed the physical changes on the earth's surface. From the first colonization of Southern Italy by the Greeks, Vesuvius afforded no other indications of its volcanic character than such as the naturalist might infer, from the analogy of its structure to other volcanos. These were recognised by Strabo, but Pliny did not include the mountain in his list of active vents. The ancient cone was of a very regular form, terminating, not as at present, in two peaks, but with a flattish summit, where the remains of an ancient crater, nearly filled up, had left a slight depression, covered in its interior by wild vines, and with a sterile plain at the bottom. On the exterior, the flanks of the mountain were covered with fertile fields richly cultivated, and at its base were the populous cities of Herculaneum and Pompeii. But the scene of repose was at length doomed to cease, and the volcanic fire was recalled to the main channel, which, at some former unknown period, had given passage to repeated streams of melted lava, sand, and scoriæ. The first symptom of the revival of the energies of this volcano was the occurrence of an earthquake in the year 63 after Christ, which did considerable injury to the cities in its vicinity. From that time to the year 79 slight shocks were frequent, and in the month of August of that

* Von Buch, Ub. einen vulcanisch. Ausbruch auf der Insel Lanzerote.

year they became more numerous and violent, till they ended at length in an eruption. The elder Pliny, who commanded the Roman fleet, was then stationed at Misenum; and in his anxiety to obtain a near view of the phenomena, he lost his life, being suffocated by sulphureous vapours. His nephew, the younger Pliny, remained at Misenum, and has given us, in his Letters, a lively description of the awful scene. A dense column of vapour was first seen rising vertically from Vesuvius, and then spreading itself out laterally, so that its upper portion resembled the head, and its lower the trunk of the pine, which characterizes the Italian landscape. This black cloud was pierced occasionally by flashes of fire as vivid as lightning, succeeded by darkness more profound than night. Ashes fell even upon the ships at Misenum, and caused a shoal in one part of the sea—the ground rocked, and the sea receded from the shores, so that many marine animals were seen on the dry sand. The appearances above described agree perfectly with those witnessed in more recent eruptions, especially those of Monte Nuovo in 1538, and of Vesuvius in 1822. In all times and countries, indeed, there is a striking uniformity in the volcanic phenomena; but it is most singular that Pliny, although giving a circumstantial detail of so many physical facts, and enlarging upon the manner of his uncle's death, and the ashes which fell when he was at Stabiæ, makes no allusion whatever to the sudden overwhelming of two large and populous cities, Herculaneum and Pompeii.

All naturalists who have searched into the memorials of the past, for records of physical events, must have been surprised at the indifference with which the most memorable occurrences are often passed by, in the works of writers of enlightened periods; as also of the extraordinary exaggeration which usually displays itself in the traditions of similar events, in ignorant and superstitious ages. But, of all omissions, the most inexplicable, perhaps, is that now under consideration; and we have no hesitation in saying, that had the buried cities never been discovered, the accounts transmitted to us of their tragical end would have been discredited by the majority, so vague and general are the other narratives, or so long subsequent to the event. Tacitus, the friend and contemporary of Pliny, when adverting in general terms to the convulsion, says merely that "cities were consumed or buried*". Suetonius, although he alludes to the eruption incidentally, is silent as to the cities. They are mentioned by Martial, in an epigram, as immersed in cinders; but the first historian who alludes to them by name is Dion Cassius†, who flourished about a century and a half after Pliny. He appears to have derived his information from the traditions of the inhabitants, and to have recorded, without discrimination, all the facts and fables which he could collect. He tells us, "that during the eruption, a multitude of men of superhuman stature, resembling giants, appeared sometimes on the mountain and sometimes in the environs—that stones and smoke were thrown out, the sun was hidden, and then the giants seemed to rise again, while the sounds of trumpets were heard, &c., &c.; and finally two entire cities, Herculaneum and Pompeii, were buried under showers of ashes, while all the people were sitting in the theatre." That many of these circumstances were invented, would have been obvious, even without the aid of Pliny's Letters; and the examination of Herculaneum and Pompeii enables us to prove, that none of the people were destroyed in the theatres, and, indeed, that there were very few of the inhabitants who did not escape from both cities. Yet some lives were lost, and there was ample foundation for the tale in its most essential particulars. This case may often serve as a caution to the geologist, who has frequent occasion to weigh, in like manner, negative evidence derived from the silence of eminent writers, against the obscure but positive testimony of popular traditions. Some authors, for example, would have us call in question the reality of the Ogygian deluge, because Homer and Hesiod say nothing of it. But they were poets, not historians, and they lived many centuries after the latest date assigned to the catastrophe. Had they even lived at the time of that flood, we might still contend that their silence ought, no more than Pliny's, to avail against the authority of tradition, however much exaggeration we may impute to the latter.

It does not appear that in the year 79 any lava flowed from Vesuvius; the ejected substances, perhaps, consisted

* Haustæ aut obrutæ urbes. Hist., lib. 1

† Hist. Rom., lib. 66.

entirely of lapilli, sand, and fragments of older lava, as when Monte Nuovo was thrown up in 1538. The first era at which we have authentic accounts of the flowing of a stream of lava, is the year 1036, which is the seventh eruption from the revival of the fires of the volcano. A few years afterwards, in 1049, another eruption is mentioned, and another in 1138 (or 1139), after which a great pause ensued of one hundred and sixty-eight years. During this long interval of repose, two minor vents opened at distant points. In the first place it is on tradition that an eruption took place from the Solfatara in the year 1198, during the reign of Frederic II., Emperor of Germany; and although no circumstantial detail of the event has reached us from those dark ages, we may receive the fact without hesitation *. Nothing more, however, can be attributed to this eruption, as Mr. Scrope observes, than the discharge of a light and scoriform trachytic lava, of recent aspect, resting upon the strata of loose tufa which covers the principal mass of trachyte †. The other occurrence is well authenticated,—the eruption, in the year 1302, of a lava-stream, from a new vent on the south-east side of the island of Ischia. During part of 1301, earthquakes had succeeded one another with fearful rapidity; and they terminated at last with the discharge of a lava-stream from a point named the Campo del Arso, not far from the town of Ischia. This lava ran quite down to the sea—a distance of about two miles: in colour it varies from iron-grey to reddish black, and is remarkable for the glassy felspars which it contains. Its surface is almost as sterile, after a period of five centuries, as if it had cooled down yesterday. A few scantlings of wild thyme, and two or three other dwarfish plants, alone appear in the interstices of the scoriæ, while the Vesuvian lava of 1767 is already covered with a luxuriant vegetation. Pontanus, whose country-house was burnt and overwhelmed, describes the dreadful scene as having lasted two months ‡. Many houses were swallowed up, and a partial emigration of the inhabitants followed. This eruption produced no

* The earliest authority, says Mr. Forbes, given for this fact, appears to be Capaccio, quoted in the Terra Tremante of Bonito. Edin. Journ. of Sci., &c. No. I., new series, p. 127, July, 1829.

† Geol. Trans., vol. ii., part 3, p. 346, second series.

‡ Lib. vi., de Bello Neap., in Grævii Thesaur.

cone, but only a slight depression, hardly deserving the name of a crater, where heaps of black and red scoriæ lie scattered around. Until this eruption, Ischia is generally believed to have enjoyed an interval of rest for about seventeen centuries; but Julius Obsequens*, who flourished A.D. 214, refers to some volcanic convulsion in the year 662, after the building of Rome. (91 B.C.) As Pliny, who lived a century before Obsequens, does not enumerate this among other volcanic eruptions, the statement of the latter author is supposed to have been erroneous; but it would be more consistent, for reasons before stated, to disregard the silence of Pliny, and to conclude that some subterranean commotion, probably of no great violence, happened at the period alluded to.

To return to Vesuvius,—the next eruption occurred in 1306; between which era and 1631 there was only one other (in 1500), and that a slight one. It has been remarked, that throughout this period Etna was in a state of such unusual activity as to lend countenance to the idea that the great Sicilian volcano may sometimes serve as a channel of discharge to elastic fluids and lava that would otherwise rise to the vents in Campania. The great pause was also marked by a memorable event in the Phlegræan Fields—the sudden formation of a new mountain in 1538, of which we have received authentic accounts from contemporary writers. Frequent earthquakes, for two years preceding, disturbed the neighbourhood of Puzzuoli; but it was not until the 27th and 28th of September, 1538, that they became alarming, when not less than twenty shocks were experienced in twenty-four hours. At length, on the night of the 29th, two hours after sunset, a gulph opened between the little town of Tripergola, which once existed on the site of the Monte Nuovo, and the baths in its suburbs, which were much frequented. This watering place contained a hospital for those who resorted thither for the benefit of the thermal springs, and it appears that there were no fewer than three inns in the principal street. A large fissure approached the town with a tremendous noise, and began to discharge pumice-stones, blocks of unmelted lava and ashes mixed with water, and occasionally flames. The ashes fell in immense

* Prodig. libell., c. 114.

quantities, even at Naples; while the neighbouring Puzzuoli was deserted by its inhabitants. The sea retired suddenly for two hundred yards, and a portion of its bed was left dry. We shall afterwards, when treating of earthquakes, show by numerous proofs derived not only from the state of the Temple of Serapis (see Frontispiece), but from many other physical phenomena, that the whole coast, from Monte Nuovo to beyond Puzzuoli, was at that time upraised to the height of many feet above the bed of the Mediterranean, and has ever since remained permanently elevated. On the 3rd of October the eruption ceased, so that the hill (fig. 1, No. 11), the great mass of which was thrown up in a day and a night, was accessible; and those who ascended reported that they found a funnel-shaped crater on its summit. (Fig. 2, No. 11.)

Monte Nuovo, formed in the Bay of Baiæ, September 29th, 1538.

1. Cone of Monte Nuovo. 2. Brim of crater of ditto.
3. Thermal spring, called Baths of Nero, or Stufe di Tritoli.

The height of Monte Nuovo has recently been determined, by the Italian mineralogist Pini, to be four hundred and forty English feet above the level of the bay; its base is about eight thousand feet, or nearly a mile and a half, in circumference. According to Pini, the depth of the crater is four hundred and twenty-one English feet from the summit of the hill, so that its bottom is only nineteen feet above the level of the sea. No lava flowed from this cavity, but the ejected matter consisted of pumiceous scoriæ and masses of trachyte, many of them schistose, and resembling clinkstone. The Monte Nuovo is declared, by the best authorities, to stand partly on the site of the Lucrine lake (fig. 4, No. 12*), which

The Phlegræan Fields.

1. Monte Nuovo. 2. Monte Barbaro.
3. Lake Avernus. 4. Lucrine Lake.
5. The Solfatara. 6. Puzzuoli.
7. Bay of Baiæ.

was nothing more than the crater of a pre-existent volcano, and was almost entirely filled during the explosion of 1538. Nothing now remains but a shallow pool, separated from the sea by an elevated beach, raised artificially.

Immediately adjoining to Monte Nuovo is the larger volcanic cone of Monte Barbaro (fig. 2, No. 12), the Gaurus inanis of Juvenal—an appellation given to it probably from its deep circular crater, which is about a mile in diameter. Large as is this cone, it was probably produced by a single eruption; and it does not, perhaps, exceed in magnitude some of the largest of those in Ischia, which there is every reason to believe to have been formed within the historical era. It is composed chiefly of indurated tufa, like Monte Nuovo, stratified conformably to its conical surface. This hill was once very celebrated for its wines, and is still covered with vineyards; but

* This representation of the Phlegræan Fields is reduced from part of Plate xxxi. of Sir William Hamilton's great work, "Campi Phlegræi," to which we refer the reader for faithful delineations of the scenery of that country.

when the vine is not in leaf it has a sterile appearance, and late in the year, when seen from the beautiful bay of Baiæ, it often contrasts so strongly in verdure with Monte Nuovo, which is always clothed with arbutus, myrtle, and other wild evergreens, that a stranger might well imagine the cone of older date to be that thrown up in the sixteenth century*. There is nothing, indeed, so calculated to instruct the geologist, as the striking manner in which the recent volcanic hills of Ischia, and that now under consideration, blend with the surrounding landscape. Nothing seems wanting or redundant; every part of the picture is in such perfect harmony with the rest, that the whole has the appearance of having been called into existence by a single effort of creative power. What other result could we have anticipated, if Nature has ever been governed by the same laws? Each new mountain thrown up—each new tract of land raised or depressed by earthquakes—should be in perfect accordance with those previously formed, if the entire configuration of the surface has been due to a long series of similar convulsions. Were it true that the greater part of the dry land originated simultaneously in its present state, and that additions were afterwards made slowly and successively; then, indeed, there might be reason to expect a strong line of demarcation between the signs of ancient and modern changes. But the continuity of the plan, and the perfect identity of the causes, are to many a source of deception, and lead them to exaggerate the energy of agents which operated in the earlier ages. In the absence of all historical information they are as unable to separate the dates of the origin of different portions of our continents, as is the stranger to determine, by their physical features alone, the distinct ages of Monte Nuovo, Monte Barbaro, Astroni, and the Solfatara.

The vast scale and violence of the volcanic operations in Campania, in the olden time, has been a theme of declamation, and has been contrasted with the comparative state of quiescence of this delightful region in the modern era. Instead of inferring, from analogy, that the ancient Vesuvius was always at rest when the craters of the Phlegræan Fields were burning,—that each

* Hamilton observes, (writing in 1770,) "the new mountain produces as yet but a very slender vegetation." This remark was not applicable in 1828.—Campi Phlegræi, p. 69.

cone rose in succession,—and that many years, and often centuries of repose intervened between each eruption—geologists seem to have conjectured that the whole group sprung up from the ground at once, like the soldiers of Cadmus when he sowed the dragon's teeth. As well might they endeavour to persuade us that on these Phlegræan Fields, as the poets feigned, the giants warred with Jove, ere yet the puny race of mortals were in being.

For nearly a century after the birth of Monte Nuovo, Vesuvius still continued in a state of tranquillity. There had then been no violent eruption for four hundred and ninety-two years; and it appears that the crater was then exactly in the condition of the present extinct volcano of Astroni, near Naples. Bracini, who visited Vesuvius not long before the eruption of 1631, gives the following interesting description of the interior. "The crater was five miles in circumference, and about a thousand paces deep; its sides were covered with brushwood, and at the bottom there was a plain on which cattle grazed. In the woody parts wild-boars frequently harboured. In one part of the plain, covered with ashes, were three small pools, one filled with hot and bitter water, another salter than the sea, and a third hot but tasteless *." But at length these forests and grassy plains were suddenly consumed —blown into the air, and their ashes scattered to the winds. In December, 1631, seven streams of lava poured at once from the crater, and overflowed several villages on the flanks and at the foot of the mountain. Resina, partly built over the ancient site of Herculaneum, was consumed by the fiery torrent. Great floods of mud were as destructive as the lava itself, as often happens during these catastrophes; for such is the violence of rains produced by the evolution of aqueous vapour, that torrents of water descend the cone, and, becoming charged with impalpable volcanic dust, roll along loose ashes, acquiring such consistency as to deserve their ordinary appellation of "aqueous lavas."

A brief period of repose ensued, which lasted only until the year 1666, from which time to the present there has

* Hamilton's Campi Phlegræi, folio, vol. i., p. 62; and Brieslak, Campanie, tome i., p. 186.

been a constant series of eruptions, with rarely an interval of rest exceeding ten years. During these three centuries no irregular volcanic agency has convulsed other points in this district. Brieslak remarked that such irregular convulsions had occurred in the Bay of Naples; in every second century, as, for example, the eruption of the Solfatara in the twelfth, of the lava of Arso, in Ischia, in the fourteenth, and of Monte Nuovo in the sixteenth; but the eighteenth has formed an exception to this rule, and this seems accounted for by the unprecedented number of eruptions of Vesuvius during that period; whereas, when the new vents opened, there had always been, as we have seen, a long intermittance of activity in the principal volcano.

CHAPTER XX.

Dimensions and structure of the cone of Vesuvius—Dikes in the recent cone, how formed—Section through Vesuvius and Somma—Vesuvian lavas and minerals—Effects on decomposition of lava—Alluvions called "aqueous lavas"—Origin and composition on the matter enveloping Herculaneum and Pompeii—Controversies on the subject—Condition and contents of the buried cities—Proofs of their having suffered by an earthquake—Small number of skeletons—State of preservation of animal and vegetable substances—Rolls of Papyrus—Probability of future discoveries of MSS—Stabiæ—Torre del Greco—Concluding remarks on the destroying and renovating agency of the Campanian volcanos.

Structure of the cone of Vesuvius.—Between the end of the eighteenth century and the year 1822, the great crater of Vesuvius had been gradually filled by lava boiling up from below, and by scoriæ falling from the explosions of minor mouths which were formed at intervals on its bottom and sides. In place of a regular cavity, therefore, there was a rough and rocky plain, covered with blocks of lava and scoriæ, and cut by numerous fissures, from which clouds of vapour were evolved. But this state of things was totally changed by the eruption of October, 1822, when violent explosions, during the space of more than twenty days, broke up and threw out all this accumulated mass, so as to leave an immense gulf or chasm, of an irregular, but somewhat elliptical shape, about three miles in circumference when measured along the very sinuous and irregular line of its extreme margin, but somewhat less than three quarters of a mile in its longest diameter, which was directed from N.E. to S.W.* The depth of this tremendous abyss has been variously estimated, for from the hour of its formation it decreased daily, by the dilapidation of its sides. It measured at first, according to the accounts of some authors, two thousand feet in depth from the extreme part of

* Account of the Eruption of Vesuvius in October, 1822, by G. P. Scrope, Esq., Journ. of Sci., &c., vol. xv., p. 175.

the existing summit *; but Mr. Scrope, when he saw it, soon after the eruption, estimated its depth at less than half that quantity. More than eight hundred feet of the cone was carried away by the explosions, so that the mountain was reduced in height from about four thousand two hundred to three thousand four hundred feet †.

As we ascend the sloping sides, the volcano appears a mass of loose materials—a mere heap of rubbish, thrown together without the slightest order; but on arriving at the brim of the crater, and obtaining a view of the interior, we are agreeably surprised to discover that the conformation of the whole displays in every part the most perfect symmetry and arrangement. The materials are disposed in regular strata slightly undulating, appearing, when viewed in front, to be disposed in horizontal planes. But as we make the circuit of the edge of the crater, and observe the cliffs by which it is encircled projecting or receding in salient or retiring angles, we behold transverse sections of the currents of lava and beds of sand and scoriæ, and recognise their true dip. We then discover that they incline outwards from the axis of the cone, at angles varying from 30° to 45°. The whole cone, in fact, is composed of a number of concentric coatings of alternating lavas, sand, and scoriæ. Every shower of ashes which has fallen from above, and every stream of lava descending from the lips of the crater, have conformed to the outward surface of the hill, so that one conical envelope may be said to have been successively folded round another, until the aggregation of the whole mountain was completed. The marked separation into distinct beds results from the different colours and degrees of coarseness in the sands, scoriæ, and lava, and the alternation of these with each other. The greatest difficulty, on the first view, is to conceive how so much regularity can be produced, notwithstanding the unequal distribution of sand and scoriæ, driven by prevailing winds in particular eruptions, and the small breadth of each sheet of lava as it first flows out from the crater. But on a closer examination we find that the appearance of extreme uniformity is delusive, for when a number of beds thin out gradually, and at different points, the eye does not without diffi-

* Mr. Forbes, Account of Mount Vesuvius, Edin. Journ. of Sci., No. xviii., p. 195, Oct., 1828.

† Ibid., p. 194.

culty recognise the termination of any one stratum, but usually supposes it continuous with some other, which at a short distance may lie precisely in the same plane. The slight undulations, moreover, produced by inequalities on the sides of the hill on which the successive layers were moulded, assists the deception. As countless beds of sand and scoriæ constitute the greater part of the whole mass, these may sometimes mantle continuously round the whole cone; and even lava-streams may be of considerable breadth when first they overflow, since in some eruptions a considerable part of the upper portion of the cone breaks down at once, and may form a sheet extending as far as the space which the eye usually takes in in a single section. The high inclination of some of the beds, and the firm union of the particles even where there is evidently no cement, is another striking feature in the volcanic tuffs and breccias, which seems at first not very easy of explanation. But the last great eruption afforded ample illustration of the manner in which these strata are formed. Fragments of lava, scoriæ, pumice, and sand, when they fall at slight distances from the summit, are only half cooled down from a state of fusion, and are afterwards acted upon by the heat from within, and by fumeroles or small crevices in the cone through which hot vapours are disengaged. Thus heated, the ejected fragments cohere together strongly; and the whole mass acquires such consistency in a few days, that fragments cannot be detached without a smart blow of the hammer. At the same time sand and scoriæ, ejected to a greater distance, remain incoherent *.

The inclined strata before mentioned, which dip outwards in all directions from the axis of the cone of Vesuvius, are intersected by veins or dikes of compact lava, for the most part in vertical position. In 1828, these were seen to be about seven in number, some of them not less than four or five hundred feet in height, and thinning out before they reached the uppermost part of the cone. Being harder than the beds through which they pass, they have resisted decomposition, and stand out in relief †.

* Monticelli and Covelli, Storia di Fenon. del Vesuv., en 1821-2-3.

† When I visited Vesuvius in November, 1828, I was prevented from descending into the crater by the constant ejections then thrown out. I only got sight of three of the dikes, but Signor Monticelli had previously had drawings made of

There can be no doubt that these dikes have been produced by the filling up of open fissures with liquid lava; but of the date of their formation we know nothing farther than that they are all subsequent to the year 79, and, relatively speaking, that they are more modern than all the lavas and scoriæ which they intersect. A considerable number of the upper strata, not traversed by them, must have been due to later eruptions if the dikes were filled from below. That the earthquakes which almost invariably precede eruptions occasion rents in the mass is well known; and, in 1822, three months before the lava flowed out, open fissures, evolving hot vapours, were numerous. It is clear that such rents must be injected with melted matter when the column of lava rises, so that the origin of the dikes is easily explained, as also the great solidity and crystalline nature of the rock composing them, which has been formed by lava cooling down slowly under great pressure.

In the annexed diagram (No. 13.) it will be seen that, on the side of Vesuvius opposite to that where a portion of the ancient cone of Somma (*a*) still remains, is a projection (*b*) called the Pedamentina, which some have supposed to be part of the circumference of the ancient crater broken down towards the sea, and over the edge of which the lavas of the modern Vesuvius have poured; the axis of the present cone of Vesuvius being, according to Visconti, precisely equidistant from the escarpment of Somma and the Pedamentina. But it has been objected (and not without reason) to this hypothesis, that if the Pedamentina and the escarpment of Somma were the remains of the original *crater*, that crater must have been many miles in diameter, and more enormous than almost any one known on the globe. It is, therefore, more probable that the ancient mountain was higher than Vesuvius (which, comparatively

the whole, which he showed me. The veins which I saw were on that side of the cone which is encircled by Somma. In the March of the year before mentioned, an eruption began at the bottom of the deep gulf formed in 1822. The ejected matter had filled up nearly one-third of the original abyss in November, and the same operation was slowly continuing, a single black cone being seen at the bottom in almost continual activity. It is clear that these ejections may continue till the throat of Vesuvius is filled up in the same manner as before 1822; and Mr. Scrope has referred the frequent occurrence of volcanic cones without craters to this cause. I found, in 1828, the lava of 1822 not yet cool on the north side of the cone, and evolving much heat and vapour from crevices.

No. 13.

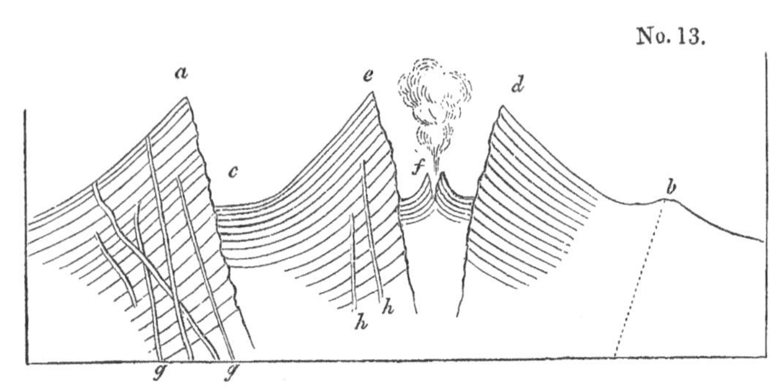

Supposed section of Vesuvius and Somma.

a. Monte Somma, or the remains of the ancient cone of Vesuvius.
b. The Pedamentina, a terrace-like projection, encircling the base of the recent cone of Vesuvius, on the south side.
c. Atrio del Cavallo*.
d. e. Crater left by eruption of 1822.
f. Small cone thrown up in 1828, at the bottom of the great crater.
g.g. Dikes intersecting Somma.
h.h. Dikes intersecting the recent cone of Vesuvius.

speaking, is a volcano of no great height); and that the explosions of the year 79 caused it not merely to disgorge the contents of its crater, which had long been choked up, but blew up a great part of the cone itself: so that the wall of Somma, and the ridge or terrace of the Pedamentina, were never the margin of a crater of eruption, but are the relics of a ruined and truncated cone. It will be seen in the diagram that the slanting beds of the cone of Vesuvius become horizontal in the Atrio del Cavallo (at *c*), where the base of the new cone meets the precipitous escarpment of Somma; for when the lava flows down to this point, as happened in 1822, its descending course is arrested, and it then runs in another direction along this small valley, circling round the base of the cone. Sand and scoriæ, also, blown by the winds, collect at the base of the cone, and are then swept away by torrents; so that there is always here a flattish plain, as represented. In the same manner the small interior cone (*f*) must be composed of sloping beds, terminating in a horizontal plain; for while this

* So called from travellers leaving their horses and mules there when they prepare to ascend the cone on foot.

monticule was gradually gaining height by successive ejections of lava and scoriæ, in 1828, it was always surrounded by a flat pool of semi-fluid lava, into which scoriæ and sand were thrown.

The escarpment of Somma exhibits a structure precisely similar to that of the cone of Vesuvius, but the beds are intersected by a much greater number of dikes. The formation of this older cone does not belong to the historical era, and we must not, therefore, enlarge upon it in this place; but we shall have occasion presently to revert to the subject, when we speak of a favorite doctrine of some modern geologists, concerning "craters of elevation" (Erhebung's Cratere), whereby, in defiance of analogy, the origin of the identical disposition of the strata and dikes in Vesuvius and Somma has been referred to a mode of operation totally dissimilar.

Vesuvian Lavas.—The modern lavas of Vesuvius are characterized by a large proportion of augite (or pyroxene). When they are composed of this mineral and felspar, they may be said to differ in no way in composition from many of the ancient volcanic rocks of Scotland. They are often porphyritic, containing disseminated crystals of augite, leucite, or some other mineral, imbedded in a more earthy base. These porphyritic lavas are often extremely compact, especially in the dikes both of Vesuvius and Somma, which, in hardness and specific gravity, are by no means inferior to ordinary veins of trap, and, like them, often preserve a remarkable parallelism in their two opposite faces for considerable distances. In regard to the structure of the Vesuvian lavas on a great scale, there are no sections of sufficient depth to enable us to draw fair comparisons between them and the products of extinct volcanos. At the fortress near Torre del Greco a section is exposed, fifteen feet in height, of a current which ran into the sea; and it evinces, especially in the lower part, a decided tendency to divide into rude columns. A still more striking example may be seen to the West of Torre del Annunziata, near Forte Scassato, where the mass is laid open by the sea to the depth of twenty feet. In both these cases, however, the rock may rather be said to be divided into numerous perpendicular fissures, than to be prismatic, although the same picturesque effect is produced. In the lava-currents of Central France (those of the Vivarais, in particular), the uppermost portion, often forty feet or more in thickness, is an amorphous mass passing downwards into lava, irregularly prismatic; and, under this, there is a foundation of regular and vertical columns, in that part of the current which must have cooled most slowly. But the lavas last mentioned are often one hundred feet or more in thickness; and we cannot expect to discover the same phenomenon in the shallow currents of Vesuvius, although it may be looked for in modern streams in Iceland, which exceed even those of ancient France in volume. Mr. Scrope* mentions, that, in the cliffs encircling the great crater of the modern cone, he saw many currents offering a columnar division, and some almost as regularly prismatic as any ranges of the older basalts; and he adds, that in some the spheroidal concretionary structure, on a large scale, was equally conspicuous. Brieslak† also informs us, that in the siliceous lava of 1737, which contains augite, leucite, and crystals of felspar, he found very regular prisms in a quarry near Torre del Greco; which observation is confirmed by modern authorities‡. The decomposition of some of the felspathic lavas, either by simple weathering, or by gaseous emanations, converts them from a hard to a soft clayey state, so that they no longer retain the smallest resemblance to rocks cooled down from a state of fusion. The exhalations of sulphuretted hydrogen and muriatic acid which are disengaged continually from the Solfatara, also produce curious changes on the trachyte of that extinct volcano: the rock is whitened and becomes porous, fissile, and honeycombed, till at length it crumbles into a white siliceous powder§. Numerous globular concretions, composed of concentric laminæ, are also formed by the same vapours in this decomposed rock‖.

They who have visited the Phlegræan Fields and the volcanic region of Sicily, and who are aware of the many problematical appearances which igneous rocks of the most modern origin assume, especially after decomposition, cannot but be

* Journ. of Sci., vol. xv., p. 177. † Voy. dans la Campanie, tome. i, p. 201.

‡ Mr. Forbes, on Mount Vesuvius, Edin. Journ. of Sci., No. xviii., Oct. 1828.

§ Daubeny on Volcanos, p. 169.

‖ Scrope, Geol. Trans., second series, vol. ii., p. 346.

astonished at the confidence with which the contending Neptunists and Vulcanists in the last century dogmatized on the igneous or aqueous origin of certain rocks of the remotest antiquity. Instead of having laboured to acquire an accurate acquaintance with the aspect of known volcanic rocks, and the transmutations which they undergo subsequently to their first consolidation, the adherents of both parties seem either to have considered themselves born with an intuitive knowledge of the effects of volcanic operations, or to have assumed that they required no other analogies than those which a laboratory and furnace might supply.

Vesuvian Minerals.—A great variety of minerals are found in the lavas of Vesuvius and Somma; for there are so many common to both, that it is unnecessary to separate them. Augite, leucite, felspar, mica, olivine, and sulphur, are most abundant. It is an extraordinary fact, that, in an area of three square miles round Vesuvius, a greater number of simple minerals have been found than in any spot of the same dimensions on the surface of the globe. Häuy only enumerated three hundred and eighty species of simple minerals as known to him, and no less than eighty-two had been found on Vesuvius before the end of the year 1828 *. Many of these are peculiar to that locality. Some mineralogists have conjectured that the greater part of these were not of Vesuvian origin, but thrown up in fragments from some older formation, through which the gaseous explosions burst. But none of the older rocks in Italy, or elsewhere, contain such an assemblage of mineral products; and the hypothesis seems to have been prompted by a disinclination to admit that, in times so recent in the earth's history, the laboratory of Nature could have been so prolific in the creation of new and rare compounds. Had Vesuvius been a volcano of high antiquity, formed when Nature

> Wanton'd as in her prime, and play'd at will
> Her virgin fancies,

it would have been readily admitted that these, or a much greater variety of substances, had been sublimed in the crevices of lava, just as several new earthy and metallic compounds are known to have been produced by fumeroles, since the

* Monticelli and Covelli, Prodrom. della Mineral. Vesuv.

eruption of 1822. But some violent hypothesis must always be resorted to, in order to explain away facts which imply the unimpaired energy of reproductive causes, in our own times.

We have hitherto described the structure of the cone; but a small part only of the ejected matter remains so near to the volcanic orifice. A large portion of sand and scoriæ is borne by the winds and scattered over the surrounding plains, or falls into the sea; and much more is swept down by torrents into the deep, during the intervals, often protracted for many centuries, between eruptions. There, horizontal deposits of tufaceous matter become intermixed with sediment of other kinds, and with shells and corals, and, when afterwards raised, form rocks of a mixed character, such as tuffs, peperinos, and volcanic conglomerates. Some of the lavas, also, of Vesuvius, reach the sea, as do those of almost all volcanos; since they are generally in islands, or bordering the coast. Here they find a bottom already rendered nearly level, for reasons before explained by us, when speaking of deltas. Instead, therefore, of being highly inclined, as around the cone, or in narrow bands as in a valley, they spread out in broad horizontal sheets so long as they retain their fluidity; and this process may probably continue for a considerable time, since, as upon the land, the upper coating of hardened lava protects the liquid and moving mass below from contact with the air, so beneath the sea the same superficial crust may prevent the great body of lava from cooling, and, being pressed upon by the weight of an increasing column of water as the current descends, it is probably squeezed down: thus the subjacent matter, still in a state of fusion, may be made to flow rapidly towards all points of the compass. This would take place the more readily if the Huttonian assumption be true, that lava cools down more slowly under the pressure of a deep sea than in the open air, which was supposed to be a corollary from Sir James Hall's experiments respecting compression, whence it was inferred that vast pressure prevented water from expanding into steam. But even if such be the case, it by no means follows that the heat of the lava could be carried off more slowly than in the air, and in seas of ordinary depth there can be no doubt that the melted matter would cool far more rapidly under water.

Besides the ejections which fall on the cone, and that

much greater mass which finds its way gradually to the neighbouring sea, there is a third portion often of no inconsiderable thickness, composed of alluvions, spread over the valleys and plains at small distances from the volcano. Immense volumes of aqueous vapour are evolved from a crater during eruptions, and often for a long time subsequently to the discharge of scoriæ and lava. These vapours are condensed in the cold atmosphere surrounding the high volcanic peak, and heavy rains are caused sometimes even in countries where, under other circumstances, such a phenomenon is entirely unknown. The floods thus occasioned sweep along the impalpable dust and light scoriæ, till a current of mud is produced, which is called, in Campania, "lava d'acqua," and is often more dreaded than an igneous stream (lava di fuoco), from the greater velocity with which it moves. So late as the 27th of October, 1822, one of these alluvions descended the cone of Vesuvius. After overspreading much cultivated soil, it flowed suddenly into the villages of St. Sebastian and Massa, and, filling the streets and interior of some of the houses, suffocated seven persons. It will therefore happen very frequently, that, towards the base of a volcanic cone, alternations will be found of lava, alluvions, and showers of ashes. To which of these two latter divisions the mass enveloping Herculaneum and Pompeii should be referred, has been a question of the keenest controversy; but the discussion might have been shortened, if the combatants had reflected that, whether volcanic sand and ashes were conveyed to the towns by running water, or through the air, during an eruption, the interior of buildings, so long as the roofs remained entire, and all underground vaults and cellars, could only be filled by an *alluvion*. We learn from history, that a heavy shower of sand, pumice, and lapilli, sufficiently great to render Pompeii and Herculaneum uninhabitable, fell for eight successive days and nights, in the year 79, accompanied by violent rains. We ought, therefore, to find a very close resemblance between the strata covering these towns, and those composing the minor cones of the Phlegræan Fields, accumulated rapidly, like Monte Nuovo, during a continued shower of ejected matter; with this difference, that the strata incumbent on the cities would be horizontal, whereas those in the cones are highly inclined, and that large angular fragments of rock, which are thrown out near the vent, would be wanting at a distance, where small lapilli only would be found. Accordingly, with these exceptions, no identity can be more perfect than the form and distribution of the matter at the base of Monte Nuovo, as laid open by the encroaching sea, and the appearance of the beds superimposed on Pompeii. That city is covered with numerous alternations of different horizontal beds of tuff and lapilli, for the most part thin, and subdivided into very fine layers. I observed the following section near the Amphitheatre, in November, 1828,—(descending series).

	Feet.	Inches.
1. Black sparkling sand from the eruption of 1822, containing minute regularly-formed crystals of augite and tourmaline from		2 to 3*
2. Vegetable mould	3	0
3. Brown incoherent tuff full of *pisolitic globules* in layers, from half an inch to 3 inches in thickness	1	6
4. Small scoriæ and white lapilli	0	3
5. Brown earthy tuff with numerous pisolitic globules .	0	9
6. Brown earthy tuff with lapilli divided into layers . .	4	0
7. Layer of whitish lapilli	0	1
8. Grey solid tuff	0	3
9. Pumice and white lapilli	0	3
	10	4

Many of the ashes in these beds are vitrified and harsh to the touch. Crystals of leucite, both fresh and farinaceous, have been found intermixed†. The depth of the bed of ashes above the houses is variable, but seldom exceeds twelve or fourteen feet, and it is said, that the higher part of the Amphitheatre always projected above the surface; though, if this were the case, it seems to be inexplicable that the city should never have been discovered till the year 1750. It will be observed, in the above section, that two of the brown half-consoli-

* The last great eruption, in 1822, only caused a covering of a few inches thick on Pompeii. Several feet are mentioned by Mr. Forbes—Ed. Journ. of Science, No. xix., p. 131, Jan. 1829; but he must have measured in spots where it had drifted. The dust and ashes were five feet thick at the top of the crater, and decreased gradually to ten inches at Torre del Annunziata. The size and weight of the ejected fragments diminished very regularly in the same continuous stratum as the distance from the centre of projection was greater.

† Forbes, Ed. Journ. of Sci., No. xix., p. 130, Jan. 1829.

dated tuffs are filled with small pisolitic globules. It is surprising that this circumstance is not alluded to in the animated controversy which the Royal Academy of Naples maintained with one of their members, Signor Lippi, as to the origin of the strata incumbent on Pompeii. The mode of aggregation of these globules has been fully explained by Mr. Scrope, who saw them formed in great numbers, in 1822, by rain falling during the eruption on fine volcanic sand, and sometimes, also, beheld them produced like hail in the air, by the mutual attraction of the minutest particles of fine damp sand. Their occurrence, therefore, agrees remarkably well with the account of heavy rain, and showers of sand and ashes, recorded in history, and is opposed to the theory of an alluvion brought from a distance by a flood of water.

Lippi entitled his work "Fu il fuoco o l' acqua che sotterrò Pompei ed Ercolano?"* and he contended that neither were the two cities destroyed in the year 79, nor by a volcanic eruption, but purely by the agency of water charged with transported matter. His Letters, wherein he endeavoured to dispense, as far as possible, with igneous agency, even at the foot of the volcano, were dedicated with great propriety to Werner, and afford an amusing illustration of the polemic style in which geological writers of that day indulged themselves. His arguments were partly of an historical nature, derived from the silence of contemporary historians, respecting the fate of the cities which, as we have already stated, is most remarkable; and were partly drawn from physical proofs. He pointed out with great clearness the resemblance of the tufaceous matter in the vaults and cellars at Herculaneum and Pompeii to aqueous alluvions, and its distinctness from ejections which had fallen through the air. Nothing, he observed, but moist, pasty matter could have received the impression of a woman's breast, which was found in a vault at Pompeii, or have given the cast of a statue discovered in the theatre at Herculaneum. It was objected to him, that the heat of the tuff in Herculaneum and Pompeii was proved by the carbonization of the timber, corn, papyrus-rolls, and other vegetable substances there discovered: but Lippi replied with truth, that the papyri would have been

* Napoli, 1816.

burnt up, if they had come in contact with fire, and that their being only carbonized, was a clear demonstration of their having been enveloped, like fossil wood, in a sediment deposited from water. The Academicians, in their report on his pamphlet, assert, that when the Amphitheatre was first cleared out, the matter was arranged, on the steps, in a succession of concave layers, accommodating themselves to the interior form of the building, just as snow would lie if it had fallen there. This observation is highly interesting, and points to the difference between the stratification of ashes in an open building, and in the interior of edifices and cellars. Nor ought we to call this allegation in question, because it could not be substantiated at the time of the controversy, when the matter was all removed; although Lippi took advantage of this removal, and met the argument of his antagonists by requiring them to prove the fact.

There is decisive evidence that no stream of lava has ever reached Pompeii since it was first built, although the foundations of the town stand upon the old leucitic lava of Somma; several of whose streams, with tuff interposed, have been cut through in excavations. At Herculaneum the case is different, although the substance which fills the interior of the houses and the vaults must have been introduced in a state of mud, like that found in similar situations in Pompeii: the superincumbent mass differs wholly in composition and thickness. Herculaneum was situated several miles nearer to the volcano, and has therefore been always more exposed to be covered, not only by showers of ashes, but by alluvions and streams of lava. Accordingly, masses of both have accumulated on each other above the city, to a depth of nowhere less than seventy, and in many places of one hundred and twelve feet*. The tuff which envelops the buildings consists of comminuted volcanic ashes, mixed with pumice. A mask imbedded in this matrix has left a cast, the sharpness of which was compared by Hamilton to those in Paris plaster; nor was the mask in the least degree scorched, as we might expect it to have been, if it had been imbedded in heated matter. This tuff is porous, and, when first excavated, is soft and easily worked, but acquires

* Hamilton's Observations on Mount Vesuvius, p. 94. London, 1774.

a considerable degree of induration on exposure to the air. Above this lowest stratum is placed, according to Hamilton, "the matter of six eruptions," each separated from the other by veins of good soil. In these soils Lippi informs us, that he collected a considerable number of land shells—an observation which is no doubt correct, for we know that in Italy several species burrow annually, in certain seasons, to the depth of five feet and more from the surface. Della Torre also informs us, that there is in one part of this superimposed mass a bed of true siliceous lava (*lava di pietra dura*); and, as no such current is believed to have flowed till near one thousand years after the destruction of Herculaneum, we must conclude, that the origin of a large part of the covering of Herculaneum was long subsequent to the first inhumation of the place. That city, as well as Pompeii, was a sea-port. Herculaneum is still very near the shore, but a tract of land, a mile in length, intervenes between the borders of the Bay of Naples and Pompeii. In both cases the gain of land is due to the filling up of the bed of the sea with volcanic matter, and not to elevation by earthquakes, for there has been no change in the relative level of land and sea. Pompeii stood on a slight eminence composed of the lavas of the ancient Vesuvius, and flights of steps led down to the water's edge. The lowermost of these steps are said to be still on an exact level with the sea.

After these observations on the nature of the strata enveloping and surrounding the cities, we may proceed to consider their internal condition and contents, so far at least as they offer facts of geological interest. Notwithstanding the much greater depth at which Herculaneum was buried, it was discovered before Pompeii, by the accidental circumstance of a well being sunk, in 1713, which came right down upon the theatre, where the statues of Hercules and Cleopatra were soon found. Whether this city or Pompeii, both of them founded by Greek colonies, was the most considerable, is not yet determined; but both are mentioned by ancient authors as among the seven most flourishing cities in Campania. The walls of Pompeii were three miles in circumference; but we have, as yet, no certain knowledge of the dimensions of Herculaneum. In the latter place the theatre alone is open for inspection; the Forum, Temple of Jupiter, and other buildings, having been filled up with rubbish as the workmen proceeded, owing to the difficulty of removing it from so great a depth below ground. Even the theatre is only seen by torch-light, and the most interesting information, perhaps, which the geologist obtains there, is the continual formation of stalactite in the galleries cut through the tuff; for there is a constant percolation of water charged with carbonate of lime mixed with a small portion of magnesia. Such mineral waters must, in the course of time, create great changes in many rocks: and we cannot but perceive the unreasonableness of the expectations of some geologists, that volcanic rocks of remote eras should accord precisely with those of modern date; since it is obvious that many of those produced in our own time will not long retain the same aspect and composition.

Both at Herculaneum and Pompeii, temples have been found with inscriptions commemorating their having been rebuilt after they were thrown down by an earthquake*. This earthquake happened in the reign of Nero, sixteen years before the inhumation of the cities. In Pompeii, one-fourth of which is now laid open to the day, both the public and private buildings bear testimony to the catastrophe. The walls are rent, and in many places traversed by fissures still open. Columns are lying on the ground only half hewn from huge blocks of travertin, and the temple for which they were designed is seen half repaired. In some few places the pavement had sunk in, but in general it was undisturbed, consisting of great flags of lava, in which two immense ruts have been worn by the constant passage of carriages through the narrow street. When the hardness of the stone is considered, the continuity of these ruts from one end of the town to the other is not a little remarkable, for there is nothing of the kind in the oldest pavements of modern cities.

A very small number of skeletons have been discovered in either city; and it is clear that the great mass of inhabitants not only found time to escape, but also to carry with them the principal part of their valuable effects. In the barracks at Pompeii were the skeletons of two soldiers chained to the stocks, and in the vaults of a country-house

* Swinburne and Lalande—Paderni, Phil. Trans., 1758, vol. 50, p. 619.

in the suburbs, were the skeletons of seventeen persons who appear to have fled there to escape from the shower of ashes. They were found inclosed in an indurated tuff, and in this matrix was preserved a perfect cast of a woman, perhaps the mistress of the house, with an infant in her arms. Although her form was imprinted on the rock, nothing but the bones remained. To these a chain of gold was suspended, and rings with jewels were on the fingers of the skeleton. Against the sides of the same vault was ranged a long line of earthen amphoræ.

The writings scribbled by the soldiers on the walls of their barracks, and the names of the owners of each house written over the doors, are still perfectly legible. The colours of fresco paintings on the stuccoed walls in the interior of buildings are almost as vivid as if they were just finished. If these artificial colours, therefore, have stood, it is not wonderful that those of shells should have remained unfaded. There are public fountains decorated with shells laid out in patterns in the same fashion as those now seen in the town of Naples; and in the room of a painter who was perhaps a naturalist, a large collection of shells was found, comprising a great variety of Mediterranean species, in as good a state of preservation as if they had remained for the same number of years in a museum. A comparison of these remains with those found so generally in a fossil state would not assist us in obtaining the least insight into the time required to produce a certain degree of decomposition or mineralization; for although, under favourable circumstances, much greater alteration might doubtless have been brought about in a shorter period, yet the example before us shows that an inhumation of seventeen centuries may sometimes effect nothing towards the reduction of shells and several other bodies to the state in which fossils are usually found.

The wooden beams in the houses at Herculaneum are black on the exterior, but when cleft open they appear to be almost in the state of ordinary wood, and the progress made by the whole mass towards the state of lignite is scarcely appreciable. Some animal and vegetable substances of more perishable kinds have of course suffered much change and decay, yet the state of conservation of these is truly remarkable. Fishing-nets are very abundant in both cities, often quite entire; and their number at Pompeii is the more interesting from the sea being now, as we stated, a mile distant. Linen has been found at Herculaneum, with the texture well defined; and in a fruiterer's shop in that city were discovered vessels full of almonds, chestnuts, walnuts, and fruit of the "carubiere," all distinctly recognizable from their shape. A loaf, also, still retaining its form, was found in a baker's shop, with his name stamped upon it thus: "Eleris Q. Crani Riser." On the counter of an apothecary was a box of pills converted into a fine earthy substance; and by the side of it a small cylindrical roll, evidently prepared to be cut into pills. By the side of these was a jar containing medicinal herbs. In 1827, moist olives were found in a square glass case, and "caviare," or roe of a fish, in a state of wonderful preservation. An examination of these curious condiments has been published by Covelli, of Naples, and they are preserved hermetically sealed in the museum there *.

There is a marked difference in the condition and appearance of the animal and vegetable substances found in Pompeii and Herculaneum; those of Pompeii being penetrated by a grey pulverulent tuff, those in Herculaneum seeming to have been first enveloped by a paste which consolidated round them, and then allowed them to become slowly carbonized. Some of the rolls of papyrus at Pompeii still retain their form; but the writing, and indeed almost all the vegetable matter, appear to have vanished and to have been replaced by volcanic tufa somewhat pulverulent. At Herculaneum the earthy matter has scarcely ever penetrated; and the vegetable substance of the papyrus has become a thin friable black matter, almost resembling in appearance the tinder which remains when stiff paper has been burnt, in which the letters may still be sometimes traced. The small bundles, composed of five or six rolls tied up together, had sometimes lain horizontally, and were pressed in that direction, but sometimes they had been placed in a vertical position. Small tickets were attached to each bundle, on which the title of the work was inscribed. In one case only have the sheets been found with writing on both sides of the pages. So numerous are the obliterations and corrections, that many must have been

* Mr. Forbes, Edin. Journ. of Sci., No. xix., p. 130, Jan., 1829.

original manuscripts. The variety of hand-writings is quite extraördinary: almost all are written in Greek, but there are a few in Latin. They were all found in the library of one private individual; and the titles of four hundred of those least injured, which have been read, are found to be unimportant works, but all entirely new, chiefly relating to music, rhetoric, and cookery. There are two volumes of Epicurus "On Nature," and the others are mostly by writers of the same school, only one fragment having been discovered, by an opponent of the Epicurean system, Crisippus *. In the opinion of some antiquaries, not one-hundredth part of the city has yet been explored; and the quarters hitherto cleared out, at great expense, are those where there was the least probability of discovering manuscripts.

As Italy could already boast splendid Roman amphitheatres and Greek temples, it was a matter of secondary interest to add to their number those in the dark and dripping galleries of Herculaneum; and having so many of the masterpieces of ancient art, we could have dispensed with the inferior busts and statues which could alone have been expected to reward our researches in the ruins of a provincial town. But from the moment that it was ascertained that rolls of papyrus preserved in this city could still be decyphered, every exertion ought to have been steadily and exclusively directed towards the discovery of other libraries. *Private dwellings* should have been searched, and no labour and expense should have been consumed in examining public edifices. A small portion of that zeal and enlightened spirit which prompted the late French and Tuscan expedition to Egypt, might, long ere this, in a country nearer home, have snatched from oblivion some of the lost works of the Augustan age, or of the most eminent Greek historians and philosophers. A single roll of papyrus might have disclosed more matter of intense interest than all that was ever written in hieroglyphics†.

* In one of the manuscripts which was in the hands of the interpreters when I visited the museum, the author indulges in the speculation that all the Homeric personages were allegorical—that Agamemnon was the ether, Achilles the sun, Helen the earth, Paris the air, Hector the moon, &c.

† During my stay at Naples, in 1828, the Neapolitan Government, after having discontinued operations for many years, cleared out a small portion of Herculaneum, near the sea, where the covering was least thick. After this expense

Besides the cities already mentioned, Stabiæ, a small town about six miles from Vesuvius, and near the site of the modern Castel-a-Mare (see map, plate 3), was overwhelmed during the eruption of 79. Pliny mentions that, when his uncle was there, he was obliged to make his escape, so great was the quantity of falling stones and ashes. In the ruins of this place, a few skeletons have been found buried in volcanic ejections, together with some antiquities of no great value, and rolls of papyrus, which, like those of Pompeii, were illegible.

Of the towns hitherto mentioned, Herculaneum alone has been overflowed by a stream of melted matter; but this did not, as we have seen, enter or injure the buildings which were previously enveloped and covered over with tuff. But burning torrents have often taken their course through the streets of Torre del Greco, and consumed or inclosed a large portion of the town in solid rock. It seems probable that the destruction of three thousand of its inhabitants, in 1631, which some accounts attribute to boiling water, was principally due to one of those alluvions which we before mentioned; but, in 1737, the lava itself flowed through the eastern side of the town, and afterwards reached the sea: and, in 1794, another current rolling over the western side, filled the street and houses, and killed more than four hundred persons. The main street is now quarried through this lava, which supplied building-stones for new houses erected where others had been annihilated. The church was half buried in a rocky mass, but the upper portion served as the foundation of a new edifice. The number of the population at present is estimated at fifteen thousand; and a satisfactory answer may readily be returned to those who inquire how the inhabitants can be so "inattentive to the voice of time and the warnings of Nature*," as to rebuild their dwellings on a spot so often devastated. No neighbouring site unoccupied by a town, or which would not be equally insecure, combines the same advantages of proximity to the

had been incurred, it was discovered that the whole of the ground had been previously examined, near a century before, by the French Prince d'Elbeuf, who had removed everything of value! The want of system with which operations have always been, and still are, carried on is such, that we may expect similar blunders to be made continually.

* Sir H. Davy, Consolations in Travel, p. 66.

capital, to the sea, and to the rich lands on the flanks of Vesuvius. If the present population were exiled, they would immediately be replaced by another, for the same reason that the Maremma of Tuscany and the Campagna di Roma will never be depopulated, although the malaria fever commits more havoc in a few years than the Vesuvian lavas in as many centuries. The district around Naples supplies one, amongst innumerable examples, that those regions where the surface is most frequently renewed, and where the renovation is accompanied, at different intervals of time, by partial destruction of animal and vegetable life, may nevertheless be amongst the most habitable and delightful on our globe. We have already made a similar remark when speaking of tracts where aqueous causes are now most active; and the observation applies as well to parts of the surface which are the abode of aquatic animals, as to those which support terrestrial species. The sloping sides of Vesuvius give nourishment to a vigorous and healthy population of about eighty thousand souls; and the surrounding hills and plains, together with several of the adjoining isles, owe the fertility of their soil to matter ejected by prior eruptions. Had the fundamental limestone of the Apennines remained uncovered throughout the whole area, the country could not have sustained a twentieth part of its present inhabitants. This will be apparent to every geologist who has marked the change in the agricultural character of the soil the moment he has passed the utmost boundary of the volcanic ejections, as when, for example, at the distance of about seven miles from Vesuvius, he leaves the plain and ascends the declivity of the Sorrentine Hills.

Yet favoured as this region has been by Nature from time immemorial, the signs of the changes imprinted on it during the period that it has served as the habitation of man, may appear in after-ages to indicate a series of unparalleled disasters. Let us suppose that at some future time the Mediterranean should form a gulf of the great ocean, and that the tidal current should encroach on the shores of Campania, as it now advances upon the eastern coast of England: the geologist will then behold the towns already buried, and many more which will inevitably be entombed hereafter, laid open in the steep cliffs, where he will discover streets superimposed above each other, with thick intervening strata of tuff or lava—some unscathed by fire, like those of Herculaneum and Pompeii, others half melted down like those of Torre del Greco, or shattered and thrown about in strange confusion like Tripergola. Among the ruins will be seen skeletons of men, and impressions of the human form stamped in solid rocks of tuff. Nor will the signs of earthquakes be wanting. The pavement of part of the Domitian Way, and the Temple of the Nymphs, submerged at high tide, will be uncovered at low water, the columns remaining erect and uninjured; while other temples which had once sunk down, like that of Serapis, will be found to have been upraised again by subsequent movements. If they who study these phenomena, and speculate on their causes, assume that there were periods when the laws of Nature differed from those established in their own time, they will scarcely hesitate to refer the wonderful monuments in question to those primeval ages. When they consider the numerous proofs of reiterated catastrophes to which the region was subject, they may, perhaps, commiserate the unhappy fate of beings condemned to inhabit a planet during its nascent and chaotic state, and feel grateful that their favoured race escaped such scenes of anarchy and misrule.

Yet what was the real condition of Campania during those years of dire convulsion? "A climate where heaven's breath smells sweet and wooingly—a vigorous and luxuriant nature unparalleled in its productions—a coast which was once the fairy land of poets, and the favourite retreat of great men. Even the tyrants of the creation loved this alluring region, spared it, adorned it, lived in it, died in it *." The inhabitants, indeed, have enjoyed no immunity from the calamities which are the lot of mankind; but the principal evils which they have suffered must be attributed to moral, not to physical causes—to disastrous events over which man might have exercised a control, rather than to the inevitable catastrophes which result from subterranean agency. When Spartacus encamped his army of ten thousand gladiators in the old extinct crater of Vesuvius, the volcano was more justly a subject of terror to Campania, than it has ever been since the rekindling of its fires.

* Forsyth's Italy, vol. ii.

CHAPTER XXI.

External physiognomy of Etna—Minor cones produced by lateral eruptions—Successive obliteration of these cones—Early eruptions of Etna—Monti Rossi thrown up in 1669—Great fissure of S. Lio—Towns overflowed by lava—Part of Catania destroyed—Mode of the advance of a current of lava—Excavation of a church under lava—Series of subterranean caverns—Linear direction of cones formed in 1811 and 1819—Flood produced in 1755 by the melting of snow during an eruption—A glacier covered by a lava-stream on Etna—Volcanic eruptions in Iceland—New island thrown up in 1783—Two lava-currents of Skaptár Jokul in the same year—Their immense volume—Eruption of Jorullo in Mexico—Humboldt's Theory respecting the convexity of the Plain of Malpais.

As we have entered into a detailed historical account of the changes in the volcanic district round Naples, our limits will only permit us to allude in a cursory manner to some of the circumstances of principal interest in the history of other volcanic mountains. After Vesuvius, our most authentic records relate to Etna, which rises near the sea in solitary grandeur to the height of nearly eleven thousand feet *, the mass being chiefly composed of volcanic matter ejected above the surface of the water. The base of the cone is almost circular, and eighty-seven English miles in circumference; but if we include the whole district over which its lavas extend, the circuit is probably twice that extent. The cone is divided by Nature into three distinct zones, called the *fertile*, the *woody*, and the *desert* regions. The first of these, comprising the delightful country around the skirts of the mountain, is well cultivated, thickly inhabited, and covered with olives, vines, corn, fruit-trees, and aromatic herbs. Higher up, the woody region encircles the mountain—an extensive forest, six or seven miles in width, affording pasturage for numerous flocks. The trees are of various species, the chestnut, oak, and pine, being most luxuriant; while, in some tracts, are groves of cork

* According to Captain Smyth (Sicily and its Islands, p. 145), its height is 10,874 feet.

and beech. Above the forest is the desert region, a waste of black lava and scoriæ; where, on a kind of plain, rises the cone to the height of about eleven hundred feet, from which sulphureous vapours are continually evolved. The most grand and original feature in the physiognomy of Etna are the multitude of minor cones which are distributed over its flanks, and which are most abundant in the woody region. These, although they appear but trifling irregularities when viewed from a distance as subordinate parts of so imposing and colossal a mountain, would, nevertheless, be deemed hills of considerable altitude in almost any other region.

Without enumerating numerous monticules of ashes thrown out at different points, there are about eighty of these secondary volcanos, of considerable dimensions; fifty-two on the west and north, and twenty-seven on the east side of Etna. One of the largest, called Monte Minardo, near Bronte, is upwards of seven hundred feet in height: and a double hill near Nicolosi called Monti Rossi, formed in 1669, is four hundred and fifty feet high, and the base two miles in circumference; so that it somewhat exceeds in size Monte Nuovo, before described. Yet it ranks only as a cone of the second magnitude amongst those produced by the lateral eruptions of Etna. On looking down from the lower borders of the desert region, these volcanos present us with one of the most beautiful and characteristic scenes in Europe. They afford every variety of height and size, and are arranged in beautiful and picturesque groups. However uniform they may appear when seen from the sea, or the plains below, nothing can be more diversified than their shape when we look from above into their craters, one side of which is generally broken down. There are, indeed, few objects in Nature more picturesque than a wooded volcanic crater. The cones situated in the higher parts of the forest zone are chiefly clothed with lofty pines; while those at a lower elevation are adorned with chestnuts, oak, beech, and holm.

The history of the eruptions of Etna, imperfect and interrupted as it is, affords, nevertheless, a full insight into the manner in which the whole mountain has successively attained its present magnitude and internal structure. The principal cone has more than once fallen in, and been reproduced. In 1444 it was three hundred and twenty feet high, and fell in after the earthquakes of 1537. In the year 1693, when a violent earthquake shook the whole of Sicily, and killed sixty thousand persons, the cone lost so much of its height, says Boccone, that it could not be seen from several places in Valdemone, whence it was before visible. The greater number of eruptions happen either from the great crater, or from lateral openings in the desert region. When hills are thrown up in the middle zone, and project beyond the general level, they gradually lose their height during subsequent eruptions; for when lava runs down from the upper parts of the mountain, and encounters any of these hills, the stream is divided, and flows round them so as to elevate the gently-sloping grounds from which they rise. In this manner a deduction is often made at once of twenty or thirty feet, or even more, from their height. Thus, one of the minor cones, called Monte Peluso, was diminished in altitude by a great lava-stream which encircled it in 1444; and another current has recently taken the same course—yet this hill still remains four or five hundred feet high. There is a cone called Monte Nucilla, near Nicolosi, round the base of which several successive currents have flowed and showers of ashes fallen since the time of history, till at last, during an eruption in 1536, the surrounding plain was so raised, that the top of the cone alone was left projecting above the general level. Monte Nero, situated above the Grotta dell' Capre, was in 1766 almost submerged by a current; and Monte Capreolo afforded, in the year 1669, a curious example of one of the last stages of obliteration; for a lava-stream descending on a high ridge which had been built up by the continued superposition of successive lavas, flowed directly into the crater, and nearly filled it. The lava, therefore, of each new lateral cone tends to detract from the relative height of lower cones above their base: so that the flanks of Etna, sloping with a gentle inclination, envelop in succession a great multitude of minor volcanos, while new ones spring up from time to time; and this has given to the older parts of the mountain, as seen in some sections two or three thousand feet perpendicular, a complex and highly interesting internal structure.

Etna appears to have been in activity from the earliest times of tradition; for Diodorus Siculus mentions an eruption which caused a district to be deserted by the Sicani before the Trojan war. Thucydides informs us *, that between the colonization of Sicily by the Greeks, and the commencement of the Peloponnesian war in the year 431 B.C., three eruptions had occurred. The last of these happened in the year 427 B.C., and ravaged the environs of Catania; and was probably that so poetically described by Pindar in his first Pythian ode.

The great eruption which happened in the year 1669 is the first to which we shall call the reader's attention. An earthquake had levelled to the ground all the houses in Nicolosi, a town situated near the lower margin of the woody region, about twenty miles from the summit of Etna, and ten from the sea at Catania. Two gulphs then opened near that town, from whence sand and scoriæ were thrown up in such quantity, that, in the course of three or four months, a double cone was formed, called Monti Rossi, about four hundred and fifty feet

No. 14.

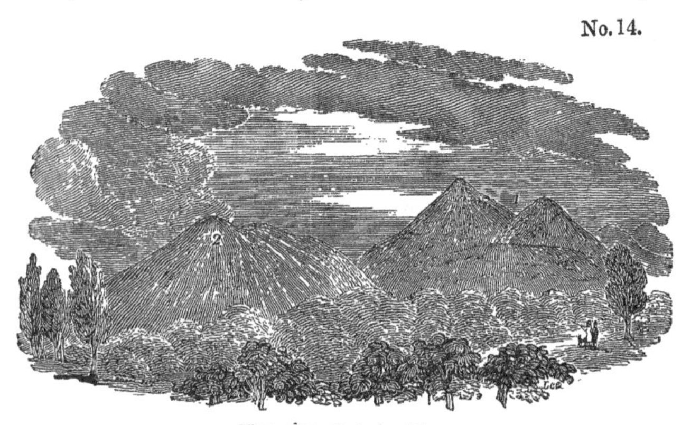

Minor cones on the flanks of Etna.

1. Monti Rossi, near Nicolosi, formed in 1669. 2. Vampeluso ? †

high. But the most extraordinary phenomenon occurred at the commencement of the convulsion in the neighbouring plain of S. Lio. A fissure six feet broad, and of unknown depth, opened with a loud crash, and ran, in a somewhat tortuous course, to within a mile of the summit of Etna. Its direction was from north to south, and its length twelve miles. It

* Book III., towards the end.

† The hill which I have here introduced was called by my guide Vampolara, but the name given in the text is the nearest to this which I find in Gemmellaro's Catalogue of Minor Cones.

emitted a most vivid light. Five other parallel fissures of considerable length afterwards opened one after the other, and emitted smoke, and gave out bellowing sounds which were heard at the distance of forty miles. This case seems to present the geologist with an illustration of the manner in which those continuous dikes of vertical porphyry were formed which are seen to traverse some of the older lavas of Etna; for the light emitted from the great rent of S. Lio appears to indicate that it was filled to a certain height with incandescent lava, probably to the height of an orifice not far distant from Monti Rossi, which at that time opened and poured out a lava-current. This lava soon reached a minor cone called Mompiliere, at the base of which it entered a subterranean grotto communicating with a suite of caverns which are common in the lavas of Etna. Here it appears to have melted down some of the vaulted foundations of the hill, so that the whole cone became slightly depressed and traversed by numerous open fissures. The lava, after overflowing fourteen towns and villages, some having a population of between three and four thousand inhabitants, arrived at length at the walls of Catania. These had been purposely raised to protect the city; but the burning flood accumulated till it rose to the top of the rampart, which was sixty feet in height, and then it fell in a fiery cascade and overwhelmed part of the city. The wall, however, was not thrown down, but was discovered long afterwards by excavations made in the rock by the Prince of Biscari; so that the traveller may now see the solid lava curling over the top of the rampart as if still in the very act of falling.

This great current had performed a course of fifteen miles before it entered the sea, where it was still six hundred yards broad and forty feet deep. It covered some territories in the environs of Catania, which had never before been visited by the lavas of Etna. While moving on, its surface was in general a mass of solid rock; and its mode of advancing, as is usual with lava-streams, was by the occasional fissuring of the solid walls. A gentleman of Catania, named Pappalardo, desiring to secure the city from the approach of the threatening torrent, went out with a party of fifty men whom he had dressed in skins to protect them from the heat, and armed with iron crows and hooks. They broke open one of the solid walls which flanked the current near Belpasso, and immediately forth issued a rivulet of melted matter which took the direction of Paternò; but the inhabitants of that town, being alarmed for their safety, took up arms and put a stop to farther operations*. As another illustration of the solidity of the walls of an advancing lava-stream, we may mention an adventure related by Recupero, who, in 1766, had ascended a small hill formed of ancient volcanic matter, to behold the slow and gradual approach of a fiery current, two miles and a half broad; when suddenly two small threads of liquid matter issuing from a crevice detached themselves from the main stream, and ran rapidly towards the hill. He and his guide had just time to escape, when they saw the hill, which was fifty feet in height, surrounded, and in a quarter of an hour melted down into the burning mass, so as to flow on with it. But it must not be supposed that this complete fusion of rocky matter coming in contact with lava is of universal, or even common occurrence. It probably happens when fresh portions of incandescent matter come successively in contact with fusible materials. In many of the dikes which intersect the tuffs and lavas of Etna, there is scarcely any perceptible alteration effected by heat on the edges of the horizontal beds, in contact with the vertical and more crystalline mass. On the site of Mompiliere, one of the towns overflowed in the great eruption above described, an excavation was made in 1704; and by immense labour the workmen reached, at the depth of thirty-five feet, the gate of the principal church, where there were three statues, held in high veneration. One of these, together with a bell, some money, and other articles, were extracted in a good state of preservation from beneath a great arch formed by the lava. It seems very extraordinary that any works of art, not encased with tuff, like those in Herculaneum, should have escaped fusion in hollow spaces left open in this lava-current, which was so hot at Catania eight years after it entered the town, that it was impossible to hold the hand in some of the crevices.

We mentioned the entrance of the lava-stream into a subterranean grotto, whereby the foundations of a hill were par-

* Ferrara, Descriz. dell' Etna, p. 108.

tially undermined. Such underground passages are among the most curious features on Etna, and appear to have been produced by the hardening of the lava, during the escape of great volumes of elastic fluids, which are often discharged for many days in succession, after the crisis of the eruption is over. Near Nicolosi, not far from Monti Rossi, one of these great openings may be seen, called the Fossa della Palomba, 625 feet in circumference at its mouth, and 78 deep. After reaching the bottom of this, we enter another dark cavity, and then others in succession, sometimes descending precipices by means of ladders. At length the vaults terminate in a great gallery ninety feet long, and from fifteen to fifty broad, beyond which there is still a passage, never yet explored; so that the extent of these caverns remains unknown*. The walls and roofs of these great vaults are composed of rough and bristling scoriæ, of the most fantastic forms.

We shall now proceed to offer some observations on the two last eruptions in 1811 and 1819. It appears, from the relation of Signor Gemmellaro, who witnessed the phenomena, that the great crater in 1811 testified, by its violent detonations, that the lava had ascended to near the summit of the mountain, by its central duct. A violent shock was then felt, and a stream broke out from the side of the cone, at no great distance from its apex. Shortly after this had ceased to flow, a second stream burst forth at another opening, considerably below the first; then a third still lower, and so on till seven different issues had been thus successively formed, all lying upon the same straight line. It has been supposed that this line was a perpendicular rent in the internal framework of the mountain, which rent was probably not produced at one shock, but prolonged successively downwards, by the lateral pressure and intense heat of the internal column of lava, as it subsided by gradual discharge through each vent†.

In 1819 three large mouths or caverns opened very near those which were formed in the eruptions of 1811, from which flames, red hot cinders, and sand, were thrown up with loud explosions. A few minutes afterwards another mouth opened below, from which flames and smoke issued;

* Ferrara, Descriz. dell' Etna. Palermo, 1818. † Scrope on Volcanos, p. 153.

and finally a fifth, lower still, whence a torrent of lava flowed which spread itself with great velocity over the valley 'del Bove.' This stream flowed two miles in the first twenty-four hours, and nearly as far in the succeeding day and night. The three original mouths at length united into one large crater, and sent forth lava, as did the four inferior apertures, so that an enormous torrent poured down the great valley 'del Bove.' When it arrived at a vast and almost perpendicular precipice, at the head of the valley of Calanna, it poured over in a cascade, and, being hardened in its descent, made an inconceivable crash as it was dashed against the bottom. So immense was the column of dust raised by the abrasion of the tufaceous hill over which the hardened mass descended, that the Catanians were in great alarm, supposing a new eruption to have burst out in the woody region, exceeding in violence that near the summit of Etna.

Of the cones thrown up during this eruption, not more than two are of sufficient magnitude to be numbered among those eighty which we before reckoned as adorning the flanks of Etna. The surface of the lava which deluged the valley 'del Bove' consists of rocky and *angular* blocks, tossed together in the utmost disorder. Nothing can be more rugged, or more unlike the smooth and even superficies which those who are unacquainted with volcanic countries may have pictured to themselves, in a mass of matter which had consolidated from a liquid state. Mr. Scrope observed this current in the year 1819, slowly progressing down a considerable slope, at the rate of about a yard an hour, nine months after its first emission. The lower stratum being arrested by the resistance of the ground, the upper or central part gradually protruded itself, and being unsupported fell down. This in its turn was covered by a mass of more liquid lava, which swelled over it from above. The current had all the appearance of a huge heap of rough and large cinders rolling over and over upon itself by the effect of an extremely slow propulsion from behind. The contraction of the crust as it solidified, and the friction of the scoriform cakes against one another, produced a crackling sound. Within the crevices a dull red heat might be seen by night, and vapour issuing in considerable quantity was visible by day*.

* Scrope, on Volcanos, p. 102.

The erosive and transporting power of running water is rarely exerted on Etna with great force, the rain which falls being immediately imbibed by the porous lavas; so that, vast as is the extent of the mountain, it feeds only a few small rivulets, and these, even, are dry throughout the greater portion of the year. The enormous rounded boulders, therefore, of trachyte and basalt, a line of which can be traced from the sea, from near Giardini, by Mascali, and Zafarana, to the valley 'del Bove,' would offer a perplexing problem to the geologist, if history had not preserved the memorials of a tremendous flood which happened in this district in the year 1755. It appears that two streams of lava flowed in that year, on the 2nd of March, from the highest crater: they were immediately precipitated upon an enormous mass of snow, which then covered the whole mountain, and was extremely deep near the summit. The sudden melting of this frozen mass, by a fiery torrent three miles in length, produced a frightful inundation, which devastated the sides of the mountain for eight miles in length, and afterwards covered the lower flanks of Etna, where they were less steep, together with the plains near the sea, with great deposits of sand, scoriæ, and blocks of lava. Many absurd stories circulated in Sicily respecting this event, such as that the water was boiling, and that it was vomited from the highest crater; that it was as salt as the sea, and full of marine shells; but these were mere inventions, to which Recupero, although he relates them as tales of the mountaineers, seems to have attached rather too much importance. Floods of considerable violence have been sometimes produced on Etna, by the fall of heavy rains, aided, probably, by the melting of snow. By this cause alone, in 1761, sixty of the inhabitants of Acicatena were killed, and many of their houses swept away*.

A remarkable discovery has lately been made on Etna of a great mass of ice, preserved for many years, perhaps for centuries from melting, by the singular event of a current of red hot lava having flowed over it. The following are the facts in attestation of a phenomenon which must at first sight appear of so paradoxical a character. The extraordinary heat experienced in the South of Europe, during the summer

* Ferrara, Descriz. dell' Etna, p. 116.

and autumn of 1828, caused the supplies of snow and ice which had been preserved in the spring of that year for the use of Catania and the adjoining parts of Sicily and the island of Malta, to fail entirely. Considerable distress was felt for the want of a commodity regarded in these countries as one of the necessaries of life rather than an article of luxury, and on the abundance of which in some large cities the salubrity of the water and the general health of the community is said in some degree to depend. The magistrates of Catania applied to Signor M. Gemmellaro, in the hope that his local knowledge of Etna might enable him to point out some crevice or natural grotto on the mountain, where drift snow was still preserved. Nor were they disappointed; for he had long suspected that a small mass of perennial ice at the foot of the highest cone was part of a larger and continuous glacier covered by a lava-current. Having procured a large body of workmen, he quarried into this ice, and proved the superposition of the lava for several hundred yards, so as completely to satisfy himself that nothing but the subsequent flowing of the lava over the ice could account for the position of the glacier. Unfortunately for the geologist, the ice was so extremely hard, and the excavation so expensive, that there is no probability of the operations being renewed. On the first of December, 1828, I visited this spot, which is on the south-east side of the cone, and not far above the Casa Inglese, but the fresh snow had already nearly filled up the new opening, so that it had only the appearance of the mouth of a grotto. I do not, however, question the accuracy of the conclusion of Signor Gemmellaro, who being well acquainted with all the appearances of drift snow in the fissures and cavities of Etna, had recognized, even before the late excavations, the peculiarity of the position of the ice in this locality. We may suppose, that, at the commencement of the eruption, a deep mass of drift snow had been covered by volcanic sand showered down upon it before the descent of the lava. A dense stratum of this fine dust mixed with scoriæ is well known to be an excellent non-conductor of heat, and may thus have preserved the snow from complete fusion when the burning flood poured over it. The shepherds in the higher regions of Etna are accustomed to provide an annual store of snow to supply their flocks with water in the summer months, by simply strewing over the snow in the spring a layer of volcanic sand a few inches thick, which effectually prevents the sun from penetrating. When lava had once consolidated over a glacier at the height of ten thousand feet above the level of the sea, we may readily conceive that the ice would endure as long as the snows of Mont Blanc, unless melted by volcanic heat from below. When I visited the great crater in the beginning of winter, (December 1st, 1828,) I found the crevices in the interior encrusted with thick ice, and in some cases hot vapours were streaming out between masses of ice and the rugged and steep walls of the crater. After the discovery of Signor Gemmellaro, it would not be surprising to find, in the cones of the Icelandic volcanos, repeated alternations of lava streams and glaciers.

Volcanic Eruptions in Iceland.—With the exception of Etna and Vesuvius, the most complete chronological records of a series of eruptions are those of Iceland: for their history reaches as far back as the ninth century of our era; and, from the beginning of the twelfth century, there is clear evidence that, during the whole period, there has never been an interval of more than forty, and very rarely one of twenty years, without either an eruption or a great earthquake. So intense is the energy of the volcanic action in this region, that some eruptions of Hecla have lasted six years without ceasing. Earthquakes have often shaken the whole island at once, causing great changes in the interior, such as the sinking down of hills, the rending of mountains, the desertion by rivers of their channels, and the appearance of new lakes *. New islands have often been thrown up near the coast, some of which still exist, while others have disappeared either by subsidences or the action of the waves.

In the interval between eruptions, innumerable hot springs afford vent to subterranean heat, and solfataras discharge copious streams of inflammable matter. The volcanos in different parts of this island are observed, like those of the Phlegræan Fields, to be in activity by turns, one vent often serving for a time as a safety-valve to the rest. Many cones

* Hoff, vol. ii., p. 393.

are often thrown up in one eruption, and in this case they take a linear direction, running generally from north-east to south-west, from the north-eastern part of the island where the volcano Krabla lies, to the promontory Reykianas.

The convulsions of the year 1783 appear to have been more tremendous than any recorded in the modern annals of Iceland; and the original Danish narrative of the catastrophe, drawn up in great detail, has since been substantiated by several English travellers, particularly in regard to the prodigious extent of country laid waste, and the volume of lava produced *. About a month previous to the eruption on the main land, a submarine volcano burst forth in the sea at the distance of thirty or forty miles in a south-west direction from Cape Reykianas, and ejected so much pumice, that the ocean was covered to the distance of one hundred and fifty miles, and ships were considerably impeded in their course. A new island was thrown up, consisting of high cliffs, within which, fire, smoke, and pumice were emitted from two or three different points. This island was claimed by his Danish Majesty, who denominated it Nyöe, or the new island; but, ere a year had elapsed, the sea resumed her ancient domain, and nothing was left but a rocky reef from five to thirty fathoms under water. Earthquakes, which had long been felt in Iceland, became violent on the 11th of June, when Skaptár Jokul, distant nearly two hundred miles from Nyöe, threw out a torrent of lava which flowed down into the river Skaptâ, and completely dried it up. The channel of the river was between high rocks, in many places from four hundred to six hundred feet in depth, and near two hundred in breadth. Not only did the lava fill up these great defiles to the brink, but it overflowed the adjacent fields to a considerable extent. The burning flood, on issuing from the confined rocky gorge, was then arrested for some

* The first narrative of the eruption was drawn up by Stephensen, then Chief Justice in Iceland, appointed Commissioner by the King of Denmark, for estimating the damage done to the country, that relief might be afforded to the sufferers. Henderson was enabled to correct some of the measurements given by Stephensen, of the depth, width, and length, of the lava currents, by reference to the MS. of Mr. Paulson, who visited the tract in 1794, and examined the lava with attention. (Journal of a Residence in Iceland, &c., p. 229.) Some of the principal facts are also corroborated by Dr. Hooker in his "Tour in Iceland," vol. ii., p. 128.

time by a deep lake, which formerly existed in the course of the river between Skaptardal and Aa, which it entirely filled. The current then proceeded again, and reaching some ancient lava full of subterraneous caverns, penetrated and melted down part of it; and in some places where the steam could not gain vent, it blew up the rock, throwing fragments to the height of more than one hundred and fifty feet. On the 18th of June, another ejection of liquid lava rushed from the volcano, which flowed down with amazing velocity over the surface of the first stream. By the damming up of the mouths of some of the tributaries of the Skaptâ, many villages were completely overflowed with water, and thus great destruction of property was caused. The lava, after flowing for several days, was precipitated down a tremendous cataract called Stapafoss, where it filled a profound abyss, which that great waterfall had been hollowing out for ages, and then the fiery current continued its course.

On the 3rd of August, fresh floods of lava still pouring from the volcano, a new branch was sent off in a different direction; for the channel of the Skaptâ was now so entirely choked up, and every opening to the west and north so obstructed, that the melted matter was forced to take a new course, and, running in a south-east direction, it discharged itself into the bed of the river Hverfisfliot, where a scene of destruction scarcely inferior to the former was occasioned. These Icelandic lavas, like the ancient streams which are met with in Auvergne, and other provinces of Central France, are stated by Stephensen to have accumulated to a prodigious depth in narrow rocky gorges, but when they came to wide alluvial plains, they spread themselves out into broad lakes of fire, sometimes from twelve to fifteen miles wide, and one hundred feet deep. When the "fiery lake" which filled up the lower portion of the valley of the Skaptâ had been augmented by new supplies, the lava flowed up the course of the river to the foot of the hills, from whence the Skaptâ takes its rise. This affords a parallel case to one which can be shewn to have happened at a remote era in the volcanic region of the Vivarais in France, when lava issued from the cone of Thueyts, and while one branch ran down, another more powerful stream flowed up the river Ardêche. The sides of the valley of the Skaptâ present superb ranges of basaltic columns of older lavas, resembling those which are laid open in the valleys descending from Mont Dor in Auvergne, where more modern lava-currents, on a scale very inferior in magnitude to those of Iceland, have also usurped the beds of the existing rivers. The eruption of Skaptár Jokul did not entirely cease till the end of two years; and when Mr. Paulson visited the tract eleven years afterwards, in 1794, he found columns of smoke still rising from parts of the lava, and several rents filled with hot water*.

Although the population of Iceland did not exceed fifty thousand, no less than twenty villages were destroyed, besides those inundated by water, and an immense number of cattle, and more than nine thousand human beings perished, partly by the depredations of the lava, partly by the noxious vapours which impregnated the air, and, in part, by the famine caused by showers of ashes throughout the island, and the desertion of the coasts by the fish.

We must now call the reader's particular attention to the extraordinary volume of melted matter produced in this eruption. Of the two branches, which flowed in nearly opposite directions, the greatest was fifty, and the lesser forty miles in length. The extreme breadth which the Skaptâ branch attained in the low countries was from twelve to fifteen miles, that of the other about seven. The ordinary height of both currents was one hundred feet, but in narrow defiles it sometimes amounted to six hundred feet. A more correct idea will be formed of the dimensions of the two streams, if we consider how striking a feature they would now form in the geology of England, had they been poured out on the bottom of the sea after the deposition, and before the elevation of our secondary and tertiary rocks. The same causes which have excavated valleys through parts of our marine strata, once continuous, might have acted with equal force on the igneous rocks, leaving, at the same time, a sufficient portion undestroyed, to enable us to discover their former extent. Let us then imagine the termination of the Skaptâ branch of lava to rest on the escarpment of the inferior and middle oolite, where it commands the vale of Gloucester. The great plateau might be one hundred feet

* Henderson's Journal, &c., p. 228.

thick, and from ten to fifteen miles broad, exceeding any which can be found in Central France. We may also suppose great tabular masses to occur at intervals, capping the summit of the Cotswold Hills between Gloucester and Oxford, by Northleach, Burford, and other towns. The wide valley of the Oxford clay would then occasion an interruption for many miles; but the same rocks might recur on the summit of Cumnor and Shotover Hills, and all the other oolitic eminences of that district. On the chalk of Berkshire, extensive plateaus, six or seven miles wide, would again be formed; and lastly, crowning the highest sands of Highgate and Hampstead, we might behold some remnants of the deepest parts of the current five or six hundred feet in thickness, rivalling or even surpassing in height Salisbury Craigs and Arthur's Seat.

The distance between the extreme points here indicated, would not exceed ninety miles in a direct line; and we might then add, at the distance of nearly two hundred miles from London, along the coast of Dorsetshire and Devonshire for example, a great mass of igneous rocks, to represent those of contemporary origin, which were produced beneath the level of the sea, where the island of Nyöe rose up. Yet, gigantic as must appear the scale of these modern volcanic operations, they are perfectly insignificant in comparison to currents of the primeval ages, if we embrace the theoretical views of some geologists of great celebrity. We are informed by Professor Brongniart, in his last work, that "aux époques géognostiques anciennes, tous les phénomènes géologiques se passoient dans des dimensions *centuples* de celles qu'ils présentent aujourd'hui*." Had Skaptár Jokul therefore been a volcano of the olden time, it would have poured forth lavas at a single eruption, a hundred times more voluminous than those which have been witnessed by the present generation. If we multiply the current before described, by a hundred, and first assume that its height and breadth remain the same, it would stretch out to the length of nine thousand miles, or about half as far again as from the pole to the equator. If, on the other hand, we suppose its length and breadth to remain the same, and multiply its height in an equal proportion, its ordinary elevation becomes

* Tableau des Terrains qui composent l'écorce du Globe, p. 52. Paris, 1829.

ten thousand feet, and its greatest more than double that of the Himalaya mountains. Amongst the ancient strata, no igneous rock of such colossal magnitude has yet been met with, nay it would be most difficult to point out a mass of igneous origin of ancient date distinctly referrible to a single eruption, which would rival in volume the matter poured out from Skaptár Jokul in 1783. It is, however, a received principle in geological reasoning, not only in France, but in England and other countries, that we ought always to assume that the energies of natural forces have been impaired and enfeebled, until the contrary can be shewn; and as we have hitherto investigated but a small part of the globe, evidence may hereafter be brought to light of the superior violence of single volcanic eruptions in remote ages. If the proofs be deficient at present in favour of the general decline of the agents of decay and renovation, we must be content with the argument of the geologist in one of Voltaire's novels, *Monsieur, on en découvrira!* *

Eruption of Jorullo in 1759.—As another example of the stupendous scale of modern volcanic eruptions, we may mention that of Jorullo in Mexico in 1759. We have already described the great region to which this mountain belongs. The plain of Malpais forms part of an elevated plateau, between two and three thousand feet above the level of the sea, and is bounded by hills composed of basalt, trachyte, and volcanic tuff, clearly indicating that the country had previously, though probably at a remote period, been the theatre of igneous action. From the era of the discovery of the New World to the middle of the last century, the district had remained undisturbed, and the space, now the site of the volcano, which is thirty-six leagues distant from the nearest sea, was occupied by fertile fields of sugar-cane and indigo, and watered by the two brooks Cuitimba and San Pedro. In the month of June, 1759, hollow sounds of an alarming nature were heard, and earthquakes succeeded each other for two months, until, in September, flames issued from the ground, and fragments of burning rocks were thrown to prodigious heights. Six volcanic cones, composed of scoriæ and fragmentary lava, were

* L'Homme aux quarante écus.

formed on the line of a chasm which ran in the direction from N.N.E. to S.S.W. The least of these cones was three hundred feet in height, and Jorullo, the central volcano, was elevated one thousand six hundred feet above the level of the plain. It sent forth great streams of basaltic lava, containing included fragments of primitive rocks, and its ejections did not cease till the month of February, 1760. Humboldt visited the country twenty years after the occurrence, and was informed by the Indians, that when they returned long after the catastrophe to the plain, they found the ground uninhabitable from the excessive heat. When the Prussian traveller himself visited the locality, there appeared, round the base of the cones, and spreading from them as from a centre over an extent of four square miles, a mass of matter five hundred and fifty feet in height in a convex form, gradually sloping in all directions towards the plain. This mass was still in a heated state, the temperature in the fissures being sufficient to light a cigar at the depth of a few inches. On this convex protuberance were thousands of flattish conical mounds, from six to nine feet high, which, as well as large fissures traversing the plain, acted as fumeroles, giving out clouds of sulphuric acid and hot aqueous vapour. The two small rivers before mentioned disappeared during the eruption, losing themselves below the eastern extremity of the plain, and reappearing as hot springs at its western limit. Humboldt attributed the convexity of the plain to inflation from below, supposing the ground, for four square miles in extent, to have risen up in the shape of a bladder, to the elevation of five hundred and fifty feet above the plain in the highest part. But this theory, which is entirely unsupported by analogy, is by no means borne out by the facts described; and it is the more necessary to scrutinize closely the proofs relied on, because the opinion of Humboldt appears to have been received as if founded on direct observation, and has been made the groundwork of other bold and extraordinary theories. Mr. Scrope has suggested that the phenomena may be accounted for far more naturally, by supposing that lava flowing simultaneously from the different orifices, and principally from Jorullo, united into a sort of pool or lake. As they were poured forth on a surface previously flat, they would, if their liquidity was not very great, remain thickest and deepest near their source, and diminish in bulk from thence towards the limits of the space which they covered. Fresh supplies were probably emitted successively during the course of an eruption which lasted a year, and some of these resting on those first emitted, might only spread to a small distance from the foot of the cone, where they would necessarily accumulate to a great height.

The showers, also, of loose and pulverulent matter from the six craters, and principally from Jorullo, would be composed of heavier and more bulky particles near the cones, and would raise the ground at their base, where, mixing with rain, they might have given rise to the stratum of black clay which is described as covering the lava. The small conical mounds (called "hornitos" or ovens) may resemble those five or six small hillocks which existed in 1823, on the Vesuvian lava, and sent forth columns of vapour, having been produced by the disengagement of elastic fluids heaving up small dome-shaped masses of lava. The fissures mentioned by Humboldt as of frequent occurrence, are such as might naturally accompany the consolidation of a thick bed of lava, contracting as it congeals; and the disappearance of rivers is the usual result of the occupation of the lower part of a valley or plain by lava, of which there are many beautiful examples in the old lava-currents of Auvergne. The heat of the "hornitos" is stated to have diminished from the first, and Mr. Bullock, who visited the spot many years after Humboldt, found the temperature of the hot spring very low, a fact which seems clearly to indicate the gradual congelation of a subjacent bed of lava, which from its immense thickness may have been enabled to retain its heat for half a century.

Another argument adduced in support of the theory of inflation from below was the hollow sound made by the steps of a horse upon the plain, which, however, proves nothing more than that the materials of which the convex mass is composed are light and porous. The sound called "rimbombo" by the Italians is very commonly returned by *made ground* when struck sharply, and has been observed not only on the sides of Vesuvius and other volcanic cones where there is a cavity below, but in plains such as the Campagna di Roma, composed in great measure of tuff and porous volcanic rocks. The reverberation, however, may, perhaps, be assisted by grottos and caverns, for these may be as numerous in the lavas of Jorullo, as in many of those of Etna; but their existence would lend no countenance to the hypothesis of a great arched cavity, or bubble, four square miles in extent, and in the centre five hundred and fifty feet high *. A subsequent eruption of Jorullo happened in 1819, accompanied by an earthquake; but unfortunately no European travellers have since visited the spot, and the only facts hitherto known are that ashes fell at the city of Guanaxuato, which is distant about one hundred and forty English miles from Jorullo, in such quantities as to lie six inches deep in the streets, and the tower of the cathedral of Guadalaxara was thrown down †.

* See Scrope on Volcanos, p. 267.

† For this information I am indebted to Captain Vetch, F.R.S.

CHAPTER XXII.

Volcanic Archipelagos—The Canaries—Eruptions of the Peak of Teneriffe—Conesthrown up in Lancerote in 1730-36—Pretended distinction between ancient and modern lavas—Recent formation of oolitic travertine in Lancerote—Grecian Archipelago—Santorin and its contiguous isles—Von Buch's Theory of "Elevation Craters" considered—New islands thrown up in the Gulf of Santorin—Supposed "Crater of Elevation" in the Isle of Palma—Description of the Caldera of Palma—Barren island in the Bay of Bengal—Origin of the deep gorge on the side of "Elevation Craters"—Stratification of submarine volcanic products—Causes of the great size of the craters of submarine volcanos—Cone of Somma, formed in the same manner as that of Vesuvius—Mineral composition of volcanic products—Speculations respecting the nature of igneous rocks produced at great depths, by modern volcanic eruptions.

In our chronological sketch of the changes which have happened within the traditionary and historical period in the volcanic district round Naples, we described the renewal of the fires of a central and habitual crater, and the almost entire cessation of a series of irregular eruptions from minor and independent vents. Some volcanic archipelagos offer interesting examples of the converse of this phenomenon, the great habitual vent having become almost sealed up, and eruptions of great violence now proceeding, either from different points in the bed of the ocean, or from adjoining islands, where, as formerly in Ischia, new cones and craters are formed from time to time. Of this state of things the Canary Islands now afford an example.

The highest crater of the Peak of Teneriffe has been in the state of a solfatara ever since it has been known to Europeans; but several eruptions have taken place from the sides of the mountain, one in the year 1430, which formed a small hill, and another in 1704 and the two following years, accompanied with great earthquakes, when the lava overflowed a town and harbour. Another eruption happened in June, 1798, not far from the summit of the peak. But these lateral emissions of lava, at distant intervals, may be considered as of a subordinate kind, and subsidiary to the great discharge which has taken place in the contiguous isles of Palma and Lancerote; and the occasional activity of the peak may be compared to the irregular eruptions before mentioned, of the Solfatara, of Arso in Ischia, and of Monte Nuovo, which have broken out since the renewal of the Vesuvian fires in 79.

We shall describe one of these insular eruptions in the Canaries, which happened in Lancerote, between the years 1730 and 1736, as the effects were remarkable; and Von Buch had an opportunity, when he visited that island in 1815, of comparing the accounts transmitted to us of the event, with the present state and geological appearances of the country*. On the 1st of September, 1730, the earth split open on a sudden two leagues from Yaira. In one night a considerable hill of ejected matter was thrown up, and a few days later, another vent opened and gave out a lava-stream, which overran Chinanfaya and other villages. It flowed first rapidly, like water, but became afterwards heavy and slow, like honey. On the 7th of September an immense rock was protruded from the bottom of the lava, with a noise like thunder, and the stream was forced to change its course, from N. to N.W., so that St. Catalina, and other villages, were overflowed. Whether this mass was protruded by an earthquake or was a mass of ancient lava, blown up like that before mentioned, in 1783, in Iceland, is not explained. On the 11th of September more lava flowed out and covered the village of Maso entirely, and, for the space of eight days, precipitated itself with a horrible roar into the sea. Dead fish floated on the waters in indescribable multitudes, or were thrown dying on the shore. After a brief interval of repose, three new openings broke forth, immediately from the site of the consumed St. Catalina. and sent out an enormous quantity of lapilli, sand, and ashes. On the 28th of October, the cattle throughout the whole country dropped lifeless to the ground, suffocated by putrid vapours, which condensed and fell down in drops. On the 1st of December, a lava-stream reached the sea, and formed an island, round which dead fish were strewed.

It is unnecessary here to give the details of the overwhelm-

* This account was principally derived by Von Buch from the MS. of Don Andrea Lorenzo Curbeto, Curate of Yaira, the point where the eruption began. Uber einen vulcanisch. Ausbruch auf der Insel Lanzerote.

ing of other places by fiery torrents, or of a storm which was equally new and terrifying to the inhabitants, as they had never known one in their country before. On the 10th of January, 1731, a high hill was thrown up, which, on the same day, precipitated itself back again into its own crater: fiery brooks of lava flowed from it to the sea. On the 3rd of February a new cone arose. Others were thrown up in March, and poured forth lava-streams. Numerous other volcanic cones were subsequently formed in succession, till at last their number amounted to about thirty. In June, 1731, during a renewal of the eruptions, all the banks and shores in the western part of the island were covered with dying fish, of different species, some of which had never before been seen. Smoke and flame arose from the sea, with loud detonations. These dreadful commotions lasted without interruption for *five successive years*, and a great emigration of the inhabitants became necessary.

As to the height of the new cones, Von Buch was assured that the formerly great and flourishing St. Catalina lay buried under hills 400 feet in height; and he observes, that the most elevated cone of the series rose 600 feet above its base, and 1378 feet above the sea, and that several others were nearly as high. The new vents were all arranged *in one line*, about two geographical miles long, and in a direction nearly east and west. If we admit the probability of Von Buch's conjecture, that these vents opened along the line of an open cleft, it seems necessary to suppose, that this subterranean fissure was only prolonged upwards to the surface by degrees, or that the rent was narrow at first, as is usually the case with fissures caused by earthquakes. Lava and elastic fluids might escape from some point on the rent where there was least resistance, till the first aperture becoming obstructed by ejections and the consolidation of lava, other orifices burst open in succession, along the line of the original fissure. Von Buch found that each crater was lowest on that side on which lava had issued; but some craters were not breached, and were without any lava-streams. In one of these were open fissures, out of which hot vapours rose, which in 1815 raised the thermometer to 145° Fahrenheit, and was probably at the boiling point lower down. The exhalations seemed to consist of aqueous vapour, yet they could not be pure steam, for the crevices were encrusted on either side by siliceous sinter, (an opal-like hydrate of silica, of a white colour,) which extended almost to the middle. This important fact attests the length of time during which chemical processes continue after eruptions, and how open fissures may be filled up laterally by mineral matter, sublimed from volcanic exhalations. The lavas of this eruption covered nearly a third of the whole island, often forming on slightly inclined planes great horizontal sheets several square leagues in area, resembling very much the basaltic plateaus of Auvergne.

One of the new lavas was observed to contain masses of olivine of an olive-green colour, resembling those which occur in one of the lavas of the Vivarais. Von Buch supposes the great crystals of olivine to have been derived from a previously existing basalt, melted up by the new volcanos, but sufficient data are not furnished for warranting such a conjecture. The older rocks of the island consist, in a great measure, of that kind of basaltic lava called dolerite, sometimes columnar, and of common basalt and amygdaloid. Some recent lavas assumed, on entering the sea, a prismatic form, and so much resembled the older lavas of the Canaries, that the only geological distinction which Von Buch appears to have been able to draw between them was, that they did not alternate with conglomerates, like the ancient basalts. Some modern writers have endeavoured to discover in the abundance of these conglomerates, a proof of the dissimilarity of the volcanic action in ancient and modern times; but this character is more probably attributable to the difference between submarine operations and those on the land. All the blocks and imperfectly rounded fragments of lava, transported, during the intervals of eruption by rivers and torrents, into the adjoining sea, or torn by the continued action of the waves from cliffs which are undermined, must accumulate in stratified breccias and conglomerates, and be covered again and again by other lavas. This is now taking place on the shores of Sicily, between Catania and Trezza, where the sea breaks down and covers the shore with blocks and pebbles of the modern lavas of Etna; and on parts of the coast of Ischia, where numerous currents of trachyte are in like manner undermined in lofty precipices. So often then as an island is raised in a volcanic archipelago, by earthquakes from the deep, the fundamental and (relatively to all above) the oldest lavas will often be distinguishable from those formed by subsequent eruptions on dry land, by their alternation with beds of sandstone and fragmentary rocks. The supposed want of identity then between the volcanic phenomena of different epochs resolves itself into the marked difference between the operations simultaneously in progress, above and below the water. Such, indeed, is the source, as we stated in our fifth chapter, of many of our strongest theoretical prejudices in geology. No sooner do we study and endeavour to explain submarine appearances, than we feel, to use a common expression, *out of our element;* and unwilling to concede, that our extreme ignorance of processes now continually going on can be the cause of our perplexity, we take refuge in a " pre-existent order of nature."

Throughout a considerable part of Lancerote, the old lavas are covered by a thin stratum of limestone, from an inch to two feet in thickness. It is of a hard stalactitic nature, sometimes oolitic, like the Jura limestone, and contains fragments of lava and terrestrial shells, chiefly helices and spiral bulimi. Von Buch imagines, that this remarkable superstratum has been produced by the furious north-west storms, which in winter drive the spray of the sea in clouds over the whole island; from whence calcareous particles may be deposited stalactitically. If this explanation be correct, and it seems highly probable, the fact is interesting, as attesting the quantity of matter held in solution by the sea-water, and ready to precipitate itself in the form of solid rock. At the bottom of such a sea, impregnated, as in the neighbourhood of all active volcanos, with mineral matter in solution, lavas must be converted into calcareous amygdaloids, a form in which the igneous rocks so frequently appear in the older European formations. We may mention that recent crevices in the rocks of Trezza, one of the Cyclopian isles at the foot of Etna, are filled with a kind of travertine, as high as the spray of the sea reaches; and in this hard veinstone, fragments, and even entire specimens of recent shells thrown up by the waves, are sometimes included.

From the year 1736 to 1815, when Von Buch visited Lancerote, there had been no eruption; but, in August, 1824, a crater opened near the port of Rescif, and formed, by its ejections, in the space of twenty-four hours, a considerable hill. Violent earthquakes preceded and accompanied this eruption*.

Grecian Archipelago.—We shall next direct our inquiry to the island of Santorin, as it will afford us an opportunity of discussing the merits of a singular theory, which has obtained no small share of popularity in modern times, respecting "craters of elevation," (Erhebungs Cratere, Cratères de soulèvement,) as they have been termed. The three islands of Santorin, Therasia, and Aspronisi surround a gulf almost circular, and above six miles in diameter. They are chiefly composed of trachytic conglomerates and tuffs, covered with pumice; but in one part of Santorin clay-slate is seen to be the fundamental rock. The beds in all these isles

No. 15.

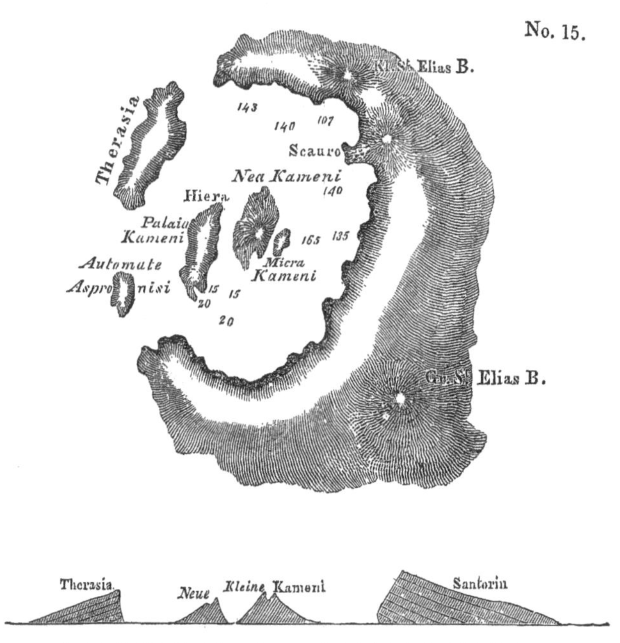

Chart and Section of Santorin and the contiguous islands in the Grecian Archipelago.

dip at a slight angle towards the exterior of the group, and lose themselves in the surrounding sea; whereas, on the con-

* Férussac, Bulletin des Sci. Nat., tome V., p. 45.—1825. The volcano was still burning when the account here cited was written.

trary, they present a high and steep escarpment towards the centre of the inclosed space. The gulf, therefore, is nearly on all sides environed by precipices; those of Santorin, which form two-thirds of the circumference, being two leagues in extent, and in some parts three hundred feet high. These rocky cliffs plunge at once into the sea, so that close to the shore soundings are only reached at a depth of eight hundred feet, and at a little distance farther at a depth of one thousand feet. In the middle of this gulf, the small isle of Hiera, now called Palaia Kameni, rose up, 144 years before the Christian era. In 1427 this isle received new accessions. In 1573 the Little Kameni was raised in the middle of the basin, its elevation being accompanied by the discharge of large quantities of pumice and a great disengagement of vapour. Lastly, in 1707 and 1709 the New Kameni was formed, which still exhales sulphureous vapours. These isles are formed of rocks of brown trachyte, which has a resinous lustre, and is full of crystals of glassy felspar. Although the birth of New Kameni was attended by an eruption, it is certain that it was upraised from a great depth by earthquakes, and was not a heap of volcanic ejections, nor of lava poured out on the spot. There were shells upon it when it first appeared; and beds of limestone and marine shells are described by several authors as entering, together with igneous rocks, into the structure of other parts of this group. In order, therefore, to explain the formation of such circular gulfs, which are common in other archipelagos, Von Buch supposes, and Humboldt adopts the same opinion, that the different beds of lava, pumice, and whatever else may be interstratified, were first horizontally disposed along the floor of the ocean. An expansive force from below then burst an opening through them, and, acting towards a central point, raised symmetrically on every side all which resisted its action, so that the uplifted strata were made to dip away on all sides from the centre outwards, as is usual in volcanic cones, while a deep hollow was left in the middle, resembling in all essential particulars an ordinary volcanic crater.

In the first instance we should inform the reader, that this theory is not founded on actual observations of analogous effects produced by the elevating forces of earthquakes, or the escape of elastic fluids in any part of the globe; for the inflation from below, of the rocks in the plain of Malpais, during the eruption of Jorullo, was, as before stated, an hypothesis proposed, long after that eruption, to account for appearances which admit of a very different explanation. Besides, in the case of Jorullo, there was no great "crater of elevation" formed in the centre. All our modern analogies, therefore, being in favour of the origin of cones and craters exclusively by *eruptions,* we are entitled to scrutinize with no small severity the new hypothesis; and we have a right to demand demonstrative evidence, that known and ordinary causes are perfectly insufficient to produce the observed phenomena. Had Von Buch and Humboldt, for instance, in the course of those extensive travels which deservedly render their opinions, in regard to all volcanic operations, of high authority, discovered a single cone composed exclusively of marine or lacustrine strata, without a fragment of any igneous rock intermixed; and in the centre a great cavity, encircled by a precipitous escarpment; then we should have been compelled at once to concede, that the cone and crater-like configuration, whatever be its mode of formation, may sometimes have no reference whatever to ordinary volcanic eruptions.

But it is not pretended that, on the whole face of the globe, a single example of this kind can be pointed out. In Europe and North America thousands of square leagues of territory have been examined, composed of marine strata, which have been elevated to various heights, sometimes to more than ten thousand feet above the level of the sea, sometimes in horizontal tabular masses; in other cases with every degree of inclination, from the horizontal to the vertical. Some have been moved without great derangement, others have been rent, contorted, or shattered with the utmost violence. Sometimes large districts, at others small spaces, appear to have changed their position. Yet, amidst the innumerable accidents to which these rocks have been subject, never have they assumed that form, exactly representing a large truncated volcanic cone, with a great cavity in the centre. Are we then called upon to believe that whenever elastic fluids generated in the subterranean regions burst through horizontal strata, so as to upheave them in the peculiar manner before adverted to, they always select, as if from choice, those spots of comparatively insignificant area, where a certain quantity of volcanic matter happens to lie, while they carefully avoid purely lacustrine and marine strata, although they often lie immediately contiguous? Why on the southern borders of the Limagne d'Auvergne, where several eruptions burst through, and elevated the horizontal marls and limestones, did these freshwater beds never acquire in any instance a conical and crateriform disposition?

But let us proceed to examine some of the most celebrated examples adduced of craters of elevation. The most perfect type of this peculiar configuration is said to be afforded by the Isle of Palma; and while we controvert Von Buch's theoretical opinions, we ought not to forget how much geology is indebted to his talents and zeal, and amongst other works for his clear and accurate description of this isle *. In the middle of Palma rises a mountain to the height of four thousand feet, presenting the general form of a great cone, the upper part of which had been truncated and replaced by an enormous funnel-shaped cavity, about four thousand feet deep; and the surrounding borders of which attain, at their highest point, an elevation of seven thousand feet above the sea. The external flanks of this cone are gently inclined, and, in part, cultivated; but the bottom and the walls of the central cavity, called by the inhabitants the Caldera, present on all sides rugged and uncultivated rocks, almost completely devoid of vegetation.

No. 16.

View of the Isle of Palma, and of the Caldera in its centre.

So steep are the sides of the Caldera, that there is no path by which they can be descended; and the only entrance is by a great ravine, which, cutting through the rocks environing the

* Physical. Besch. der Canarischen Inseln. Berlin, 1825.

circus, runs down to the sea. The sides of this gorge are jagged, broken, and precipitous. In the mural escarpments surrounding the Caldera are seen nothing but beds of basalt, and conglomerates composed of broken fragments of basalt, which dip away with the greatest regularity, from the centre to the circumference of the cone. Now, according to the theory of "elevation craters," we are called upon to suppose that, in the first place, a series of horizontal beds of volcanic matter accumulated over each other, to the enormous depth of more than four thousand feet—a circumstance which alone would imply the proximity, at least, of a vent from which immense quantities of igneous rocks had proceeded. After the aggregation of the mass, the expansive force was directed on a given point with such extraordinary energy, as to lift up bodily the whole mass, so that it should rise to the height of seven thousand feet above the sea, leaving a great gulf or cavity in the middle. Yet, notwithstanding this prodigious effort of gaseous explosions, concentrated on so small a point, the beds, instead of being shattered, contorted, and thrown into the utmost disorder, have acquired that gentle inclination, and that regular and symmetrical arrangement, which characterize the flanks of a large cone of eruption, like Etna! We admit that earthquakes, when they act on extensive tracts of country, may elevate and depress them without deranging, considerably, the relative position of hills, valleys, and ravines. But is it possible to conceive that elastic fluids could break through a mere point as it were of the earth's crust, and that too where the beds were not composed of soft, yielding clay, or incoherent sand, but of solid basalt, thousands of feet thick, and that they could inflate them, as it were, in the manner of a bladder? Would not the rocks, on the contrary, be fractured, fissured, thrown into a vertical, and often into a reversed position; and, ere they attained the height of seven thousand feet, would they not be reduced to a mere confused and chaotic heap?

The Great Canary is an island of a circular form, analogous to that of Palma. Barren Island, also, in the Bay of Bengal, is proposed as a striking illustration of the same phenomenon; and here it is said we have the advantage of being able to contrast the ancient crater of elevation with a cone and crater of eruption in its centre. When seen from the ocean, this island presents, on almost all sides, a surface of bare rocks, which rise up with a moderate declivity towards the interior; but at one point there is a narrow cleft, by which we can penetrate into the centre, and there discover that it is occupied by a great

No. 17.

Cone and Crater of Barren Island, in the Bay of Bengal.

circular basin, filled by the waters of the sea, bordered all around by steep rocks, in the midst of which rises a volcanic cone, very frequently in eruption. The summit of this cone is 1690 French feet in height, corresponding to that of the circular border which incloses the basin; so that it can only be seen from the sea through the ravine, which precisely resembles the deep gorge by which we penetrate into the Caldera of the Isle of Palma, and of which an equivalent, more or less decided in its characters, is said to occur in all elevation craters.

The cone of the high peak of Teyda, in Teneriffe, is also represented as rising out of the middle of a crater of elevation, standing like a tower surrounded by its foss and bastion; the foss being the remains of the ancient gulf, and the bastion the escarpment of the circular inclosure. So that Teneriffe is an exact counterpart of Barren Island, except that one is raised to an immense height, while the other is still on a level with the sea, and in part concealed beneath its waters.

Now, without enumerating more examples, let us consider what form the products of submarine volcanos may naturally be expected to assume. There is every reason to conclude, from the few accounts which we possess of eruptions at the bottom of the sea, that they take place in the same manner there as on the open surface of a continent*. That the volcanic phenomena, if they are ever developed at unfathomable depths, may

* Scrope on Volcanos, p. 171.

be extremely different, is very possible; but when they have been witnessed by the crews of vessels casually passing, the explosions of aëriform fluids beneath the waters have closely resembled those of volcanos on the land. Rocky fragments, ignited scoriæ, and comminuted ashes, are thrown up, and in several cases conical islands have been formed, which afterwards disappeared; as when, in 1691 and 1720, small isles were thrown up off St. Michael in the Azores, or as Sabrina rose in 1811 near the same spot, and, in 1783, Nyöe, off the coast of Iceland. Where the cones have disappeared, they probably consisted of loose matters, easily reduced by the waves and currents to a submarine shoal. When islands have remained firm, as in the case of Hiera, and the New and Little Kameni in the Gulf of Santorin (see wood-cut No. 15), they have consisted in part of solid lava. Whatever doubts might have been entertained as to the action of volcanos entirely submarine, yet it must always have been clear, that in those numerous cases where they just raise their peaks above the waves, the ejected sand, scoriæ, and fragments of rock, must accumulate round the vent into a cone with a central crater, while the lighter will be borne to a distance by tides and currents, as by winds during eruptions in open air. The lava which issues from the crater spreads over the subaqueous bottom, seeking the lowest levels, or accumulating upon itself, according to its liquidity, volume, and rapidity of congelation; following, in short, the same laws as when flowing in the atmosphere *.

But we may next enquire, what characters may enable a geologist to distinguish between cones formed entirely, or in great part beneath the waters of the sea, and those formed on land. In the first place, large beds of shells and corals often grow on the sloping sides of submarine cones, particularly in the Pacific, and these often become interstratified with lavas. Instead of alluvions containing land-shells, like some of those which cover Herculaneum, great beds of tufaceous sand and conglomerate, mixed with marine remains, might be expected on such parts of the flanks of a volcano like Stromboli as are submerged beneath the waters. The pressure of a column of

* Scrope on Volcanos, p. 173.

water exceeding many times that of the atmosphere, must impede the escape of the elastic fluids and of lava, until the resistance is augmented in the same proportion; hence the explosions will be more violent, and when a cone is formed it will be liable to be blown up and truncated at a lower level than in shallower water or in the open air. Add to this, that when a submarine volcano has repaired its cone, it is liable to be destroyed again by the waves, as in several cases before adverted to. The vent will then become choked up with strata of sand and fragments of rock, swept in by the tides and currents. These materials are far more readily consolidated under water than in the air, especially as mineral matter is so copiously introduced by the springs which issue from the ground in all volcanic regions hitherto carefully investigated. Beds of solid travertin, also, and in hot countries coral reefs, must often, during long intervals of quiescence, obstruct the vent, and thus increase the repressive force and augment the violence of eruptions. The probabilities, therefore, in a submarine volcano, of the destruction of a larger part of the cone, and the formation of a more extensive crater, are obvious; nor can the dimensions of "craters of elevation," if referred to such operations, surprise us. During an eruption in 1444, accompanied by a tremendous earthquake, the summit of Etna was destroyed, and an enormous crater was left, from which lava flowed. The segment of that crater may still be seen near the Casa Inglese, and, when complete, it must have measured several miles in diameter. The cone was afterwards repaired; but this would not have happened so easily had Etna been placed like Stromboli in a deep sea, with its peak exposed to the fury of the waves. Let us suppose the Etnean crater of 1444 to have been filled up with beds of coral and conglomerate, and that during succeeding eruptions these were thrown out by violent explosions, so that the cone became truncated down to the upper margin of the woody region, a circular basin would then be formed thirty Italian miles in circumference, exceeding by five or six miles the circuit of the Gulf of Santorin. Yet we know by numerous sections that the strata of trachyte, basalt, and trachytic breccia, would, in that part of the great cone of Etna, dip on all sides off from the centre at a gentle angle to every point of the compass, except where irregularities were occasioned, at points where the small buried cones before mentioned occurred. If this gulf were then again choked up, and the vent obstructed, so that new explosions of great violence should truncate the cone once more down to the inferior border of the forest zone of Etna, the circumference of the gulf would then be fifty Italian miles *. Yet even then the ruins of the cone of Etna might form a circular island entirely composed of volcanic rocks, sloping gently outwards on all sides at a very slight angle; and this island might be between seventy and eighty English miles in its exterior circuit, while the circular bay within might be between forty and fifty miles round. In fertility it would rival the Island of Palma; and the deep gorge which leads down from the Valley of Calanna to Zafarana, might well serve as an equivalent to the grand défile which leads into the Caldera.

It is most probable, then, that the exterior inclosure of Barren Island, *c d*, in the annexed diagram, is nothing more than the remains of the truncated cone *c a b d*, a great portion of which has been carried away, partly by the action of the

No. 18.

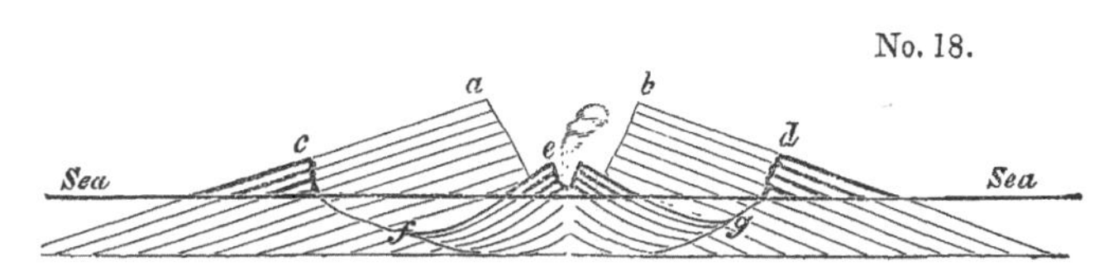

Supposed Section of Barren Island, in the Bay of Bengal.

waves, and partly by explosions which preceded the formation of the new interior cone, *f e g*. Whether the outer and larger cone has in this particular case, together with the bottom of the ocean on which it rests, been upheaved, or whether it originally projected in great part like Stromboli above the level of the sea, may, probably, be determined by geological investigations; for, in the former case, some beds replete with marine remains may be interstratified with volcanic ejections.

Some of the accounts transmitted to us by eye-witnesses, of the gradual manner in which New Kameni first rose covered with living shells in the Gulf of Santorin, appear, certainly,

* For the measurements of different parts of the cone of Etna, see Trattato dei boschi dell' Etna, Scuderi, Acti. dell' Acad. Gion. de Catan., vol. i.

to establish the possibility of the elevation of small masses from a depth of several hundred feet during an eruption, and during the emission of lava. But the protrusion of isolated masses, under such circumstances, affords no analogy to the supposed action of the expansive force in the formation of craters of elevation. It is hardly necessary, after the observations now made, to refer the reader again to our section of Somma and Vesuvius, and to say that we ascribed the formation of the ancient and the modern cone to operations precisely analogous.

M. Necker * long ago pointed out the correspondence of their structure, and explained most distinctly the origin of the form of Somma; and his views were afterwards confirmed by Mr. Scrope. But, notwithstanding the juxta-position of the entire and the ruined cone, the identity of the slope and quâquâ-versal dip of the beds, the similarity of their mineral composition, and the intersection of both cones by porphyritic dikes, the defenders of the "elevation" theory have declared that the lavas and breccias of Somma were once horizontal, and were afterwards raised into a conical mass, while they admit that those in Vesuvius have always been as highly inclined as they are now.

In controverting Von Buch's theory, we might have adduced as the most conclusive argument against it, that it would lead its advocates, if consistent with themselves, to the extravagant conclusion, that the two cones of Vesuvius had derived their form from very distinct causes. But as these geologists are not afraid to follow their system into all its consequences, and have even appealed to Somma as confirmatory of their views, it would be vain to hope, by pointing out the closest analogies between the effects of ordinary volcanic action and "craters of elevation," to induce them to abandon their hypothesis.

The marine shelly strata, interstratified with basalt, through which the great cone of Etna rises, are also said to have constituted an ancient crater of elevation; but when we allude more particularly to the geology of Sicily, it will appear that the strata in question do not dip so as to countenance in the least degree such an hypothesis. The nearest approach, perhaps, to the production of a conical mass by elevation from

* Mémoire sur le Mont Somma, Mém. de la Soc. de Phys. et d'Hist. Nat. de Genève, tom. ii., part I., p. 155.

below, is in the Cantal in Central France. The volcanic eruptions which produced at some remote period the volcanic mountain called the Plomb du Cantal, broke up through fresh-water strata, which must have been deposited originally in an horizontal position, on rocks of granitic schist. During the gradual formation of the great cone, beds of lava and tuff, thousands of feet in thickness, were thrown out from one or more central vents, so as to cover great part of the lacustrine strata, and these at the same time were traversed by dikes, and in parts lifted up together with the subjacent granitic rocks; so that if the igneous products could now be removed, and the marls, limestones, and fundamental schists, supported at their present elevation, they would form a kind of dome-shaped protuberance. But the outline of this shattered mass would be very unlike that of a regular cone, and the dip of the beds would be often horizontal, as near Aurillac, often vertical, often reversed, nor would there be in the centre any great cavity or crater of elevation *. On the other hand, the *volcanic* beds of the Plomb du Cantal are arranged in a conical form, like those of Etna, not by elevation from below, but because they flowed down during successive eruptions *from above*.

We may observe that the Fossa Grande on Vesuvius, a deep ravine washed out by the winter-torrents which descend from the Atrio del Cavallo, may represent, on a small scale, the Valley of Calanna, and its continuation, the Valley of St. Giacomo on Etna. In the Fossa Grande, a small body of water has cut through tuff, and in some parts solid beds of lava of considerable thickness; and the channel, although repeatedly blocked up by modern lavas, has always been re-excavated. It is natural that on one side of every large hollow, such as the crater of a truncated cone, there should be a channel to drain off the water; and this becoming in the course of ages a deep ravine, may have caused such gorges as exist in Palma and other isles of similar conformation.

Mineral Composition of Volcanic Products.—The mineral called felspar, forms in general more than half of the mass of

* See a Memoir by Messrs. Murchison and Lyell, Sur les Dépôts Lacustres Tertiaires du Cantal, &c., Ann. des Sci. Nat., October, 1829.

modern lavas. When it is in great excess, lavas are called trachytic; when augite (or pyroxene) predominates, they are termed basaltic. But lavas of composition, precisely intermediate, occur, and from their colour have been called graystones. A great abundance of quartz characterizes the granitic and other ancient rocks, now generally considered by geologists as of igneous origin, whereas that mineral, which is nothing more than silex crystallized, is rare in recent lavas, although silex enters largely into their composition. Hornblende, which is so common in ancient rocks, is rare in modern lava, nor does it enter largely into rocks of any age in which augite abounds. Mica occurs plentifully in some recent trachytes, but is rarely present where augite is in excess. We must beware, however, not to refer too hastily to a difference of era, characters which may, in truth, belong to the different circumstances under which the products of fire originate.

When we speak of the igneous rocks of our own times, we mean that small portion which happens in violent eruptions to be forced up by elastic fluids to the surface of the earth. We merely allude to the sand, scoriæ, and lava, which cool in the open air; but we cannot obtain access to that which is congealed under the pressure of many hundred, or many thousand atmospheres. We may, indeed, see in the dikes of Vesuvius rocks consolidated from a liquid state, under a pressure of perhaps a thousand feet of lava, and the rock so formed is more crystalline and of greater specific gravity than ordinary lavas. But the column of melted matter raised above the level of the sea during an eruption of Vesuvius must be more than three thousand feet in height, and more than ten thousand feet in Etna; and we know not how many miles deep may be the ducts which communicate between the mountain and those subterranean lakes or seas of burning matter which supply for thousands of years, without being exhausted, the same volcanic vents. The continual escape of hot vapours from many craters during the interval between eruptions, and the chemical changes which are going on for ages in the fumeroles of volcanos, prove that the volcanic foci retain their intense heat constantly, nor can we suppose it to be otherwise; for as lava-currents of moderate thickness require many years to cool down in the open air, we must suppose the great reservoirs of melted matter at vast depths in the nether regions to preserve their high temperature and fluidity for thousands of years.

During the last century, about fifty eruptions are recorded of the five European volcanos, Vesuvius, Etna, Volcano, Santorin, and Iceland, but many beneath the sea in the Grecian Archipelago and near Iceland may doubtless have passed unnoticed. If some of them produced no lava, others on the contrary, like that of Skaptár Jokul in 1783, poured out melted matter for five or six years consecutively, which cases, being reckoned as single eruptions, will compensate for those of inferior strength. Now, if we consider the active volcanos of Europe to constitute about a fortieth part of those already known on the globe, and calculate, that, one with another, they are about equal in activity to the burning mountains in other districts, we may then compute that there happen on the earth about two thousand eruptions in the course of a century, or about twenty every year.

However inconsiderable, therefore, may be the superficial rocks which the operations of fire produce on the surface, we must suppose the subterranean changes now constantly in progress to be on the grandest scale. The loftiest volcanic cones must be as insignificant, when contrasted to the products of fire in the nether regions, as are the deposits formed in shallow estuaries when compared to submarine formations accumulating in the abysses of the ocean. In regard to the characters of these volcanic rocks, formed in our own times in the bowels of the earth, whether in rents and caverns, or by the cooling of lakes of melted lava, we may safely infer that the rocks are heavier and less porous than true lavas, and more crystalline, although composed of the same mineral ingredients. As the hardest crystals produced artificially in the laboratory, require the longest time for their formation, so we must suppose that where the cooling down of melted matter takes place by insensible degrees, in the course of ages a variety of minerals will be produced far harder than any formed by natural processes within the short period of human observation.

These subterranean volcanic rocks, moreover, cannot be stratified in the same manner as sedimentary deposits from water, although it is evident that when great masses consolidate from a state of fusion, they may separate into natural divisions; for this is seen to be the case in many lava-currents. We may also expect that the rocks in question will often be rent by earthquakes, since these are common in volcanic regions, and the fissures will be often injected with similar matter, so that dikes of crystalline rock will traverse masses of similar composition. It is also clear that no organic remains can be included in such masses, unless where sedimentary strata have subsided to great depths, and in this case the fossil substances will probably be so acted upon by heat, that all signs of organization will be obliterated. Lastly, these deep-seated igneous formations must underlie all the strata containing organic remains, because the heat proceeds from below upwards, and the intensity required to reduce the mineral ingredients to a fluid state must destroy all organic bodies in rocks either subjacent or included in the midst of them. If, by a continued series of elevatory movements, such masses shall hereafter be brought up to the surface, in the same manner as sedimentary marine strata have, in the course of ages, been upheaved to the summit of the loftiest mountains, it is not difficult to foresee what perplexing problems may be presented to the geologist. He may then, perhaps, study in some mountain chain the very rocks produced at the depth of several miles beneath the Andes, Iceland, or Java, in the time of Leibnitz, and draw from them the same conclusion which that philosopher derived from certain igneous products of high antiquity; for he conceived our globe to have been, for an indefinite period, in the state of a comet, without an ocean, and uninhabitable alike by aquatic or terrestrial animals.

CHAPTER XXIII.

Earthquakes and their effects—Deficiency of ancient accounts—Ordinary atmospheric phenomena—Changes produced by earthquakes in modern times considered in chronological order—Earthquake in Murcia, 1829—Bogota in 1827—Chili in 1822—Great extent of country elevated—Aleppo in 1822—Ionian Isles in 1820—Island of Sumbawa in 1815—Town of Tomboro submerged—Earthquake of Cutch in 1819—Subsidence of the delta of the Indus—Earthquake of Caraccas in 1812—South Carolina in 1811—Geographical changes in the valley of the Mississippi—Volcanic convulsions in the Aleutian Islands in 1806—Reflections on the earthquakes of the eighteenth century—Earthquake in Quito, 1797—Cumana, 1797—Caraccas, 1790—Sicily, 1790—Java, 1786—Sinking down of large tracts.

We have already stated, in our sketch of the geographical boundaries of volcanic regions, that, although the points of eruption are but thinly scattered, and form mere spots on the surface of those vast districts, yet the subterranean movements extend, simultaneously, over immense areas. We shall now proceed to consider the changes which these movements have been observed to produce on the surface, and in the internal structure of the earth's crust.

It is only within the last century and a half, since Hooke first promulgated his views respecting the connexion between geological phenomena and earthquakes, that the permanent changes effected by these convulsions have excited attention. Before that time, the narrative of the historian was almost exclusively confined to the number of human beings who perished, the number of cities laid in ruins, the value of property destroyed, or certain atmospheric appearances which dazzled or terrified the observers. The creation of a new lake, the engulphing of a city, or the raising of a new island, are sometimes, it is true, adverted to, as being too obvious, or of too much geographical interest, to be passed over in silence. But no researches were made expressly with a view of ascertaining the precise amount of depression or elevation of the ground, or the particular alterations in the relative position of sea and land; and very little distinction was made between the raising of soil by volcanic ejections, and the upheaving of it by forces acting from below. The same remark applies to a very large proportion of modern accounts; and how much reason we have to regret this deficiency of information is apparent from the fact, that in every instance where a spirit of scientific inquiry has animated the eye-witnesses of these events, facts calculated to throw much light on former modifications of the earth's structure have been recorded.

As we shall confine ourselves almost entirely, in our notice of certain earthquakes, to the changes brought about by them in the configuration of the earth's crust, we may mention, generally, some accompaniments of these terrible events which are almost uniformly commemorated in history, that it may be unnecessary to advert to them again. Irregularities in the seasons precede or follow the shocks; sudden gusts of wind, interrupted by dead calms; violent rains, in countries or at seasons when such phenomena are unusual or unknown; a reddening of the sun's disk, and a haziness in the air, often continued for months; an evolution of electric matter, or of inflammable gas from the soil, with sulphureous and mephitic vapours; noises underground, like the running of carriages, or the discharge of artillery, or distant thunder; animals utter cries of distress, and evince extraordinary alarm, being more sensitive than men of the slightest movement; a sensation like sea-sickness, and a dizziness in the head, are experienced:—these, and other phenomena which do not immediately bear on our present subject, have recurred again and again at distant ages, and in all parts of the globe.

We shall now begin our enumeration with the latest authentic narratives of earthquakes, and so carry back our survey retrospectively, that we may bring before the reader, in the first place, the minute and circumstantial details of modern times, and enable him, by observing the extraordinary amount of change within the last hundred and fifty years, to perceive how great must be the deficiency in the meagre annals of earlier eras.

Murcia, 1829.—The first event which presents itself in our chronological order, is the earthquake which happened in the south of Spain on the 21st of March, 1829. It appears, by the narrative of M. Cassas, the French Consul at Alicant, that the accounts of the catastrophe were generally much exaggerated. The district violently agitated was only about four square miles in area, being the basin of the river Segura between Orihuela and the sea. All the villages in this tract were thrown down by a vertical movement, the soil being traversed by innumerable crevices four or five inches broad. In the alluvial plain, especially that part near the sea, small circular apertures were formed, out of which black mud, salt-water, and marine shells were vomited; and in other places fine yellowish-green micaceous sand, like that on the beach at Alicant, was thrown up in jets. No crater sent forth lava, as was asserted in several Spanish journals *.

Bogota, 1827.—On the 16th November, 1827, the plain of Bogota was convulsed by an earthquake, and a great number of towns were thrown down. Torrents of rain swelled the Magdalena, sweeping along vast quantities of mud and other substances, which emitted a sulphureous vapour and destroyed the fish. Popayan, which is distant two hundred geographical miles S.S.W. of Bogota, suffered greatly. Wide crevices appeared in the road of Guanacas, leaving no doubt that the whole of the Cordilleras sustained a powerful shock. Other fissures opened near Costa, in the plains of Bogota, into which the river Tunza immediately began to flow †. In such cases, we may observe, the ancient gravel bed of a river is deserted, and a new one formed at a lower level; so that, a want of relation in the position of alluvial beds to the existing water-courses may be no test of the high antiquity of such deposits to a geologist, in countries habitually convulsed by earthquakes. Extraordinary rains accompanied the shocks before mentioned, and two volcanos are said to have been in eruption in the mountain-chain nearest to Bogota.

Chili, 1822.—On the 19th of November, 1822, the coast of Chili was visited by a most destructive earthquake. The shock was felt simultaneously throughout a space of one thousand two hundred miles from north to south. St. Jago, Valparaiso, and some other places, were greatly injured. When the district

* Férussac, Bulletin des Sci. Nat., November, 1829, p. 203.
† Phil. Mag., July, 1828, p. 37.

round Valparaiso was examined on the morning after the shock, it was found that the whole line of coast for the distance of above one hundred miles was raised above its former level *. At Valparaiso the elevation was three feet, and at Quintero about four feet. Part of the bed of the sea remained bare and dry at high water, "with beds of oysters, muscles, and other shells adhering to the rocks on which they grew, the fish being all dead, and exhaling most offensive effluvia †." An old wreck of a ship, which before could not be approached, became accessible from the land; although its distance from the original sea-shore had not altered. It was observed, that the water-course of a mill, at the distance of about a mile from the sea, gained a fall of fourteen inches, in little more than one hundred yards; and from this fact it is inferred, that the rise in some parts of the inland country was far more considerable than on the coast ‡. Part of the coast thus elevated consisted of granite, in which parallel fissures were caused, some of which were traced for a mile and a half inland. Cones of earth, about four feet high, were thrown up in several districts, by the forcing up of water mixed with sand, through funnel-shaped hollows—a phenomenon very common in Calabria, and the explanation of which will hereafter be considered. Those houses in Chili, of which the foundations were on rock, were less damaged than such as were built on alluvial soil. The area over which this permanent alteration of level extended, was estimated at one hundred thousand square miles. The whole country, from the foot of the Andes to a great distance under the sea, is supposed to have been raised, the greatest rise being at the distance of about two miles from the shore. "The rise upon the coast was from two to four feet:—at the distance of a mile inland it must have been from five to six, or seven feet §." The soundings in the harbour of Valparaiso have been materially changed by this shock, and the bottom has become shallower. The shocks continued up to the end of September, 1823: even then, forty-eight hours seldom passed without one, and sometimes two or three were felt during twenty-four hours. Mrs.

* See Geol. Trans., vol. i., second series; and also Journ. of Sci., 1824, vol. xvii., p. 40.

† Geol. Trans., vol. i., second series, p. 415.

‡ Journ. of Sci., vol. xvii., p. 42.

§ Ibid., pp. 40, 45.

Graham observed, after the earthquake of 1822, that, besides the beach newly raised above high-water mark, there were several older elevated lines of beach one above the other, consisting of shingle mixed with shells, extending in a parallel direction to the shore, to the height of fifty feet above the sea*.

Aleppo, 1822.—In 1822 Aleppo was destroyed by an earthquake, and alterations are said to have been caused in the level of the land; but of these we have no exact details. At the same time two rocks were reported by the captain of a French vessel to have risen from the sea, in the neighbourhood of Cyprus, an island well known to be subject to subterranean movements, and almost under the same latitude as Aleppo†. In these and similar instances, where there is no evidence of a submarine eruption, it is not the magnitude of the masses lifted above the sea which are of importance, but the indication apparently afforded by them, that a submarine tract, of which they merely form the highest points, has undergone some change of level.

Ionian Isles, 1820.—In the year 1820, from the 15th of February to the 6th of March, Santa Maura, one of the Ionian isles, experienced a succession of destructive earthquakes. Immediately afterwards a rocky island was observed not far from the coast, which had never been known before‡. No indications of a submarine eruption were observed on this spot: it is, therefore, most probable that this rock was elevated by the earthquake; but an examination of its structure is much to be desired.

Island of Sumbawa, 1815.—In April, 1815, one of the most frightful eruptions recorded in history occurred in the mountain Tomboro, in the island of Sumbawa. It began on the 5th of April, and was most violent on the 11th and 12th, and did not entirely cease till July. The sound of the explosions was heard in Sumatra, at the distance of nine hundred

* Geol. Trans., vol. i., second series, p. 415.

† Journ. of Sci., vol. xiv., p. 450.

‡ Allgemeine Zeitung, 1820, No. 146. Verneul, Journal des Voyages, vol. vi., p. 383; cited by Von Hoff, vol. ii., p. 180.

and seventy geographical miles in a direct line; and at Ternate in an opposite direction, at the distance of seven hundred and twenty miles. Out of a population of twelve thousand, only twenty-six individuals survived on the island. Violent whirlwinds carried up men, horses, cattle, and whatever else came within their influence, into the air, tore up the largest trees by the roots, and covered the whole sea with floating timber*. Great tracts of land were covered by lava, several streams of which, issuing from the crater of the Tomboro mountain, reached the sea. So heavy was the fall of ashes, that they broke into the Resident's house at Bima, forty miles east of the volcano, and rendered it, as well as many other dwellings in the town, uninhabitable. On the side of Java, the ashes were carried to the distance of three hundred miles, and two hundred and seventeen towards Celebes, in sufficient quantity to darken the air. The floating cinders to the westward of Sumatra formed on the 12th of April a mass two feet thick, and several miles in extent, through which ships with difficulty forced their way. The darkness occasioned in the daytime by the ashes in Java was so profound, that nothing equal to it was ever witnessed in the darkest night. Although this volcanic dust when it fell was an impalpable powder, it was of considerable weight, when compressed, a pint of it weighing twelve ounces and three quarters. Along the sea-coast of Sumbawa, and the adjacent isles, the sea rose suddenly to the height of from two to twelve feet, a great wave rushing up the estuaries, and then suddenly subsiding. Although the wind at Bima was still during the whole time, the sea rolled in upon the shore, and filled the lower parts of the houses with water a foot deep. Every prow and boat was forced from the anchorage, and driven on shore.

On the 19th of April, says one of Raffles's correspondents, "we grounded on the bank of Bima town. The anchorage at Bima must have altered considerably, as where we grounded the Ternate cruiser lay at anchor in six fathoms a few months before." Unfortunately no facts are stated by which we may judge with certainty whether this shoal, implying a change of depth of more than thirty feet, was caused by an accumulation of ashes, or by an upheaving of the bottom of the sea. It is

* Raffles's Java, vol. i., p. 28.

stated, however, that the surrounding country was covered with ashes. On the other hand, the town called Tomboro, on the west side of the volcano, was overflowed by the sea, which encroached upon the shore at the foot of the volcano, so that the water remained permanently eighteen feet deep in places where there was land before. Here we may observe, that the amount of subsidence of land was very apparent *in spite of the ashes*, which would naturally have caused the limits of the coast to be extended.

The area over which tremulous noises and other volcanic effects extended, was one thousand English miles in circumference, including the whole of the Molucca islands, Java, a considerable portion of Celebes, Sumatra, and Borneo. In the island of Amboyna, in the same month and year, the ground opened, threw out water, and then closed again *. We may conclude, by reminding the reader, that but for the accidental presence of Sir Stamford Raffles, then governor of Java, we should scarcely have heard in Europe of this tremendous catastrophe. He required all the residents in the various districts under his authority to send in a statement of the circumstances which occurred within their own knowledge; but, valuable as were their communications, they are often calculated to excite rather than to satisfy the curiosity of the geologist. They mention that similar effects, though in a less degree, had about seven years before accompanied an eruption of Carang Assam, a volcano in the island of Bali, west of Sumbawa; but no particulars of this catastrophe are recorded †.

Cutch, 1819.—A violent earthquake occurred at Cutch, in Bombay, on the 16th of June, 1819. The principal town, Bhooi, was converted into a heap of ruins, and its stone buildings thrown down. The shock extended to Ahmedhabad, where it was very destructive; and at Poonah, four hundred miles farther, it was feebly felt. At the former city, the great mosque erected by Sultan Ahmed nearly four hundred and fifty years before, fell to the ground, attesting how long a period had elapsed since a shock of similar violence had visited that point. At Anjar, the fort, with its towers and guns, was hurled to the ground in

* Raffles's Hist. of Java, vol. i., p. 25.—Ed. Phil. Journ., vol. iii., p. 389.

† Life and Services of Sir Stamford Raffles, p. 241. London, 1830.

one common mass of ruin. The shocks continued some days until the 20th, when, thirty miles from Bhooi, a volcano burst out in eruption, and the convulsions ceased. Although the ruin of towns was great, the face of Nature in the inland country, says Captain Macmurdo, was not visibly altered. In the hills some large masses only of rock and soil were detached from the precipices; but the eastern and almost deserted channel of the Indus, which bounds the province of Cutch, was greatly changed. This estuary or inlet of the sea was, before the earthquake, fordable at Luckput, being only about a foot deep when the tide was at ebb, and at flood tide never more than six feet; but it was deepened at the fort of Luckput, after the shock, to more than *eighteen feet at low water**. On sounding other parts of the channel, it was found, that where previously the depth of the water at flood never exceeded one or two feet, it had become from four to ten feet deep; and this increase of depth extended from Cutch to the Sindh shore, a distance of three or four miles. The channel of the Runn, which extends from Luckput round the north of the province of Cutch, was sunk so much, that, instead of being dry as before during that period of the year, it was no longer fordable, except at one spot only. By these remarkable changes of level, a part of the inland navigation of that country, which had been closed for centuries, became again practicable.

The fort and village of Sindree, situated where the Runn joins the Indus, was overflowed; and, after the shock, the tops of the houses and wall were alone to be seen above the water, for the houses, although submerged, were not cast down. Had they been situated in the interior, where so many forts were levelled o the ground, their site would perhaps have been regarded as having remained comparatively unmoved. From this circumstance we may feel assured that great permanent upheavings and depressions of soil may be the result of earthquakes, without the inhabitants being in the least degree conscious of any change of level.

Cones of sand, six or eight feet in height, were thrown out of the lands near the Runn. Somewhat farther to the east of the line of this earthquake lies Oojain (called Ozene in the *Peryplus Maris Erythr).* Ruins of an old town are there

* Ed. Phil. Journ., vol. iv., p. 106.

found, a mile north of the present, sunk in the earth to the depth of from fifteen to sixteen feet, which sinking is known to have been the consequence of a tremendous catastrophe in the time of the Rajah Viermaditya.

Caraccas, 1812.—On the 26th of March, 1812, several violent shocks of an earthquake were felt in Caraccas. The surface undulated like a boiling liquid, and terrific sounds were heard underground. The whole city with its splendid churches was in an instant a heap of ruins, under which ten thousand of the inhabitants were buried. On the 5th of April, enormous rocks were detached from the mountains. It was believed that the mountain Silla lost from three hundred to three hundred and sixty feet of its height by subsidence; but this was an opinion not founded on any measurement. On the 27th of April, a volcano in St. Vincent's threw out ashes; and on the 30th, lava flowed from its crater into the sea, while its explosions were heard at a distance equal to that between Vesuvius and Switzerland, the sound being transmitted, as Humboldt supposes, through the ground. During the earthquake which destroyed Caraccas, an immense quantity of water was thrown out at Valecillo near Valencia, as also at Porto Cabello, through openings in the earth; and in the Lake Maracaybo the water sank *.

Although the great change of level in the mountain Silla was not distinctly proved, the opinion of the inhabitants deserves attention, because we shall afterwards have to mention some well-authenticated alterations in the same district during preceding earthquakes. Humboldt observed that the Cordilleras, composed of gneiss and mica slate, and the country immediately at their foot, were more violently shaken than the plains.

South Carolina, 1811.—Previous to the destruction of Laguira and Caraccas in 1811, South Carolina was convulsed by earthquakes, and the shocks continued till those cities were destroyed. The valley also of the Mississippi, from the village of New Madrid to the mouth of the Ohio in one direction, and to the St. Francis in another, was convulsed to such

* Humboldt's Pers. Nar., vol. iv., p. 12; and Ed. Phil. Journ., vol. i., p. 272. 1819.

a degree as to create lakes and islands. Flint, the geographer, who visited the country seven years after the event, informs us, that a tract of many miles in extent, near the Little Prairie, became covered with water three or four feet deep; and when the water disappeared, a stratum of sand was left in its place. Large lakes of twenty miles in extent were formed in the course of an hour, and others were drained. The grave-yard at New Madrid was precipitated into the bed of the Mississippi. The inhabitants related that the earth rose in great undulations; and when these reached a certain fearful height, the soil burst, and vast volumes of water, sand, and pit-coal, were discharged as high as the tops of the trees. Flint saw hundreds of these deep chasms remaining in a tender alluvial soil seven years after. The people in the country, although inexperienced in such convulsions, had remarked that the chasms in the earth were in a direction from S.W. to N.E.; and they accordingly felled the tallest trees, and, laying them at right angles to the chasms, stationed themselves upon them. By this invention, when chasms opened more than once under these trees, several persons were prevented from being swallowed up *. At one period during this earthquake, the ground not far below New Madrid swelled up so as to arrest the Mississippi in its course, and to cause a temporary reflux of its waves. The motion of some of the shocks was horizontal, and of others perpendicular; and the vertical movement is said to have been much less desolating than the horizontal. If this be often the case, those shocks which injure cities least may often produce the greatest alteration of level.

Aleutian Islands, 1806.—In the year 1806 a new island, in the form of a peak, with some low conical hills upon it, rose from the sea among the Aleutian Islands, north of Kamtschatka. According to Langsdorf †, it was four geographical miles in circumference; and Von Buch infers from its magnitude, and from its not having again subsided below the level of the sea, that it did not consist merely of ejected matter, like Monte Nuovo, but of solid rock upheaved *. Another extraordinary eruption happened in the spring of the year 1814, in

* Silliman's Journ., Jan., 1829.

† Bemerkungen, auf einer Reise um die Welt., bd. ii., s. 209.

the sea near Unalaschka, in the same archipelago. A new isle was then produced of considerable size, and with a peak three thousand feet high, which remained standing for a year afterwards, though with somewhat diminished height.

Although it is not improbable that the earthquakes accompanying the tremendous eruptions above mentioned may have heaved up part of the bed of the sea, yet we must wait for fuller information before we assume this as a fact. The circumstance of these islands not having disappeared like Sabrina, may have arisen from the emission of lava. If Jorullo, for example, in 1759, had risen from a shallow sea to the height of one thousand seven hundred feet, instead of attaining that elevation above the Mexican plateau, the massive current of basaltic lava which poured out from its crater would have enabled it to withstand, for a long period, the action of a turbulent sea.

We are now about to pass on to the events of the eighteenth century; but, before we leave the consideration of those already enumerated, let us pause for a moment, and reflect how many remarkable facts of geological interest are afforded by the earthquakes above described, though they constitute but a small part of the convulsions even of the last thirty years. New rocks have risen from the waters; the coast of Chili for one hundred miles has been permanently elevated; part of the delta of the Indus has sunk down, and some of its shallow channels have become navigable; the town of Tomboro has been submerged, and twelve thousand of the inhabitants of Sumbawa have been destroyed. Yet, with a knowledge of these terrific catastrophes, witnessed during so brief a period by the present generation, will the geologist declare with perfect composure that the earth has at length settled into a state of repose? Will he continue to assert that the changes of relative level of land and sea, so common in former ages of the world, have now ceased? If, in the face of so many striking facts, he persists in maintaining this favorite dogma, it is in vain to hope that, by accumulating the proofs of similar convulsions during a series of antecedent ages, we shall shake his tenacity of purpose,

Si fractus illabatur orbis
Impavidum ferient ruinæ.

* Neue Allgem. Geogr. Ephemer. bd. iii., s. 348.

Quito, 1797.—On the morning of February 4th, 1797, the volcano of Tunguragua in Quito, and the surrounding district, for forty leagues, from south to north, and twenty leagues from west to east, experienced an undulating movement, which lasted four minutes. The same shock was felt over a tract of one hundred and seventy leagues from south to north, from Piura to Popayan; and one hundred and forty from west to east, from the sea to the river Napo. In the smaller district, first mentioned, every town was levelled to the ground; and Riobamba, Quero, and other places, were buried under masses detached from the mountains. At the foot of Tunguragua the earth was rent open in several places; and streams of water and fetid mud, called "moya," poured out, overflowing and wasting everything. In valleys one thousand feet broad, the water of these floods reached to the height of six hundred feet; and the mud deposit barred up the course of the river, so as to form lakes, which in some places continued for more than eighty days. Flames and suffocating vapours escaped from the lake Quilotoa, and killed all the cattle on its shores. The shocks continued all February and March, and on the 5th of April they recurred with almost as much violence as at first. We are told that the form of the surface in the district most shaken was entirely altered, but no exact measurements are given whereby we may estimate the degree of elevation or subsidence *. Indeed it would be difficult, except in the immediate neighbourhood of the sea, to obtain any certain standard of comparison, if the levels were really as much altered as the narrations seem to imply.

Cumana, 1797.—In the same year, on the 14th of December, the small Antilles experienced subterranean movements, and four-fifths of the town of Cumana was shaken down by a vertical shock. The form of the shoal of Mornerouge, at the mouth of the river Bourdones, was changed by an upheaving of the ground†.

* Cavanilles, Journ. de Phys. tome xlix., p. 230. Gilberts, Annalen, bd. vi., p. 67. Humboldt's Voy., p. 317.

† Humboldt's Voy., Relat. Hist., part i., p. 309.

Caraccas, 1790.—In the Caraccas, near where the Caura joins the Orinoco, between the towns San Pedro de Alcantara and San Francisco de Aripao, an earthquake on St. Matthew's Day, 1790, caused a sinking in of the granitic soil, and left a lake eight hundred yards in diameter, and from eighty to one hundred in depth. It was a portion of the forest of Aripao which subsided, and the trees remained green for several months under water*.

Sicily, 1790.—On the 18th of March in the same year, at S. Maria di Niscemi, some miles from Terranuova, near the south coast of Sicily, the ground gradually sunk down for a circumference of three Italian miles, during seven shocks; and, in one place, to the depth of thirty feet. It continued to subside to the end of the month. Several fissures sent forth sulphur, petroleum, steam, and hot water; and a stream of mud which flowed for two hours, and covered a space sixty feet long, and thirty broad. This happened far from both the ancient and modern volcanic district, in a group of strata, consisting chiefly of blue clay†.

Java, 1786.—About the year 1786 an earthquake was felt at intervals, for the period of four months, in the neighbourhood of Batur, in Java, and an eruption followed. Various rents were formed which emitted a sulphureous vapour; separate tracts sunk away, and were swallowed by the earth. Into one of these the rivulet Dotog entered, and afterwards continued to follow a subterraneous course. The village of Jampang was buried in the ground, with thirty-eight of its inhabitants, who had not time to escape. We are indebted to Dr. Horsfield for having verified the above mentioned facts‡.

* Humboldt's Voy., Relat. Hist., part ii., p. 632. † Ferrara, Campi. fl., p. 51.
‡ Batav. Trans., vol. viii., p. 141.

CHAPTER XXIV.

Earthquake in Calabria, February 5th, 1783—Shocks continued to the end of the year 1786—Authorities—Extent of the area convulsed—Geological structure of the district—Difficulty of ascertaining changes of relative level even on the sea-coast—Subsidence of the quay at Messina—Shift or fault in the Round Tower of Terranuova—Movement in the stones of two obelisks—Alternate opening and closing of fissures—Cause of this phenomenon—Large edifices engulphed—Dimensions of new caverns and fissures—Gradual closing in of rents—Bounding of detached masses into the air—Landslips—Buildings transported entire, to great distances—Formation of fifty new lakes—Currents of mud—Small funnel-shaped hollows in alluvial plains—Fall of cliffs along the sea-coast—Shore near Scilla inundated—State of Stromboli and Etna during the shocks—Illustration afforded by this earthquake of the mode in which valleys are formed.

Of the numerous earthquakes which have occurred in different parts of the globe, during the last hundred years, that of Calabria, in 1783, is the only one of which the geologist can be said to have such a circumstantial account as to enable him fully to appreciate the changes which this cause is capable of producing in the lapse of ages. The shocks began in February, 1783, and lasted for nearly four years, to the end of 1786. Neither in duration, nor in violence, nor in the extent of territory moved, was this convulsion remarkable, when contrasted with many experienced in other countries, both during the last and present century; nor were the alterations which it occasioned in the relative level of hill and valley, land and sea, so great as those effected by some subterranean movements in South America, in our own times. The importance of the earthquake in question arises from the circumstance, that Calabria is the only spot hitherto visited, both during and after the convulsions, by men possessing sufficient leisure, zeal, and scientific information, to enable them to collect and describe with accuracy the physical facts which throw light on geological questions.

Among the numerous authorities, Vivenzio, physician to the

King of Naples, transmitted to the court a regular statement of his observations during the continuance of the shocks; and his narrative is drawn up with care and clearness * Francesco Antonio Grimaldi, then secretary of war, visited the different provinces at the king's command, and published a most detailed description of the permanent changes in the surface †. He measured the length, breadth, and depth of the different fissures and gulphs which opened, and ascertained their number in many provinces. His comments, moreover, on the reports of the inhabitants, and his explanations of their relations, are judicious and instructive. Pignataro, a physician residing at Monteleone, a town placed in the very centre of the convulsions, kept a register of the shocks, distinguishing them into four classes, according to their degree of violence. From his work, it appears that, in the year 1783, the number was nine hundred and forty-nine, of which five hundred and one were shocks of the first degree of force; and in the following year there were one hundred and fifty-one, of which ninety-eight were of the first magnitude. Count Ippolito, also, and many others, wrote descriptions of the earthquake; and the Royal Academy of Naples, not satisfied with these and other observations, sent a deputation from their own body into Calabria, before the shocks had ceased, who were accompanied by artists instructed to illustrate by drawings the physical changes of the district, and the state of ruined towns and edifices. Unfortunately these artists were not very successful in their representations of the condition of the country, particularly when they attempted to express, on a large scale, the extraordinary revolutions which many of the great and minor river-courses underwent. But many of the plates published by the Academy are valuable; and we shall frequently avail ourselves of them to illustrate the facts about to be described ‡. In addition to these Neapolitan sources of information, our countryman, Sir William Hamilton, surveyed the district, not without some personal risk, before the shocks had ceased; and his sketch, published in the Philosophical Transactions, supplies many facts that would otherwise have been lost. He has explained

* Istoria de' Tremuoti della Calabria, del 1783.

† Descriz. de' Tremuoti Accad. nelle Calabria nel 1783. Napoli, 1784.

‡ Istoria de' Fenomeni del Tremoto, &c. nell' an. 1783, posta in luce dalla Real. Accad., &c., di Nap. Napoli, 1784, fol.

in a rational manner many events which, as related in the language of some eye-witnesses, appeared marvellous and incredible. Dolomieu also examined Calabria, soon after the catastrophe, and wrote an account of the earthquake, correcting a mistake into which Hamilton had fallen, who supposed that a part of the tract shaken had consisted of volcanic tuff. It is, indeed, a circumstance which enhances the geological interest of the commotions which so often modify the surface of Calabria, that they are confined to a country where there are neither ancient nor modern rocks of igneous origin; so that at some future time, when the era of disturbance shall have passed by, the cause of former revolutions will be as latent as in parts of Great Britain now occupied exclusively by ancient marine formations.

The convulsion of the earth, sea, and air, extended over the whole of Calabria Ultra, the south-east part of Calabria Citra, and across the sea to Messina and its environs—a district lying between the 38th and 39th degrees of latitude. The concussion was perceptible over a great part of Sicily, and as far north as Naples; but the surface over which the shocks acted so forcibly as to excite intense alarm, did not generally exceed five hundred square miles in circumference. The soil of that part of Calabria is composed chiefly, like the southern part of Sicily, of calcareo-argillaceous strata of great thickness, containing marine shells. This clay is sometimes associated with beds of sand and limestone. For the most part these formations resemble in appearance and consistency the Subapennine marls, with their accompanying sands and sandstones; and the whole group bears considerable resemblance, in the yielding nature of its materials, to most of our tertiary deposits in France and England. Chronologically considered, however, the Calabrian formations are comparatively of very modern date, and abound in fossil shells referrible to species now living in the Mediterranean.

We learn from Vivenzio, that on the 20th and 26th of March, 1783, earthquakes occurred in the islands of Zante, Cephalonia, and St. Maura; and in the last-mentioned isle several public edifices and private houses were overthrown, and many people destroyed. We have already shown that the Ionian Isles fall within the line of the same great volcanic region as Calabria; so that both earthquakes were probably derived from a common source, and it is not improbable that the bed of the whole intermediate sea was convulsed.

If the city of Oppido, in Calabria, be taken as a centre, and round that centre a circle be described with a radius of twenty-two miles, this space will comprehend the surface of the country which suffered the greatest alteration, and where all the towns and villages were destroyed. But if we describe the circle with a radius of seventy-two miles, this will then comprehend the whole country that had any permanent marks of having been affected by the earthquake. The first shock, of February 5th, 1783, threw down, in two minutes, the greater part of the houses in all the cities, towns, and villages, from the western flanks of the Apennines in Calabria Ultra, to Messina in Sicily, and convulsed the whole surface of the country. Another occurred on the 28th of March, with almost equal violence. The granitic chain which passes through Calabria from north to south, and attains the height of many thousand feet, was shaken but slightly; but it is said that a great part of the shocks which were propagated with a wave-like motion through the recent strata from west to east, became very violent when they reached the point of junction with the granite, as if a reaction was produced where the undulatory movement of the soft strata was suddenly arrested by the more solid rocks. The surface of the country often heaved like the billows of a swelling sea, which produced a swimming in the head like sea-sickness. It is particularly stated, in almost all the accounts, that just before each shock the clouds appeared motionless; and although no explanation is offered of this phenomenon, it is obviously the same as that observed in a ship at sea when it pitches violently. The clouds seem arrested in their career as often as the vessel rises in a direction contrary to their course; so that the Calabrians must have experienced precisely the same motion on the land.

We shall first consider that class of physical changes produced by the earthquake, which are connected with changes in the relative level of the different parts of the land; and afterwards describe those which are more immediately connected with the derangement of the regular drainage of the country, and where the force of running water co-operated with that of the earthquake.

In regard to alterations of relative level, none of the accounts establish that they were on a considerable scale; but it must always be remembered, that in proportion to the area moved is the difficulty of proving that the general level has undergone any change, unless the sea-coast happens to have participated in the principal movement. Even then it is often impossible to determine whether an elevation or depression even of several feet has occurred, because there is nothing novel in a band of sand and shingle of unequal breadth above the level of the sea, marking the point reached by the waves during spring-tides or the most violent tempests. The scientific investigator has not sufficient topographical knowledge to discover whether the extent of beach has diminished or increased; and he who has the necessary local information feels no interest in ascertaining the amount of the rise or fall of the ground. Add to this the great difficulty of making correct observations, in consequence of the enormous waves which roll in upon a coast during an earthquake, and efface every landmark near the shore.

It is evidently in sea-ports alone that we can look for very accurate indications of slight changes of level; and when we find them, we may presume that they would not be rare at other points, if equal facilities of comparing relative altitudes were afforded. Grimaldi states (and his account is confirmed by Hamilton and others) that at Messina in Sicily the shore was rent; and the soil along the port, which before the shock was perfectly level, was found afterwards to be inclined towards the sea, the sea itself near the "Banchina" becoming deeper, and its bottom in several places disordered. The quay also sank down about fourteen inches below the level of the sea, and the houses in its vicinity were much fissured*. Among various proofs of partial elevation and depression in the interior, the Academicians mention, in their Survey, that the ground was sometimes on the same level on both sides of new ravines and fissures, but sometimes there had been a considerable shifting, either by the upheaving of one side or the subsidence of the other. Thus, on the sides of long rents in the territory of Soriano, the stratified masses had altered their relative position to the extent of from eight to fourteen palms (six to ten and a half feet). Similar shifts in the strata are alluded to in the territory of Polistena, where there appeared

* Phil. Trans., 1783.

innumerable fissures in the earth. One of these was of great length and depth; and in parts, the level of the corresponding

Deep fissure near Polistena, caused by the earthquake of 1783.

sides was greatly changed. In the town of Terranuova, some houses were seen uplifted above the common level, and others adjoining sunk down into the earth. In several streets, the soil appeared thrust up, and abutted against the walls of houses; a large circular tower of solid masonry, which had withstood the general destruction, was divided by a vertical rent, and one side was upraised, and the foundations heaved out of the

Shift or "fault" in the round tower of Terranuova in Calabria, occasioned by the earthquake of 1783.

ground. It was compared by the Academicians to a great tooth half extracted from the alveolus, with the upper part of the fangs exposed. (See cut No. 20.)

Along the line of this shift, or "fault" as it would be termed technically by miners, the walls were found to adhere firmly to each other, and to fit so well, that the only signs of their having been disunited was the want of correspondence in the courses of stone on either side of the rent.

In some walls which had been thrown down, or violently shaken, in Monteleone, the separate stones were parted from the mortar so as to leave an exact mould where they had rested, whereas in other cases the mortar was ground to dust between the stones.

It appears that the wave-like motions, and those which are called vorticose or whirling in a vortex, often produced effects of the most capricious kind. Thus, in some streets of Monteleone, every house was thrown down but one; in others, all but two; and the buildings which were spared were often scarcely in the least degree injured.

In many cities of Calabria, all the most solid buildings were thrown down, while those which were slightly built, escaped; but at Rosarno, as also at Messina, in Sicily, it was precisely the reverse, the massive edifices being the only ones that stood.

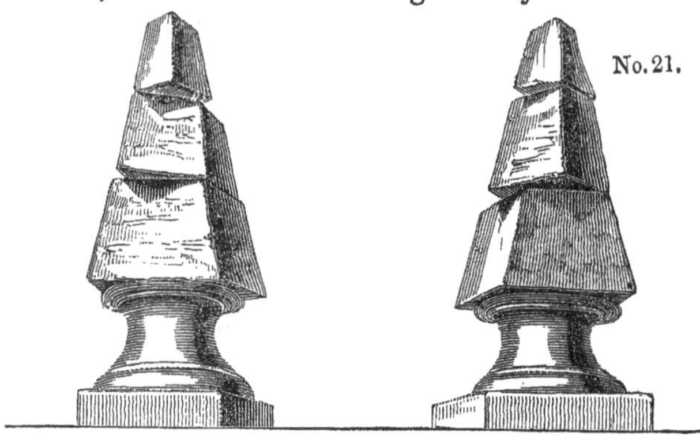

No. 21.

Two obelisks (No. 21) placed at the extremities of a magnificent façade in the convent of S. Bruno, in a small town called Stefano del Bosco, were observed to have undergone a movement of a singular kind. The shock which agitated the building is described as having been horizontal and vorticose. The pedestal of each obelisk remained in its original place; but the separate stones above were turned partially round, and removed sometimes nine inches from their position, without falling.

It appears evident that a great part of the rending and fissuring of the ground was the effect of a violent motion from below upwards; and in a multitude of cases where the rents and chasms opened and closed alternately, we must suppose that the earth was by turns heaved up, and then let fall again. We may conceive the same effect to be produced on a small scale, if, by some mechanical force, a pavement composed of large flags of stone should be raised up and then allowed to fall suddenly, so as to resume its original position. If any small pebbles happened to be lying on the line of contact of two flags, they would fall into the opening when the pavement rose, and be swallowed up, so that no trace of them would appear after the subsidence of the stones. In the same manner, when the earth was upheaved, large houses, trees, cattle, and men were engulphed in an instant in chasms and fissures; and when the ground sank down again, the earth closed upon them, so that no vestige of them was discoverable on the surface. In many instances, individuals were swallowed up by one shock, and then thrown out again alive, together with large jets of water, by the shock which immediately succeeded.

At Jerocarne, a country which, according to the Academicians, was *lacerated* in a most extraordinary manner, the fissures ran in every direction like cracks on a broken pane

No. 22.

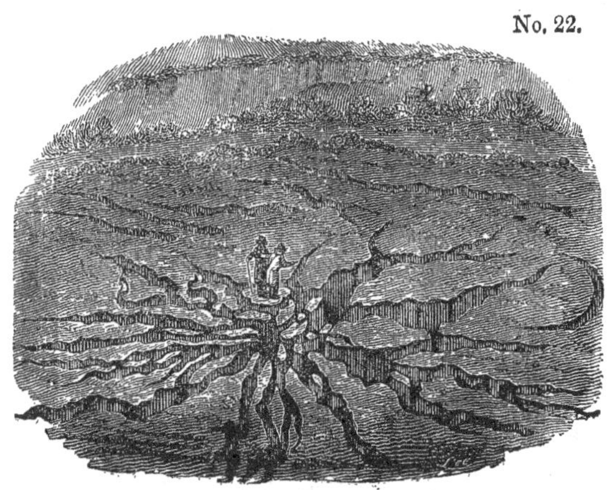

Fissures near Jerocarne, in Calabria, caused by the earthquake of 1783.

of glass (see cut No. 22); and, as a great portion of them remained open after the shocks, it is very possible that this country was permanently upraised.

In the vicinity of Oppido, the central point from which the earthquake diffused its violent movements, many houses were swallowed up by the yawning earth, which closed immediately over them. In the adjacent district also of Cannamaria, four farm-houses, several oil-stores, and some spacious dwelling-houses were so completely engulphed in one chasm, that not a vestige of them was afterwards discernible. The same phenomenon occurred at Terranuova, S. Christina, and Sinopoli. The Academicians state particularly that when deep abysses had opened in the argillaceous strata of Terranuova, and houses had sunk into them, the sides of the chasms closed with such violence, that, on excavating afterwards to recover articles of value, the workmen found the contents and detached parts of the buildings jammed together so as to become one compact mass. It is unnecessary to accumulate examples of similar occurrences; but so many are well authenticated during this earthquake in Calabria, that we may, without hesitation, yield assent to the accounts of catastrophes of the same kind repeated again and again in history, where whole towns are declared to have been engulphed, and nothing but a pool of water or tract of sand left in their place.

On the sloping side of a hill near Oppido, a great chasm opened, and, although a large quantity of soil was precipitated

No. 23.

Chasm formed by the earthquake of 1783, near Oppido, in Calabria.

into the abyss, together with a considerable number of olive-trees and part of a vineyard, a great gulph remained after the shock in the form of an amphitheatre, five hundred feet long and two hundred feet deep (see cut No. 23).

According to Grimaldi, many fissures and chasms, formed by the first shock of February 5th, were greatly widened, lengthened, and deepened by the violent convulsions of March 28th. In the territory of San Fili, this observer found a new ravine, half a mile in length, two feet and a half broad, and twenty-five feet deep; and another of similar dimensions in the territory of Rosarno. A ravine *nearly a mile long, one hundred and five feet broad*, and thirty feet deep, opened in the district of Plaisano, where, also, two gulphs were caused—one in a place called Cerzulle, three quarters of a mile long, *one hundred and fifty feet broad*, and above *one hundred feet deep*, and another at La Fortuna, nearly a quarter of a mile long, above thirty feet in breadth, and no less than *two hundred and twenty-five feet deep*. In the district of Fosolano three gulphs opened: one of these measured three hundred feet square, and above thirty feet deep; another was nearly half a mile long, fifteen feet broad, and above thirty feet deep; the third was seven hundred and fifty feet square. Lastly, a calcareous mountain, called Zefirio, at the southern extremity of the Italian peninsula, was cleft in two for the length of nearly half a mile, and an irregular breadth of many feet. Some of these chasms were in the form of a crescent. The annexed cut (No. 24) repre-

No. 24.

Chasm in the hill of St. Angelo, near Soriano, in Calabria; caused by the earthquake of 1783.

sents one by no means remarkable for its dimensions, which remained open by the side of a small pass over the hill of St. Angelo, near Soriano. The small river Mesima is seen in the foreground.

In the vicinity of Seminara, a lake was suddenly formed by the opening of a great chasm, from the bottom of which water issued. This lake was called Lago del Tolfilo. It extended 2380 palms in length, by 1250 in breadth, and 70 in depth. The inhabitants, dreading the miasma of this stagnant pool, endeavoured, at great cost, to drain it by canals, but without success, as it was fed by springs issuing from the bottom of the deep chasm. A small circular subsidence occurred not far from Polistena, of which a representation is given in the annexed cut.

No. 25.

Circular pond near Polistena, in Calabria, caused by the earthquake in 1783.

Sir W. Hamilton was shown several deep fissures in the vicinity of Mileto, which, although not one of them was above a foot in breadth, had opened so wide during the earthquake as to swallow up an ox and near one hundred goats. The Academicians also found, on their return through districts which they had passed at the commencement of their tour, that many rents had in that short interval gradually closed in, so that their width had diminished several feet, and the opposite walls had sometimes nearly met. It is natural that this should happen in argillaceous strata, while in more solid rocks we may expect that fissures will remain open for ages. Should this be ascertained to be a general fact in countries convulsed by earthquakes, it would afford a satisfactory explanation of a common phenomenon in mineral veins. Such veins often retain their full size so long as the rocks consist of limestone, granite, or other indurated materials; but they contract their dimensions, become mere threads, or are even entirely cut off, where masses of an argillaceous nature are interposed. If we suppose the filling up of fissures with metallic and other ingredients to be a process requiring ages for its completion, it is obvious that the opposite walls of rents, where strata consist of yielding materials, must collapse or approach very near to each other before sufficient time is allowed for the accretion of a large quantity of veinstone.

It is stated by Grimaldi that the thermal waters of St. Euphemia, in Terra di Amato, which first burst out during the earthquake of 1638, acquired, in February 1783, an augmentation both in quantity and degree of heat. This fact appears to indicate a connexion between the heat of the interior and the fissures caused by the Calabrian earthquakes, notwithstanding the absence of volcanic rocks either ancient or modern in that district.

The violence of the movement of the ground upwards was singularly illustrated by what the Academicians call the "sbalzo," or bounding into the air, to the height of several yards, of masses slightly adhering to the surface. In some towns a great part of the pavement-stones were thrown up, and found lying with their lower sides uppermost. In these cases we must suppose that they were propelled upwards by the momentum which they had acquired, and that the adhesion of one end of the mass being greater than that of the other, a rotatory motion had been communicated to them. When the stone was projected to a sufficient height to perform somewhat more than a quarter of a revolution in the air, it pitched down on its edge and fell with its lower side uppermost.

The next class of effects to be considered, are those more immediately connected with the formation of valleys, in which the action of water was often combined with that of the earthquake. The country agitated was composed, as we before stated, chiefly of argillaceous strata, intersected by deep narrow valleys, sometimes from five to six hundred feet deep. As the boundary cliffs were in great part vertical, it will readily be conceived that, amidst the various movements of the earth, the precipices overhanging the rivers, being without support on one side, were often thrown down. We find, indeed, that inundations produced by obstructions in river-courses are among the most disastrous consequences of great earthquakes in all parts of the world; for the alluvial plains in the bottoms of valleys are usually the most fertile and well peopled parts of the whole country, and whether the site of a town is above or below a temporary barrier in the channel of a river, it is exposed to injury by the waters either of a lake or a flood.

From each side of the deep valley or ravine of Terranuova, enormous masses of the adjoining flat country were detached and cast down into the course of the river, so as to give rise to great lakes. Oaks, olive-trees, vineyards, and corn, were often seen growing at the bottom of the ravine, as little injured as their companions from which they were separated in the plain above at least five hundred feet higher, and at the distance of about three-quarters of a mile. In one part of this ravine was an enormous mass, two hundred feet high, and about four hundred feet in diameter at its basis, which had been detached by some former earthquake. It is well attested that this mass travelled down the ravine near four miles, having been put in motion by the earthquake of the 5th of February. Hamilton, after examining the locality, declared that this phenomenon might be accounted for by the declivity of the valley, the great abundance of rain which fell, and the great weight of the alluvial matter which pressed behind it. The momentum of the "terre movitine," or lavas, as the flowing mud is called in the country, is no doubt very great; but the transportation of masses that might be compared to small hills, for a distance of several miles at a time, is an effect which could never have been anticipated: and the fact should serve as a hint to those geologists who are fond of appealing to alluvial phenomena as proofs of the superior violence of aqueous causes in former ages.

The first account sent to Naples of the two great slides or landslips above alluded to, which caused a great lake near Terranuova, was couched in these words:—"Two mountains on the opposite sides of a valley walked from their original position until they met in the middle of the plain, and there joining together, they intercepted the course of a river, &c." The expressions here used resemble singularly those applied to phenomena, probably very analogous, which are said to have occurred at Fez, during the great Lisbon earthquake, as also in Jamaica and Java at other periods.

Not far from Soriano, which was levelled to the ground by the great shock of February the 5th, a small valley, containing a beautiful olive-grove, called Fra Ramondo, underwent a most extraordinary revolution. Innumerable fissures first traversed the river-plain in all directions, and absorbed the water until the argillaceous substratum became soaked, and a great part of it was reduced to a state of fluid paste. Strange alterations in the outline of the ground were the consequence, as the soil to a great depth was easily moulded into any form. In addition to this change, the ruins of the neighbouring hills were precipitated into the hollow; and while many olives were

No. 26.

Changes of the surface at Fra Ramondo, near Soriano, in Calabria.

1. Portion of a hill covered with olives thrown down.
2. New bed of the river Caridi. 3. Town of Soriano.

uprooted, others remained growing on the fallen masses, and inclined at various angles (see cut No. 26). The small river Caridi was entirely concealed for many days; and when at length it reappeared, it had shaped for itself an entirely new channel.

Near Seminara, an extensive olive-ground and orchard were hurled to a distance of two hundred feet, into a valley sixty feet in depth. At the same time a deep chasm was riven in another part of the high plateau from which the orchard had been detached, and the river immediately entered the fissure, leaving its former bed completely dry. A small inhabited house, standing on the mass of earth carried down into the valley, went along with it entire, and without injury to the inhabitants. The olive-trees, also, continued to grow on the land which had slid into the valley, and bore the same year an abundant crop of fruit.

Two tracts of land on which a great part of the town of Polistena stood, consisting of some hundreds of houses, were detached into a contiguous ravine, and nearly across it about half a mile from their original site; and, what is most extraordinary, several of the inhabitants were dug out from the ruins alive, and unhurt.

Two tenements, near Mileto, called the Macini and Vaticano, about a mile long, and half a mile broad, were carried for a mile down a valley. A thatched cottage, together with large olive and mulberry-trees, most of which remained erect, were carried uninjured to this extraordinary distance. According to Hamilton, the surface removed had been long undermined by rivulets, which were afterwards in full view on the bare spot deserted by the tenements. The earthquake seems to have opened a passage in the adjoining argillaceous hills, by which water charged with loose soil had suddenly taken its course into the subterranean channels of the rivulets immediately under the tenements, so that the entire piece of ground was floated off. Another example of subsidence, where the edifices were not destroyed, is mentioned by Grimaldi, as having taken place in the city of Catanzaro, the capital of the province of that name. The houses in the quarter called San Giuseppe subsided with the ground to various depths from two to four feet, but the buildings remained uninjured.

It would be tedious, and our space would not permit us, to follow the different authors through their local details of landslips produced in numerous minor valleys; but they are highly interesting, as showing to how great an extent the power of rivers to widen valleys, and to carry away large portions of soil towards the sea, is increased where earthquakes are of

No. 27.

Landslips near Cinquefrondi, caused by the earthquake of 1783.

periodical occurrence. Among other territories, that of Cinquefrondi was greatly convulsed, various portions of soil being raised or sunk, and innumerable fissures traversing the country in all directions (see cut No. 27). Along the flanks of a small valley in this district there appears to have been an almost uninterrupted line of landslips.

Vivenzio states, that near Sitizzano a valley was very nearly filled up to a level with the high grounds on each side, by the enormous masses detached from the boundary hills, and cast down into the course of two streams. By this barrier a lake was formed of great depth, about two miles long and a mile broad. The same author mentions that upon the whole, there were fifty lakes occasioned during the convulsions, and he assigns localities to all of these. The government surveyors enumerated two hundred and fifteen lakes, but they included in this number many small and insignificant ponds.

Near S. Lucido, among other places, the soil is described as having been "dissolved," so that large torrents of mud inundated all the low grounds, like lava. Just emerging from this mud, the tops only of trees and of the ruins of farm-houses, were seen. Two miles from Laureana the swampy soil in two ravines became filled with calcareous matter, which oozed out from the ground immediately before the first great shock. This mud, rapidly accumulating, began, ere long, to roll onward like a flood of lava into the valley, where the two streams uniting, moved forward with increased impetus from east to west. It now presented a breadth of three hundred palms by twenty in depth, and before it ceased to move, covered a surface equal in length to an Italian mile. In its progress it overwhelmed a flock of thirty goats, and tore up by the roots many olive and mulberry-trees, which floated like ships upon its surface. When this calcareous lava had ceased to move, it gradually became dry and hard, during which process the mass was lowered ten palms. It contained fragments of earth of a ferruginous colour, and emitting a sulphureous smell.

Many of the appearances exhibited in the alluvial plains indicate clearly the alternate rising and sinking of the ground. The first effect of the more violent shocks was usually to dry up the rivers, but they immediately afterwards overflowed their banks. Along the alluvial plains, and in marshy places, an immense number of cones of sand were thrown up. These appearances Hamilton explains, by supposing that the first movement raised the fissured plain from below upwards, so that the rivers and stagnant waters in bogs sank down, or at least were not upraised with the soil. But when the ground

No. 28.

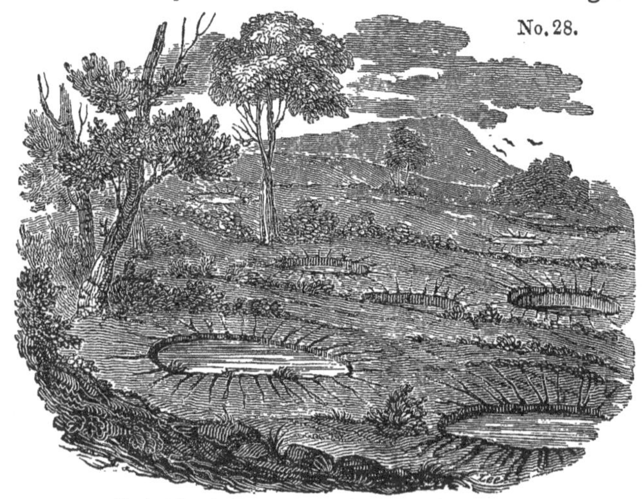

Circular hollows in the plain of Rosarno, formed by the earthquake of 1783.

returned with violence to its former position, the water was thrown up in jets through fissures *.

In the Report of the Academy, we find that some plains were covered with circular hollows, for the most part about the size of carriage-wheels, but often somewhat larger or smaller. When filled with water to within a foot or two of the surface, they appeared like wells; but, in general, they were filled with dry sand, sometimes with a concave surface, and at other times convex. On digging down, they found them to be funnel-shaped, and the moist loose sand in the centre marked the tube up which the water spouted. The annexed cut represents a section of one of these inverted cones when the water had disappeared, and nothing but dry micaceous sand remained.

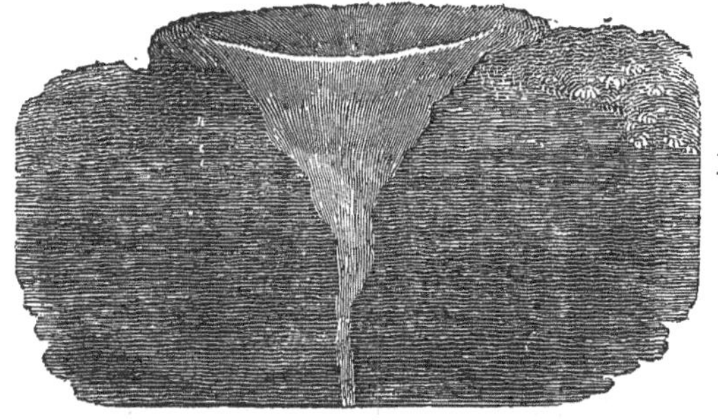

No. 29.

Along the sea-coast of the straits of Messina, near the celebrated rock of Scilla, the fall of huge masses detached from the bold and lofty cliffs overwhelmed many villas and gardens. At Gian Greco a continuous line of cliff, for a mile in length, was thrown down. Great agitation was frequently observed in the bed of the sea during the shocks, and, on those parts of the coast where the movement was most violent, all kinds of fish were taken in greater abundance, and with much greater facility. Some rare species, as that called Cicirelli, which usually lie buried in the sand, were taken on the surface of the waters in great quantity. The sea is said to have boiled up near Messina, and to have been agitated as if by a copious discharge of vapours from its bottom. The Prince of Scilla

* Phil. Trans., vol. lxxiii., p. 180.

had persuaded a great part of his vassals to betake themselves to their fishing-boats for safety, and he himself had gone on board. On the night of the 5th of February, when some of the people were sleeping in the boats, and others on a level plain slightly elevated above the sea, the earth rocked, and suddenly a great mass was torn from the contiguous Mount Jaci, and thrown down with a dreadful crash upon the plain. Immediately afterwards, the sea rising thirty palms above the level of this low tract, rolled foaming over it, and swept away the multitude. It then retreated, but soon rushed back again with greater violence, bringing with it some of the people and animals it had carried away. At the same time every boat was sunk or dashed against the beach, and some of them were swept far inland. The aged Prince, with one thousand four hundred and thirty of his people, was destroyed. The number of persons who perished during the earthquake in the two Calabrias and Sicily is estimated by Hamilton at about forty thousand, and about twenty thousand more died by epidemics which were caused by insufficient nourishment, exposure to the atmosphere, and malaria, arising from the new stagnant lakes and pools. By far the greater number were buried under the ruins of their houses; while some were burnt to death in the conflagrations which almost invariably followed the shocks, and consumed immense magazines of oil and other provisions. A small number were engulphed in chasms and fissures, and their skeletons are perhaps buried in the earth to this day, at the depth of several hundred feet, for such was the profundity of some of the openings which did not close in again.

The inhabitants of Pizzo remarked, that on the 5th of February, 1783, when the first great shock afflicted Calabria, the volcano of Stromboli, which is in full view of that town, and at the distance of about fifty miles, smoked less, and threw up a less quantity of inflamed matter, than it had done for some years previously. On the other hand, the great crater of Etna is said to have given out a considerable quantity of vapour towards the beginning, and Stromboli towards the close of the commotions. But as no eruption happened from either of these great vents during the whole earthquake, the sources of the Calabrian convulsions, and of the volcanic fires of Etna and Stromboli, appear to be very independent of each other;

unless, indeed, they have the same mutual relation as Vesuvius and the volcanos of the Phlegræan Fields and Ischia, a violent disturbance in one district serving as a safety-valve to the other, and both never being in full activity at once.

It is impossible for the geologist to consider attentively the effect of this single earthquake of 1783, and to look forward to the alterations in the physical condition of the country to which a continued series of such movements will hereafter give rise, without perceiving that the formation of valleys by running water can never be understood, if we consider the question independently of the agency of earthquakes. Rivers do not begin to act, as some seem to imagine, when a country is already elevated far above the level of the sea, but while it is *rising* or *sinking* by successive movements. Whether Calabria is now undergoing any considerable change of relative level, in regard to the sea, or is, upon the whole, nearly stationary, is a question which our observations, confined almost entirely to the last half century, cannot possibly enable us to determine. But we know that strata, containing species of shells identical with those now living in the contiguous parts of the Mediterranean, have been raised in this country, as they have in Sicily, to the height of several thousand feet. Now those geologists who merely grant that the present course of Nature, in the inanimate world, has been unchanged since the existing species of animals were in being, will not feel surprise that the Calabrian streams and rivers have cut out of such comparatively modern strata a great system of valleys varying in depth from fifty to six hundred feet, and often several miles wide, when they consider how numerous must have been the earthquakes which lifted those recent marine strata to so prodigious a height. Some speculators, indeed, who disregard the analogy of existing Nature, and who are as prodigal of violence as they are thrifty of time, may suppose that Calabria "rose like an exhalation" from the deep, after the manner of Milton's Pandemonium. But such an hypothesis will deprive them of that peculiar removing force required to form a regular system of deep and wide valleys, for *time* is essential to the operation. Landslips must be cleared away in the intervals between subterranean movements, otherwise fallen masses will serve as buttresses to the precipitous cliffs bordering a valley, so that the succeeding earthquake will be unable to exert its full power. Barriers must be worn through and swept away, and steep or overhanging cliffs again left without support, before another shock can take effect in the same manner.

If a single convulsion be too violent, and agitate at once an entire hydrographical basin, or if the shocks follow each other too rapidly, the previously-existing valleys will be annihilated, instead of being modified and enlarged. Every stream will be compelled to begin its operations anew, and to open for itself a passage through strata before undisturbed, instead of continuing to deepen and widen channels already in great part excavated. On the other hand, if, consistently with all that is known from observation of the laws which regulate subterranean movements, we consider their action to have been intermittent— if sufficient periods have always intervened between the severer shocks to allow the drainage of the country to be nearly restored to its original state, then are both the kind and degree of force supplied which may enable running water to hollow out a valley of any depth and size consistent with the degree of elevation above the sea which the district in question may happen at any time to have attained during a succession of physical revolutions.

Nothwithstanding the great derangement caused by violent earthquakes, there is an evident tendency in running water to remain constant to the same connected series of valleys. The softening of the soil is invariably greatest in the channels of rivers and in alluvial plains. The water is absorbed in an infinite number of rents, and when the ground is swelled with water it is reduced almost to a state of mud by the vehement agitation of the ground in every direction, and often for several years consecutively. The erosive and transporting action of running water is, therefore, facilitated in the tracts already excavated.

When we read of the drying up and desertion of the channels of rivers, the accounts most frequently refer to their deflection into some other part of the same alluvial plain, perhaps several miles distant. Under certain circumstances a change of level may undoubtedly force the water to flow over into some distinct hydrographical basin; but even then it will fall immediately into valleys already formed. Provided, therefore, we suppose the elevation and subsidence of mountain-chains to be a gradual process, there is no difficulty in explaining how the rivers draining our continents have converted ravines into valleys, and enlarged and deepened valleys to an enormous extent. On the contrary, the signs of slow and gradual action so manifest in the sinuosities and other characters of valleys are admirably reconcileable with the great width and depth of the excavations, if we are content not only to suppose a great succession of ordinary earthquakes, but also the usual intervals of time between the shocks.

We may observe that earthquakes alone could never give rise to a regular system of valleys ramifying from a main trunk like the veins from the great arteries of the human body. On the contrary, they would, in the course of time, destroy every system of valleys on the globe, were it not for the agency of aqueous causes. We learn from history that ever since the first Greek colonists, the Bruttii, settled in Calabria, that region has been subject to devastation by earthquakes, and, for the last century and a half, ten years have seldom elapsed without a shock; but the severer convulsions have not only been separated by intervals of twenty, fifty, or one hundred years, but have not affected precisely the same points when they recurred. Thus the earthquake of 1783, although confined within the same geographical limits as that of 1638, and not very inferior in violence, visited, according to Grimaldi, very different localities. The points where the local intensity of the force is developed, being thus perpetually varied, more time is allowed for the removal of separate mountain masses thrown into river channels by each shock.

When chasms and deep hollows open at the bottom of valleys, they must often be filled with those "mud lavas" before described; and these must be extremely analogous to the enormous ancient deposits of mud which are seen in many countries, as in the basin of the Tay, Isla, and North Esk rivers, for example, in Scotland—alluvions hundreds of feet thick, which are neither stratified nor laminated like the sediment which subsides from water. Whenever a landslip blocks up a river, these currents of mud will be arrested, and accumulate to an enormous depth.

The transportation for several miles at a time, of masses as large as great edifices by the momentum of these floods of mud combined with the motion of the earthquake, and the enveloping of land animals, together with many other facts mentioned in the Calabrian account, cannot but excite in the mind of every geologist a strong desire to become more acquainted with the changes now in progress in those vast regions of the globe which are habitually devastated by earthquakes. To our extreme ignorance of this important class of phenomena we may probably refer the obscurity of many of the appearances of superficial alluvions throughout the greater part of Europe, as well as the diversity of opinion relating to them, and the extravagant theories which have passed current.

The portion of the Calabrian valleys formed within the last three thousand years, must, undoubtedly, be inconsiderable in amount, compared to that previously formed, just as the lavas which have flowed from Etna since the historical era constitute but a small proportion of the whole cone. But as a continued series of such eruptions as man has witnessed would reproduce another cone like Etna, so a sufficient number of earthquakes like that of 1783 would enable torrents and rivers to re-excavate all the Calabrian valleys if they were now to be entirely obliterated. It must be evident that more change is effected in two centuries in the width and depth of the valleys of that region, than in many thousand years in a country as undisturbed by earthquakes as Great Britain. For the same reason, therefore, that he who desires to comprehend the volcanic phenomena of Central France will repair to Vesuvius, Etna, or Hecla, so they who aspire to explain the mode in which valleys are formed must visit countries where earthquakes are of frequent occurrence. For we may be assured, that the power which uplifted our more ancient tertiary strata of marine origin to more than a thousand feet above the level of the sea, co-operated at some former epoch with the force of rivers in the removal of large portions of rock and soil, just as the elevatory power which has upraised newer strata to the height of several thousand feet in the south of Italy has caused those formations to be already intersected by deep valleys and ravines.

He who studies the hydrographical basin of the Thames, and compares its present state with its condition when it was a Roman province, may have good reason to declare that if that river and its tributaries had since their origin been always as inactive, and as impotent as they are now, they could never, not even in millions of years, have excavated the valleys through which they flow: but, if he concludes from these premises, that the valleys in this basin were not formed by ordinary causes, he reasons like one, who having found a solfatara which for many centuries has thrown out nothing more than vapour and a few handfuls of sand and scoriæ, infers that a lofty cone, composed of successive streams of lava and ejections, can no longer be produced by volcanic agency.

CHAPTER XXV.

Earthquakes of the eighteenth century, *continued*—Java, 1772—Truncation of a lofty cone—Caucasus, 1772—Java, 1771—Colombia, 1766—Chili, 1760—Azores, 1757—Lisbon, 1755—Sinking down of the quay to the depth of six hundred feet—Shocks felt throughout Europe, Northern Africa, and the West Indies—Great wave—Shocks felt at sea—St. Domingo, 1751—Conception Bay, 1750—Permanent elevation of the bed of the sea to the height of twenty-four feet—Peru, 1746—Kamtschatka, 1737—Martinique, 1727—Iceland, 1725—Teneriffe, 1706—Java, 1699—Landslips obstruct the Batavian and Tangaran rivers—Quito, 1698—Sicily, 1693—Subsidence of land—Moluccas, 1693—Jamaica, 1692—Large tracts engulphed—Portion of Port Royal sunk from twenty to fifty feet under water—The Blue Mountains shattered—Reflections on the amount of change in the last one hundred and forty years—Proofs of elevation and subsidence of land on the coast of the Bay of Baiæ—Evidence of the same afforded by the present state of the Temple of Serapis.

In the preceding chapters we have considered a small part of those earthquakes only which have occurred during the last fifty years, of which accurate and authentic descriptions happen to have been recorded. We shall next proceed to examine some of earlier date, respecting which information of geological interest has been obtained.

Java, 1772.—In the year 1772, Papandayang, formerly one of the loftiest volcanos in the island of Java, was in eruption. Before all the inhabitants on the declivities of the mountain could save themselves by flight, the ground began to give way, and a great part of the volcano fell in and disappeared. It is estimated that an extent of ground of the mountain itself and its immediate environs fifteen miles long and full six broad, was by this commotion swallowed up in the bowels of the earth. Forty villages were destroyed, some being engulphed and some covered by the substances thrown out on this occasion, and two thousand nine hundred and fifty-seven of the inhabitants perished. A proportionate number of cattle were also killed, and most of the plantations of cotton, indigo, and coffee in the adjacent districts were buried under the volcanic matter. This catastrophe appears to have resembled, although on a grander scale, that of the ancient Vesuvius in the year 79. The cone was reduced in height from nine thousand to about five thousand feet, and, as vapours still escape from the crater on its summit, a new cone may one day rise out of the ruins of the ancient mountain, as the modern Vesuvius has risen from the remains of Somma*.

Caucasus, 1772.—About the year 1772, an earthquake convulsed the ground in the province of Beshtau, in the Caucasus, so that part of the hill Metshuka sunk into an abyss†.

Java, 1771.—By an earthquake in the year 1771, several tracts of ground were upraised in Java, and a new bank made its appearance opposite the mouth of the river of Batavia‡.

Colombia, 1766.—On the 21st of October, 1766, the ground was agitated at once at Cumana, at Caraccas, at Maracaybo, and on the banks of the rivers Casanare, the Meta, the Orinoco, and the Ventuario. These districts were much fissured, and great fallings in of the earth took place in the mountain Paurari; Trinidad was violently shaken. A small island in the Orinoco, near the rock Aravacoto, sunk down and disappeared§. At the same time the ground was raised in the sea near Cariaco, where the Point Del Gardo was enlarged. A rock also rose up in the river Guarapica, near the village of Maturin‖. The shocks continued in Colombia hourly for fourteen months.

Chili, 1760.—In 1760, the volcano Peteroa, in Chili, was in eruption, and formed a new crater. A fissure, several miles in

* Dr. Horsfield, Batav. Trans., vol. viii., p. 26. Dr. H. informs me that he has seen this truncated mountain, and though he did not ascend it, he has conversed with those who have examined it. Raffles's account (History of Java, vol. i.) is derived from Horsfield.

† Pallas's Travels in Southern Russia.

‡ Raffles's History of Java, vol. ii., p. 232.

§ Humboldt's Personal Narrative, vol. iv., p. 45, and Saggio di Storia Americana, vol. ii., p. 6.

‖ Humboldt, Voy. Relat. Hist., part i., p. 307, and part ii., p. 23.

length, opened in a neighbouring hill, and a great landslip obstructed the river Lontue for ten days, giving rise to a considerable lake.

Azores, 1757.—In the year 1757, the island of St. George was struck by an earthquake, and eighteen small islets rose at the distance of about two hundred yards from the shore. These may possibly have been produced by a submarine eruption.

Lisbon, 1755.—In no part of the volcanic region of southern Europe has so tremendous an earthquake occurred in modern times as that which began on the 1st of November, 1755, at Lisbon. A sound of thunder was heard under ground, and immediately afterwards a violent shock threw down the greater part of that city. In the course of about six minutes, sixty thousand persons perished. The sea first retired and laid the bar dry; it then rolled in, rising fifty feet or more above its ordinary level. The mountains of Arrabida, Estrella, Julio, Marvan, and Cintra, being some of the largest in Portugal, were impetuously shaken, as it were, from their very foundations; and most of them opened at their summits, which were split and rent in a wonderful manner, huge masses of them being thrown down into the subjacent valleys *. Flames are related to have issued from these mountains, which are supposed to have been electric; they are also said to have smoked; but vast clouds of dust seem to have given rise to this appearance. The most extraordinary incident which occurred at Lisbon during the catastrophe was the subsidence of a new quay, built entirely of marble, at an immense expense. A great concourse of people had collected there for safety, as a spot where they might be beyond the reach of falling ruins; but, suddenly, the quay sank down with all the people on it, and not one of the dead bodies ever floated to the surface. A great number of boats and small vessels anchored near it, all full of people, were swallowed up, as in a whirlpool †. No fragments of these

* Hist. and Philos. of Earthquakes, p. 317.

† Rev. C. Davy's Letters, vol. ii., Letter ii., p. 12, who was at Lisbon at the time, and ascertained that the boats and vessels said to have been swallowed were missing.

wrecks ever rose again to the surface, and the water in the place where the quay had stood is stated, in many accounts, to be unfathomable; but, Whitehurst * says, he ascertained it to be one hundred fathoms.

In this case, we must either suppose that a certain tract sank down into a subterranean hollow which would cause a "fault" in the strata to the depth of six hundred feet, or we may infer, as some have done, from the entire disappearance of the substances engulphed, that a chasm opened and closed again. Yet, in adopting this latter hypothesis, we must suppose that the upper part of the chasm, to the depth of one hundred fathoms, remained open.

The great area over which this Lisbon earthquake extended is very remarkable. The movement was most violent in Spain, Portugal, and the north of Africa; but nearly the whole of Europe, and even the West Indies, felt the shock on the same day. A sea-port, called St. Eubals, about twenty miles south of Lisbon, was engulphed. At Algiers and Fez, in Africa, the agitation of the earth was equally violent, and at the distance of eight leagues from Morocco, a village, with the inhabitants to the number of about eight or ten thousand persons, together with all their cattle, were swallowed up. Soon after the earth closed again over them. A great wave swept over the coast of Spain, and is said to have been sixty feet high at Cadiz. At Tangier, in Africa, it rose and fell eighteen times on the coast. At Funchal, in Madeira, it rose full fifteen feet perpendicular above high-water mark, although the tide which ebbs and flows there seven feet was then at half ebb. Besides entering that city, and committing great havoc, it overflowed other sea-ports in the island. At Kinsale, in Ireland, a body of water rushed into the harbour, whirled round several vessels, and poured into the market place.

The shock was felt at sea, on the deck of a ship to the west of Lisbon, and produced very much the same sensation as on dry land. Off St. Lucar, the captain of the Nancy frigate felt his ship so violently shaken that he thought he had struck the ground; but, on heaving the lead, found he was in a great depth of water. Captain Clark from Denia, in north latitude

* On the Formation of the Earth, p. 55.

36° 24′, between nine and ten in the morning, had his ship shaken and strained as if she had struck upon a rock, so that the seams of the deck opened, and the compass was overturned in the binnacle. Another ship forty leagues west of St. Vincent experienced so violent a concussion, that the men were thrown a foot and a half perpendicularly up from the deck. In Antigua and Barbadoes, as also in Norway, Sweden, Germany, Holland, Corsica, Switzerland, and Italy, tremors and slight oscillations of the ground were felt.

The agitation of lakes, rivers, and springs, in Great Britain was remarkable. At Loch Lomond in Scotland, for example, the water, without the least apparent cause, rose against its banks, and then subsided below its usual level. The greatest perpendicular height of this swell was two feet four inches. It is said that the movement of this earthquake was undulatory, and that it travelled at the rate of twenty miles a minute, its velocity being calculated by the intervals between the time when the first shock was felt at Lisbon, and its time of occurrence at other distant places *.

St. Domingo, 1751.—On the 15th of September, 1751, a shock began to be experienced in several of the West India Islands, and on the 21st of November, a violent one destroyed the capital of St. Domingo, Port au Prince. Part of the coast twenty leagues in length sank down and has ever since formed a bay of the sea †.

Conception, 1750.—On the 24th of May 1750, the ancient town of Conception, otherwise called Penco, in Chili, was totally destroyed by an earthquake and the sea rolled over it. The ancient port was rendered entirely useless, and the inhabitants built another town ten miles from the sea-coast, in order to be beyond the reach of similar inundations. During a late survey of Conception Bay, Captains Beechey and Belcher discovered that the ancient harbour, which formerly admitted all large merchant vessels which went round the Cape, is now occupied by a reef of sandstone, certain points of which pro-

* Michell on the Cause and Phenomena of Earthquakes, Phil. Trans., vol. li. p. 566. 1760.

† Hist. de l'Acad. des Sciences. 1752. Paris.

ject above the sea at low-water, the greater part being very shallow. A tract of a mile and a half in length, where, according to the report of the inhabitants, the water was formerly four or five fathoms deep, is now a shoal. The correctness of this statement of the original depth may be concluded from the circumstance, that the large trading vessels which formerly frequented the port could not have anchored in less than four fathoms water. Our hydrographers found the reef to consist of hard sandstone, so that it cannot be supposed to have been formed by recent deposits of the river Biobio, an arm of which carries down loose micaceous sand into the same side of the bay. Besides it is a well known fact, that ever since the shock of 1750, no vessels have been able to approach within a mile and a half of the ancient port of Penco. That shock, therefore, uplifted the bed of the sea to the height of twenty-four feet at the least, and most probably the adjoining coast shared in the elevation, for an enormous bed of shells of the same species as those now living in the bay, are seen raised above high-water mark along the beach, filled with micaceous sand like that which the Biobio now conveys to the bay. These shells, as well as others which cover the adjoining hills of mica-schist to the height of from one thousand to one thousand five hundred feet, have lately been examined by experienced conchologists in London, and identified with those taken at the same time in a living state from the Bay and its neighbourhood *.

Ulloa, therefore, was perfectly correct in his statement, that at various heights above the sea between Talcaguana and Conception, "mines were found of various sorts of shells used for lime of the very same kinds as those found in the adjoining sea." Among them, he mentions the great mussel called Choros, and two others which he describes. Some of these, he says, are entire, and others broken; they occur at the bottom of the sea, in four, six, ten, or twelve fathom water, where they adhere to a sea-plant called Cochayuyo. They are taken in dredges, and have no resemblance to those found on the shore or in shallow water, yet beds of them occur at various heights on the hills. "I was the more pleased with the sight," he adds, "as it

* Captain Belcher has shewn me these shells, and the collection has been examined by Mr. Broderip.

appeared to me a convincing proof of the universality of the deluge, although I am not ignorant that some have attributed their position to other causes; but an unanswerable confutation of their subterfuge is, that the various sorts of shells which compose these strata, both in the plains and mountains, are the very same with those found in the bay *." Perhaps the diluvian theory of this distinguished navigator, the companion of Condamine, may account for his never having recorded even reports of changes in the relative level of land and sea on the shores of South America. He could not, however, have given us a relation of the rise of the reef above alluded to, for the destruction of Penco happened a few years after the publication of his Voyages. If we duly consider these facts so recently brought to light, as well as the elevation before mentioned of the coast at Valparaiso in 1822, we shall be less sceptical than Raspe, in regard to an event for which Hooke had cited Purchas's Travels. In that passage it was stated, "that a certain sea-coast in a province of South America called Chili, was, during a violent earthquake, propelled upwards with such force and velocity, that some ships on the sea were grounded in it, and the sea receded to a distance." Raspe, being himself of opinion that all the continents had been upraised gradually by earthquakes from the sea, admitted that the circumstance was not impossible, but he complains that Purchas had interpolated the account of the earthquake (which happened probably at the close of the seventeenth century) into Da Costa's History of the West Indies †.

Peru, 1746.—Peru was visited on the 28th of October, 1746, by an earthquake, which is declared to have been more tremendous and extensive than even that of Lisbon in 1755. In the first twenty-four hours, two hundred shocks were experienced. The ocean twice retired and returned impetuously upon the land: Lima was destroyed, and part of the coast near Callao was converted into a bay; four other harbours, among which were Cavalla and Guanape, shared the same fate. There were twenty-three ships and vessels great and small in the

* Ulloa's Voyage to South America, vol. ii., Book 8, chap. 6.

† De Novis Insulis, p. 120. 1753.

harbour of Callao, of which nineteen were sunk, and the other four, among which was a frigate called St. Fermin, were carried by the force of the waves to a great distance up the country. The number of the inhabitants in this city amounted to four thousand. Two hundred only escaped, twenty-two of whom were saved on a small fragment of the fort of Vera Cruz, which remained as the only memorial of the site of the town after this dreadful inundation.

A volcano in Lucanas burst forth the same night, and such quantities of water descended from the cone, that the whole country was overflowed; and in the mountain near Patao, called Conversiones de Caxamarquilla, three other volcanos burst out, and frightful torrents of water swept down their sides *.

Kamtschatka, 1737.—The eastern side of the peninsula of Kamtschatka, at Awatchka bay, was shaken by an earthquake on October the 6th, 1737. The sea was violently agitated, and overflowed the land to an immense height, and then withdrew so far as to lay bare its bottom between the first and second of the Kurile Isles. The shape of the ground was greatly changed. Several plains were uplifted and formed hills, and on the other hand many subsidences occasioned inland lakes and new bays on the coast †.

Martinique, 1727.—In the year 1727, a hill sunk down in Martinique during an earthquake ‡.

Iceland, 1725.—In Iceland during the eruption of the volcano Leirhnukur, in 1725-6, a tract of high land sunk down, and formed a lake, and half a mile from the same place a hill rose in a lake and converted it into dry land §.

Teneriffe, 1706.—May 5th, 1706, a lateral eruption of Teneriffe took place south of the harbour of Garachico, which was overwhelmed with lava. Many springs disappeared, and

* Ulloa's Voyage, vol. ii., Book 7, chap. 7.

† Kracheninikon by Chappe d'Auteroche, p. 337.

‡ Geog. of America, Schlözer, Part II., p. 554.

§ Dureau de la Malle, Géog., de la Mer Noire, p. 203.

there were such changes of level as to alter the whole face of the country, hills having risen up where there were plains before *.

Java, 1699.—On the 5th of January, 1699, a terrible earthquake visited Java, and no less than two hundred and eight considerable shocks were reckoned. Many houses in Batavia were overturned, and the flame and noise of a volcanic eruption were seen and heard at that city, which were afterwards found to proceed from Mount Salak †, a volcano six days' journey distant. Next morning the Batavian river, which has its rise from that mountain, became very high and muddy, and brought down abundance of bushes and trees, half burnt. The channel of the river being stopped up, the water overflowed the country round, the gardens about the town, and some of the streets, so that fishes lay dead in them. All the fish in the river, except the carps, were killed by the mud and turbid water. A great number of drowned buffaloes, tigers, rhinoceroses, deer, apes, and other wild beasts were brought down by the current, and "notwithstanding," observes one of the writers, "that a crocodile is amphibious, several of them were found dead among the rest ‡." It is stated, that seven hills bounding the river sank down, by which is merely meant, as by similar expressions in the description of the Calabrian earthquakes, seven great landslips. These hills, descending some from one side of the valley and some from the other, filled the channel, and the waters then finding their way under the mass, flowed out thick and muddy. The Tangaran river was also dammed up by nine hills, and in its channel were large quantities of drift trees. Seven of its tributaries also are said to have been "covered up with earth." A high tract of forest land, between the two great rivers before mentioned, is described as having been changed into an open country, destitute of trees, the surface being spread over with a fine red clay. This part of the account may, perhaps, merely refer to the sliding down of woody tracts into the valleys, as happened to so many extensive vineyards and olive-grounds in Calabria,

* Humboldt and Bonpland, Voy. Relat. Hist., Part I., p. 177.

† Misspelt Sales in Hooke's account.

‡ Hooke's Posthumous Works, p. 437, 1705.

in 1783. The close packing of large trees in the Batavian river is represented as very remarkable, and it attests in a striking manner the destruction of soil bordering the valleys which had been caused by floods and landslips*.

Quito, 1698.—In Quito, on the 19th of July, 1698, during an earthquake, a great part of the crater and summit of the volcano Carguairazo fell in, and a stream of water and mud issued from the broken sides of the hill †.

Sicily, 1693.—Shocks of earthquakes spread over all Sicily in 1693, and on the 11th of January the city of Catania and forty nine other places were levelled to the ground, and about one hundred thousand people killed. The bottom of the sea, says Vicentino Bonajutus, sank down considerably both in ports, inclosed bays, and open parts of the coast, and water bubbled up along the shores. Numerous long fissures of various breadths were caused, which threw out sulphureous water, and one of them, in the plain of Catania (the delta of the Simeto), at the distance of four miles from the sea, sent forth water as salt as the sea. The stone buildings of a street in the city of Noto, for the length of half a mile, sank into the ground, and remained hanging on one side. In another street, an opening large enough to swallow a man and horse appeared ‡.

Moluccas, 1693.—The small isle of Sorea, which consists of one great volcano, was in eruption in the year 1693. Different parts of the cone fell one after the other into a deep crater, until almost half the space of the island was converted into a fiery lake. Most of the inhabitants fled to Banda, but great pieces of the mountain continued to fall down, so that the lake became wider, and finally the whole population was compelled to emigrate. It is stated, that in proportion as the lake of lava increased in size, the earthquakes were less vehement §.

Jamaica, 1692.—In the year 1692 the island of Jamaica was visited by a violent earthquake, the ground swelled and

* Phil. Trans., 1700. † Humboldt, Atl. Pit., p. 106.
‡ Phil. Trans., 1693—4. § Ibid. 1693.

heaved like a rolling sea, and was traversed by numerous cracks, two or three hundred of which were often seen at a time opening and then closing rapidly again. Many people were swallowed up in these rents; some the earth caught by the middle and squeezed to death; the heads of others only appeared above ground, and some were first engulphed and then cast up again with great quantities of water. Such was the devastation, that even at Port Royal, then the capital, where more houses are said to have been left standing than in the whole island beside, three quarters of the buildings, together with the ground they stood on, sank down with their inhabitants entirely under water. The large store-houses on the harbour side subsided, so as to be twenty-four, thirty-six, and forty-eight feet under water; yet many of them appeared to have remained standing, for it is stated that, after the earthquake, the mast-heads of several ships wrecked in the harbour, together with the chimney-tops of houses, were seen just projecting above the waves. A tract of land round the town, about a thousand acres in extent, sank down in less than one minute, during the first shock, and the sea immediately rolled in. The Swan frigate, which was repairing in the wharf, was driven over the tops of many buildings, and then thrown upon one of the roofs, through which it broke. The breadth of one of the streets is said to have been doubled by the earthquake. At several thousand places in Jamaica the earth is related to have opened. On the north of the island several plantations, with their inhabitants, were swallowed up, and a lake appeared in their place, covering above a thousand acres, which afterwards dried up, leaving nothing but sand and gravel, without the least sign that there had ever been a house or tree there. Several tenements at Yallowes were buried under landslips; and one plantation was removed half a mile from its place, the crops continuing to grow upon it uninjured. Between Spanish town and Sixteen-mile-walk the high and perpendicular cliffs bounding the river fell in, stopped the passage of the river, and flooded the latter place for nine days, so that the people "concluded it had been sunk as Port Royal was." But the flood at length subsided, for the river had found some new passage at a great distance.

The Blue and other of the highest mountains are declared

to have been strangely torn and rent. They appeared shattered and half-naked, no longer affording a fine green prospect, as before, but stripped of their woods and natural verdure. The rivers on these mountains first ceased to flow for about twenty-four hours, and then brought down into the sea at Port Royal and other places, several hundred thousand tons of timber which looked like floating islands on the ocean. The trees were in general barked, most of their branches having been torn off in the descent. It is particularly remarked in this, as in the narratives of so many earthquakes, that fish were taken in great numbers on the coast during the shocks. The correspondents of Sir Hans Sloane, who collected with care the accounts of eye-witnesses of the catastrophe, refer constantly to *subsidences*, and some supposed the whole of Jamaica to have sunk down*.

We have now only enumerated the earthquakes of the last hundred and forty years, respecting which, facts illustrative of geological inquiries are on record. Even if our limits permitted, it would be a tedious and unprofitable task to examine all the obscure and ambiguous narratives of similar events of earlier epochs, although, if the localities were now examined by geologists well practised in the art of interpreting the monuments of physical changes, many events which have happened within the historical era might still be determined with precision. The reader must not imagine, that in our sketch of the occurrences in the short period above alluded to, we have given an account of all, or even the greater part of the mutations which the earth has undergone, by the agency of subterranean movements. Thus, for example, the earthquake of Aleppo, in the present century, and of Syria in the middle of the eighteenth, would doubtless have afforded numerous phenomena of great geological importance, had those catastrophes been described by scientific observers. The shocks in Syria in 1759, were protracted for three months, throughout a space of ten thousand square leagues, an area compared to which that of the Calabrian earthquake, of 1783, was insignificant. Accon, Saphat, Balbeck, Damascus, Sidon, Tripoli, and many other places, were almost entirely levelled to the ground. Many thousands of the inhabitants perished in each, and in the valley

* Phil. Trans., 1694.

of Balbeck alone twenty thousand men are said to have been victims to the convulsion. It would be as irrelevant to our present purpose to enter into a detailed account of such calamities, as to follow the track of an invading army, to enumerate the cities burnt or rased to the ground, and reckon the number of individuals who perished by famine or the sword. If such then be the amount of ascertained changes in the last one hundred and forty years, notwithstanding the extreme deficiency of our records during that brief period, how important must we presume the physical revolutions to have been in the course of thirty or forty centuries, during which, some countries habitually convulsed by earthquakes have been peopled by civilized nations! Towns engulphed during one earthquake may, by repeated shocks, have sunk to enormous depths beneath the surface, while their ruins remain as imperishable as the hardest rocks in which they are inclosed. Buildings and cities submerged for a time beneath seas or lakes, and covered with sedimentary deposits, must, in some places, have been re-elevated to considerable heights above the level of the ocean. The signs of these events have probably been rendered visible by subsequent mutations, as by the encroachments of the sea upon the coast, by deep excavations made by torrents and rivers, by the opening of new ravines and chasms, and other effects of natural agents, so active in districts agitated by subterranean movements. If it be asked why if such wonderful monuments exist, so few have hitherto been brought to light—we reply—because they have not been searched for. In order to rescue from oblivion the memorials of former occurrences, we must know what we may reasonably expect to discover; and under what peculiar local circumstances. The inquirer, moreover, must be acquainted with the action and effects of physical causes, in order to recognise, explain, and describe, correctly, the phenomena when they present themselves.

The best known of the great volcanic regions of which we sketched the boundaries, in the eighteenth chapter, is that which includes Southern Europe, Northern Africa, and Central Asia, yet nearly the whole even of this region must be laid down in a geological map as "Terra Incognita." Even Calabria may be regarded as unexplored, as also Spain, Portugal, the Barbary states, the Ionian Isles, the Morea, Asia Minor, Cyprus, Syria, and the countries between the Caspian and Black Seas. We are, in truth, beginning to obtain some insight into one small spot of that great zone of volcanic disturbance, the district around Naples, a tract by no means remarkable for the violence of the earthquakes which have convulsed it.

If, in this part of Campania, we are enabled to establish, that considerable changes in the relative level of land and sea have taken place since the Christian era, it is all that we could have expected, and it is to recent antiquarian and geological research, not to history, that we are principally indebted for the information. We shall proceed to lay before the reader some of the results of modern investigations in the Bay of Baiæ and the adjoining coast.

Temple of Jupiter Serapis.—This celebrated monument of antiquity affords, in itself alone, unequivocal evidence, that the relative level of land and sea has changed twice at Puzzuoli, since the Christian era, and each movement both of elevation and subsidence has exceeded twenty feet. Before examining these proofs we may observe, that a geological examination of the coast of the Bay of Baiæ, both on the north and south of Puzzuoli, establishes in the most satisfactory manner an elevation at no remote period, of more than twenty feet, and the evidence of this change would have been complete even if the temple had to this day remained undiscovered. If we coast along the shore from Naples to Puzzuoli we find, on approaching the latter place, that the lofty and precipitous cliffs of indurated tuff, resembling that of which Naples is built, retire slightly from the sea, and that a low level tract of fertile land, of a very different aspect, intervenes between the present sea-beach, and what was evidently the ancient line of coast. The inland cliff is in many parts eighty feet high near Puzzuoli, and as perpendicular as if it was still undermined by the waves. At its base, the new deposit attains a height of about twenty feet above the sea, and as it consists of regular sedimentary deposits, containing marine shells, its position proves that since its formation there has been a change of more than twenty feet in the relative level of land and sea.

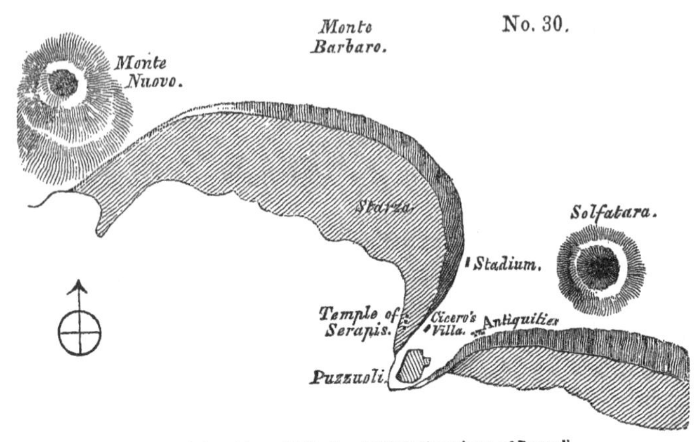

Ground plan of the coast of the Bay of Baiæ in the environs of Puzzuoli.

The sea encroaches on these new incoherent strata, and as the soil is valuable, a wall has been built for its protection ; but when I visited the spot in 1828, the waves had swept away part of this rampart, and exposed to view a regular series of strata of tuff, more or less argillaceous, alternating with beds

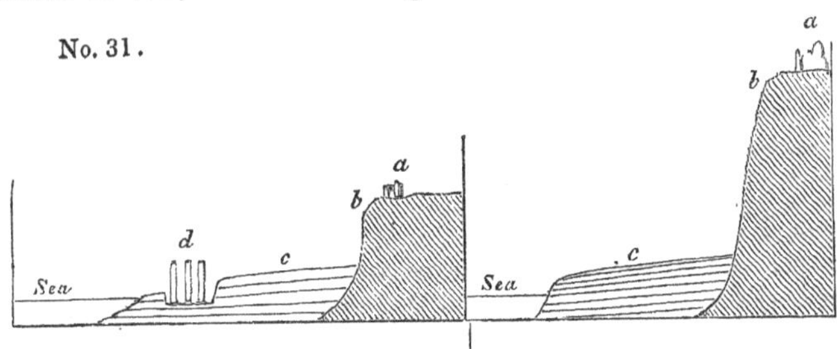

a. Remains of Cicero's villa, N. side of Puzzuoli *.	*a.* Antiquities on hill S.E. of Puzzuoli.
b. Ancient cliff now inland.	*b.* Ancient cliff now inland.
c. Terrace composed of recent submarine deposit.	*c.* Terrace composed of recent submarine deposit.
d. Temple of Serapis.	

of pumice and lapilli, and containing great abundance of marine shells, of species now common on this coast, and amongst them Cardium rusticum, Ostrea edulis, Donax trunculus (Lam.) and

* The spot here indicated on the summit of the cliff, is that from which Hamilton's view, plate 26, Campi Phlegræi is taken, and on which he observes Cicero's villa called the Academia anciently stood.

others. The strata vary from about a foot to a foot and half in thickness, and one of them contains abundantly remains of works of art, tiles, squares of mosaic pavement of different colours, and small sculptured ornaments, perfectly uninjured. Intermixed with these I collected some teeth of the pig and ox. These fragments of building occur below as well as above strata containing marine shells.

If we then pass to the north of Puzzuoli and examine the coast between that town and Monte Nuovo, we find a repetition of analogous phenomena. The sloping sides of Monte Barbaro slant down within a short distance of the coast, and terminate in an inland cliff of moderate elevation, to which the geologist perceives at once, that the sea must, at some former period, have extended. Between this cliff and the sea is a low plain or terrace, called La Starza, corresponding to that before described on the south-east of the town; and, as the sea encroaches rapidly, fresh sections of the strata may readily be obtained, of which the annexed is an example.

Section on the shore north of the town of Puzzuoli.

	Ft.	*In.*
1. Vegetable soil	1	0
2. Horizontal beds of pumice and scoriæ, with broken fragments of unrolled bricks, bones of animals, and marine shells	1	6
3. Beds of lapilli, containing abundance of marine shells, principally Cardium rusticum, Donax trunculus Lam., Ostrea edulis, Triton cutaceum, Lam. and Buccinum serratum, Brocchi, the beds varying in thickness from one to eighteen inches	10	0
4. Argillaceous tuff containing bricks and fragments of buildings not rounded by attrition	1	6

The thickness of many of these beds varies greatly as we trace them along the shore, and sometimes the whole group rises to a greater height than at the point above described. The surface of the tract which they compose appears to slope gently upwards towards the base of the old cliffs. Puzzuoli itself stands chiefly on a promontory of the older tufaceous formation, which cuts off the new deposit, although I detected a small patch of the latter in a garden under the town.

Now if these appearances presented themselves on the eastern or southern coast of England, a geologist would naturally endeavour to seek an explanation in some local depression of high water-mark, in consequence of a change in the set of the tides and currents: for towns have been built, like ancient Brighton, on sandy tracts intervening between the old cliff and the sea, and in some cases they have been finally swept away by the return of the ocean. On the other hand, the inland cliff at Lowestoff, in Suffolk, remains, as we stated in the fifteenth chapter, at some distance from the shore, and the low green tract called the Ness may be compared to the low flat called La Starza, near Puzzuoli. But there are no tides in the Mediterranean; and to suppose that sea to have sunk generally from twenty to twenty-five feet since the shores of Campania were covered with sumptuous buildings, is an hypothesis obviously untenable. The observations, indeed, made during modern surveys on the moles and cothons (docks) constructed by the ancients in various ports of the Mediterranean, have proved that there has been no sensible variation of level in that sea during the last two thousand years. A very slight change would have been perceptible; and had any been ascertained to have taken place, and had it amounted only to a difference of a few feet, it would not have appeared very extraordinary, since the equilibrium of the Mediterranean is only restored by a powerful current from the Atlantic*.

Thus we arrive, without the aid of the celebrated temple, at the conclusion that the recent marine deposit at Puzzuoli was upraised in modern times above the level of the sea, and that not only this change of position, but the accumulation of the modern strata, was posterior to the destruction of many edifices, of which they contain the imbedded remains. If we now examine the evidence afforded by the temple itself, it appears, from the most authentic accounts, that the three pillars now standing erect, continued, down to the middle of the last century, half buried in the new marine strata before described. The upper part of the columns, being concealed by bushes, had not attracted the notice of antiquaries; but, when the soil was removed in 1750, they were seen to form part of the remains of a splendid edifice, the pavement of which was still preserved, and upon it lay a number of columns of African breccia and of granite. The original plan of the building

* Captain W. H. Smyth, R.N. obtained, during his survey, numerous proofs of the permanency of the level of the Mediterranean from a remote historical period.

could be traced distinctly; it was of a quandrangular form, seventy feet in diameter, and the roof had been supported by forty-six noble columns, twenty-four of granite, and the rest of marble. The large court was surrounded by apartments, supposed to have been used as bathing-rooms; for a thermal spring, still used for medicinal purposes, issues now just behind the building, and the water, it is said, of this spring, was conveyed by marble ducts into the chambers. Many antiquaries have entered into elaborate discussions as to the deity to which this edifice was consecrated; but Signor Carelli, who has written the last able treatise on the subject *, endeavours to show that all the religious edifices of Greece were of a form essentially different—that the building, therefore, could never have been a temple—that it corresponded to the public bathing-rooms at many of our watering-places, and, lastly, that if it had been a temple, it could not have been dedicated to Serapis,—the worship of the Egyptian god being strictly prohibited at the time when this edifice was in use, by the senate of Rome.

It is not for the geologist to offer an opinion on these topics, and we shall, therefore, designate this valuable relic of antiquity by its generally received name, and proceed to consider the memorials of physical changes, inscribed on the three standing columns in most legible characters by the hand of nature. (See Frontispiece †.) The pillars are forty-two feet in height; their surface is smooth and uninjured to the height of about twelve feet above their pedestals. Above this, is a zone, twelve feet in height, where the marble has been pierced by a species of marine perforating bivalve—Lithodomus, Cuv. ‡. The holes of these animals are pear-shaped, the external opening being minute, and gradually increasing downwards. At the bottom of the cavities, many shells are still found, notwithstanding the great numbers that have been taken out by visitors. The perforations are so considerable in depth and size, that they manifest a long continued abode of the Lithodomi in the

* Dissertazione esergetica sulla sagra Architettura degli Antichi.

† The representation of the present state of the temple in the frontispiece has been carefully reduced from that given by the Canonico Andrea de Jorio, Ricerche sul Tempio di Serapide, in Puzzuoli. Napoli, 1820.

‡ Modiola lithophaga, Lam. Mytilus lithophagus, Linn.

columns; for, as the inhabitant grows older and increases in size, it bores a larger cavity, to correspond with the increasing magnitude of its shell. We must, consequently, infer a long continued immersion of the pillars in sea-water, at a time when the lower part was covered up and protected by strata of tuff and the rubbish of buildings, the highest part at the same time projecting above the waters, and being consequently weathered, but not materially injured. On the pavement of the temple, lie some columns of marble, which are perforated in the same manner in certain parts, one, for example, to the length of eight feet, while, for the length of four feet, it is uninjured. Several of these broken columns are eaten into, not only on the exterior, but on the cross fracture, and, on some of them, other marine animals have fixed themselves *. All the granite pillars are untouched by Lithodomi. The platform of the Temple is at present about one foot below high-water mark, (for there are small tides in the Bay of Naples,) and the sea, which is only one hundred feet distant, soaks through the intervening soil. The upper part of the perforations then are at least twenty-three feet above high-water mark, and it is clear, that the columns must have continued for a long time in an erect position, immersed in salt-water. After remaining for many years submerged, they must have been upraised to the height of about twenty-three feet above the level of the sea.

So far the information derived from the Temple corroborates that before obtained from the new strata in the plain of La Starza, and proves nothing more. But as the temple could not have been built originally at the bottom of the sea, it must have first sunk down below the waves, and afterwards have been elevated. Of such subsidences there are numerous independent proofs in the Bay of Baiæ. Not far from the shore, to the north-west of the Temple of Serapis, are the ruins of a Temple of Neptune, and a Temple of the Nymphs, now under water. These buildings probably participated in the movement which raised the Starza, but, either they were deeper under water than the Temple of Serapis, or they were not raised up again to so great a height. There are also two Roman roads under water in the

* Serpula contortuplicata, Linn., and Vermilia triquetra, Lam. These species, as well as the Lithodomus, are now inhabitants of the neighbouring sea.

Bay, one reaching from Puzzuoli towards the Lucrine Lake, which may still be seen, and the other near the Castle of Baiæ. The ancient mole too, which exists at the Port of Puzzuoli, and which is commonly called that of Caligula, has the water up to a considerable height of the arches; whereas Brieslak * justly observes, it is next to certain, that the piers must formerly have reached the surface before the springing of the arches. A modern writer also reminds us, that these effects are not so local as some would have us believe; for on the opposite side of the Bay of Naples, on the Sorrentine coast, which, as well as Puzzuoli, is subject to earthquakes, a road, with some fragments of Roman buildings, is covered to some depth by the sea. In the island of Capri, also, which is situated some way at sea, in the opening of the Bay of Naples, one of the palaces of Tiberius is now covered with water †. They who have attentively considered the effects of earthquakes before enumerated by us during the last one hundred and forty years, will not feel astonished at these signs of alternate elevation and depression of the bed of the sea and the adjoining coast during the course of eighteen centuries, but, on the contrary, they will be very much astonished if future researches fail to bring to light similar indications of change in all regions of volcanic disturbances. That buildings should have been submerged, and afterwards upheaved, without being entirely reduced to a heap of ruins, will appear no anomaly, when we recollect that in the year 1819, when the delta of the Indus sank down, the houses within the fort of Sindree subsided beneath the waves without being overthrown. In like manner, in the year 1692, the buildings around the harbour of Port Royal, in Jamaica, descended suddenly to the depth of between thirty and fifty feet under the sea without falling. Even on small portions of land, transported to a distance of a mile, down a declivity, tenements like those near Mileto, in Calabria, were carried entire. At Valparaiso, buildings were left standing when their foundations, together with a long tract of the Chilian coast, were permanently upraised to the height of several feet in 1822. It is true that, in the year 1750, when the bottom of the sea

* Voy. dans la Campanie, tome ii., p. 162.

† Mr. Forbes, Physical Notices of the Bay of Naples. Ed. Journ. of Sci., No. 2, new series, p. 280. October, 1829.

in the harbour of Penco was suddenly uplifted to the extraordinary elevation of twenty-four feet above its former level, the buildings of that town were thrown down; but we might still suppose that a great portion of them would have escaped, had the walls been supported on the exterior and interior with a deposit, like that which surrounded and filled to the height of ten or twelve feet the Temple of Serapis at Puzzuoli.

The next subject of inquiry, is the era when these remarkable changes took place in the Bay of Baiæ. It appears, that in the Atrium of the Temple of Serapis, inscriptions were found in which Septimus Severus and Marcus Aurelius record their labours in adorning it with precious marbles *. We may, therefore, conclude, that it existed at least down to the third century of our era in its original position. On the other hand, we have evidence that the marine deposit forming the flat land called La Starza was still covered by the sea in the year 1530, or just eight years anterior to the tremendous explosion of Monte Nuovo. Mr. Forbes † has lately pointed out the distinct testimony of an old Italian writer Loffredo, in confirmation of this important point. Writing in 1580, Loffredo declares that fifty years previously, the sea washed the base of the hills which rise from the flat land before alluded to, and at that time he expressly tells us that a person *might have fished* from the site of those ruins which are now called the Stadium. (See wood cut, No. 30.) Hence it follows, that the subsidence of the ground on which the Temple stood, happened at some period between the third century and the beginning of the sixteenth century. Now in this interval the only two events which are recorded in the imperfect annals of the dark ages, are the eruption of the Solfatara in 1198, and an earthquake in 1488 by which Puzzuoli was ruined. It is at least highly probable, that earthquakes, which preceded the eruption of the Solfatara, which is very near the Temple, (see wood cut, No. 30) caused a subsidence, and the pumice and other matters ejected from that volcano might have fallen in heavy showers into the sea, and would thus immediately have covered up the lower part of the columns. The action of the waves might afterwards have thrown down many pillars, and formed strata of broken fragments of the

* Brieslak, Voy. dans la Campanie, tom. ii., p. 167.
† Ed. Journ. of Science, new series, No. II., p. 281.

building intermixed with volcanic ejections, before the Lithodomi had time to perforate the lower part of the columns. In like manner, the sea acting on other submerged buildings, would naturally have caused a similar stratum, containing works of art and shells for several miles along the coast.

Now it is perfectly evident from Loffredo's statement, that the re-elevation of the low tract called La Starza took place after the year 1530, and long before the year 1580; and from this alone we might confidently conclude that the change happened in the year 1538 when Monte Nuovo was formed. But fortunately we are not left in the slightest doubt that such was the date of this remarkable event. Sir William Hamilton * has given us two original letters describing the eruption of 1538, the first of which by Falconi, dated 1538, contains the following passages. "It is now two years since there have been frequent earthquakes at Puzzuoli, Naples, and the neighbouring parts. On the day and in the night before the eruption (of Monte Nuovo), above twenty shocks great and small were felt.—The next morning (after the formation of Monte Nuovo) the poor inhabitants of Puzzuoli quitted their habitations, &c., some with their children in their arms, some with sacks full of their goods, others carrying quantities of birds of various sorts that had fallen dead at the beginning of the eruption, others again with fish which they had found, and which were to be met with in plenty on the shore, the sea having *left them dry for a considerable time.*—I accompanied Signor Moramaldo to behold the wonderful effects of the eruption. The sea had retired on the side of Baiæ, *abandoning a considerable tract*, and the shore appeared almost entirely dry from the quantity of ashes and broken pumice-stones thrown up by the eruption. I saw two springs *in the newly discovered ruins*, one before the house that was the Queen's, of hot and salt-water, &c." So far Falconi—the other account is by Pietro Giacomo di Toledo, which begins thus: "It is now two years since this province of Campagna has been afflicted with earthquakes, the country about Puzzuoli much more so than any other parts: but the 27th and the 28th of the month of September last, the earthquakes did not cease day or night in the town of Puzzuoli;

* Campi Phlegræi, p. 70.

that plain which lies between lake Avernus, the Monte Barbaro and the sea was *raised a little*, and many cracks were made in it, from some of which issued water; at the same time the sea immediately adjoining the plain *dried up about two hundred paces*, so that the fish were left on the sand a prey to the inhabitants of Puzzuoli. At last, on the 29th of the same month, about two o'clock in the night, the earth opened, &c." Now both these accounts, written immediately after the birth of Monte Nuovo, agree in expressly stating, that the sea retired, and one mentions that its bottom was upraised. To this elevation we have already seen that Hooke, writing at the close of the seventeenth century, alludes as to a well known fact*. The preposterous theories, therefore, that have been advanced in order to dispense with the elevation of the land, in the face of all this historical and physical evidence, are not entitled to a serious refutation. The flat land, when first upraised, must have been more extensive than now, for the sea encroaches somewhat rapidly, both to the north and south-east of Puzzuoli. The coast has of late years given way more than a foot in a twelvemonth, and I was assured by fishermen in the bay, that it has lost ground near Puzzuoli, to the extent of thirty feet, within their memory. It is, probably, this gradual encroachment which has led many authors to imagine that the level of the sea is slowly rising in the Bay of Baiæ, an opinion by no means warranted by such circumstances. In the course of time the whole of the low land will, perhaps, be carried away, unless some earthquake shall remodify the surface of the country, before the waves reach the ancient coast-line; but the removal of this narrow tract will by no means restore the country to its former state, for the old tufaceous hills and the interstratified current of trachytic lava which has flowed from the Solfatara, must have participated in the movement of 1538; and these will remain upraised even though the sea may regain its ancient limits.

In 1828 excavations were made below the marble pavement of the Temple of Serapis, and another costly pavement of mosaic was found, at the depth of five feet or more below the other. The existence of these two pavements at different levels seems clearly to imply some subsidence previously to all the

* Ante, p. 34.

changes already alluded to, which had rendered it necessary to construct a new floor at a higher level. But to these and other circumstances bearing on the history of the Temple antecedently to the revolutions already explained, we shall not refer at present, trusting that future investigations will set them in a clearer light.

In concluding this subject, we may observe, that the interminable controversies to which the phenomena of the Bay of Baiæ gave rise, have sprung from an extreme reluctance to admit that the land rather than the sea is subject alternately to rise and fall. Had it been assumed that the level of the ocean was invariable, on the ground that no fluctuations have as yet been clearly established, and that, on the other hand, the continents are inconstant in their level, as has been demonstrated by the most unequivocal proofs again and again, from the time of Strabo to our own times, the appearances of the temple at Puzzuoli could never have been regarded as enigmatical. Even if contemporary accounts had not distinctly attested the upraising of the coast, this explanation should have been proposed in the first instance as the most natural, instead of being now adopted unwillingly when all others have failed. To the strong prejudices still existing in regard to the mobility of the land, we may attribute the rarity of such discoveries as have been recently brought to light in the Bay of Baiæ and the Bay of Conception. A false theory it is well known may render us blind to facts, which are opposed to our prepossessions, or may conceal from us their true import when we behold them. But it is time that the geologist should in some degree overcome those first and natural impressions which induced the poets of old to select the rock as the emblem of firmness—the sea as the image of inconstancy. Our modern poet, in a more philosophical spirit, saw in the latter "The image of Eternity," and has finely contrasted the fleeting existence of the successive empires which have flourished and fallen, on the borders of the ocean, with its own unchanged stability.

> ——— Their decay
> Has dried up realms to deserts:—not so thou,
> Unchangeable, save to thy wild waves' play:
> Time writes no wrinkle on thine azure brow;
> Such as creation's dawn beheld, thou rollest now.
>
> CHILDE HAROLD, Canto iv.

CHAPTER XXVI.

Magnitude of the subterranean changes produced by earthquakes at great depths below the surface—Obscurity of geological phenomena no proof of want of uniformity in the system, because subterranean processes are but little understood—Reasons for presuming the earthquake and volcano to have a common origin—Probable analogy between the agency of steam in the Icelandic geysers, and in volcanos during eruptions—Effects of hydrostatic pressure of high columns of lava—Of the condensation of vapours in the interior of the earth—That some earthquakes may be abortive eruptions—Why all volcanos are in islands or maritime tracts—Gases evolved from volcanos—Regular discharge of heat and of gaseous and earthy matter from the subterranean regions—Cause of the wave-like motion and of the retreat of the sea during earthquakes—Difference of circumstances of heat and pressure at great depths—Inferences from the superficial changes brought about by earthquakes—In what matter the repair of land destroyed by aqueous causes takes place—Proofs that the sinking in of the earth's crust somewhat exceeds the forcing out by earthquakes—Geological consequences of this hypothesis, that there is no ground for presuming that the degree of force exerted by subterranean movements in a given time has diminished—Concluding remarks.

WHEN we consider attentively the changes brought about by earthquakes during the last century, and reflect on the light which they already throw on the ancient history of the globe, we cannot but regret that investigations into the effects of this powerful cause have hitherto been prosecuted with so little zeal. The disregard of this important subject may be attributed to the general persuasion, that former revolutions of the earth were not brought about by causes now in operation,—a theory which, if true, would fully justify a geologist in neglecting the study of such phenomena. We may say of the superficial alterations arising from subterranean movements, as we have already declared of the visible effects of active volcanos, that, important as they are in themselves, they are still more so as indicative of far greater changes in the interior of the earth's crust. That both the chemical and mechanical changes in the subterranean regions must often be value of some terms. So if a student of Nature, who, when he first examines the monuments of former changes upon our globe, is acquainted only with one-tenth part of the processes now going on upon or far below the surface, or in the depths of the sea, should still find that he comprehends at once the import of the signs of all, or even half the changes that went on in the same regions some hundred or thousand centuries ago, he might declare without hesitation that the ancient laws of nature have been subverted. Even after toiling for centuries, and learning more both of the present and former state of things, he must never expect to gain a perfect insight into all that formerly happened, so long as his acquaintance is very limited in regard to much that is now going on. So completely has the force of this line of argument been overlooked, that when any one has ventured to presume that all former changes were simply the result of causes now in operation, they have invariably been called upon to explain every obscure phenomenon in geology, and if they failed, it was considered as conclusive against their assumption. Whereas, in truth, there is no part of the evidence in favour of the uniformity of the system, more cogent than the fact, that with much that is intelligible, there is still more which is yet novel, mysterious, and inexplicable in the monuments of ancient mutations in the earth's crust.

Before the immense depth of the sources of volcanic fire was generally admitted, the causes of subterranean movements were sought in peculiar states of the atmosphere. These were imagined to afford not only prognostics of the convulsions, but to have considerable influence in their production. But the supposed signs of approaching earthquakes were of a most uncertain and contradictory character. Aristotle, Pliny, and Seneca, taught that earthquakes were preceded by a serene state of the air; whereas several modern writers have been of opinion that a cloudy sky and sudden storms are the forerunners of these commotions. That there is an intimate connexion between subterranean convulsions and particular states of the weather is unquestionable; but as Michell truly remarked, "it is more probable that the air should be affected by the causes of earthquakes, than that the earth should be affected in so extraordinary a manner, and to so great a depth, by a cause residing in the air."

value of some terms. So if a student of Nature, who, when he first examines the monuments of former changes upon our globe, is acquainted only with one-tenth part of the processes now going on upon or far below the surface, or in the depths of the sea, should still find that he comprehends at once the import of the signs of all, or even half the changes that went on in the same regions some hundred or thousand centuries ago, he might declare without hesitation that the ancient laws of nature have been subverted. Even after toiling for centuries, and learning more both of the present and former state of things, he must never expect to gain a perfect insight into all that formerly happened, so long as his acquaintance is very limited in regard to much that is now going on. So completely has the force of this line of argument been overlooked, that when any one has ventured to presume that all former changes were simply the result of causes now in operation, they have invariably been called upon to explain every obscure phenomenon in geology, and if they failed, it was considered as conclusive against their assumption. Whereas, in truth, there is no part of the evidence in favour of the uniformity of the system, more cogent than the fact, that with much that is intelligible, there is still more which is yet novel, mysterious, and inexplicable in the monuments of ancient mutations in the earth's crust.

Before the immense depth of the sources of volcanic fire was generally admitted, the causes of subterranean movements were sought in peculiar states of the atmosphere. These were imagined to afford not only prognostics of the convulsions, but to have considerable influence in their production. But the supposed signs of approaching earthquakes were of a most uncertain and contradictory character. Aristotle, Pliny, and Seneca, taught that earthquakes were preceded by a serene state of the air; whereas several modern writers have been of opinion that a cloudy sky and sudden storms are the forerunners of these commotions. That there is an intimate connexion between subterranean convulsions and particular states of the weather is unquestionable; but as Michell truly remarked, "it is more probable that the air should be affected by the causes of earthquakes, than that the earth should be affected in so extraordinary a manner, and to so great a depth, by a cause residing in the air."

After violent earthquakes the regular drainage of a country is obstructed; lakes and pools are caused by local subsidences or landslips, and the evaporation of an extensive surface of shallow water produces unseasonable rains. Fogs proceed from the damp soil which is traversed by numerous rents and crevices filled with water. In addition to these circumstances, the electrical effect produced by the movement and friction of great masses of rock against each other may cause lightning, gusts of wind, luminous exhalations, and other atmospheric phenomena. Rains, moreover, are sometimes derived from volcanic eruptions accompanying earthquakes; for eruptions, as we before stated, are attended with a copious discharge of aqueous vapour.

Before we attempt to enquire farther into the true causes of earthquakes, we shall briefly recapitulate our reasons for considering them as originating from the same sources as volcanic phenomena. In the first place, the regions convulsed by violent earthquakes include within them the site of all the active volcanos. Earthquakes, sometimes local, sometimes extending over vast areas, precede volcanic eruptions. Both the subterranean movement and the eruption return again and again, at unequal intervals of time, and with unequal degrees of force, to the same places. The duration of both may continue for a few hours, or for several consecutive years. Paroxysmal convulsions of both kinds are usually followed by long periods of tranquillity. Thermal springs, and those containing abundance of mineral matter in solution, are characteristic of countries where active volcanos or earthquakes are frequent. In districts considerably distant from volcanic vents, the temperature of hot springs has been sometimes raised by subterranean movements. In addition to these signs of relation and analogy, we may observe, that it is not very easy to conceive how columns of melted matter can be raised to such great heights, as we know them to attain in volcanos, without exerting an hydrostatic pressure capable of moving enormous masses of land; nor can we be surprised that elastic fluids capable of forcing up so great a weight of rock in fusion, and of projecting large stones to immense heights in the air, should also cause tremors, vibrations, and violent movements in the solid crust of the earth. The volcano of Cotopaxi has thrown a mass of rock, about one hundred cubic yards in volume, to the distance of eight or nine miles, and we may well conceive that the slightest obstruction to the escape of such an expansive force may convulse a considerable tract in South America. "If these vapours," says Michell, "when they find a vent are capable of shaking a country to the distance of ten or twenty miles round the volcano, what may we not expect from them when they are confined?" As there is no doubt that aqueous vapour constitutes the most abundant of the aëriform products of volcanic eruptions, it may be well to consider attentively a case in which steam is exclusively the moving power—the Geysers of Iceland. These intermittent hot springs rise from a large tract, covered to a considerable depth by a stream of lava; and where thermal waters, and apertures evolving steam, are very common. The great Geyser rises out of a spacious basin at the summit of a circular mound, composed of siliceous incrustations deposited from the spray of its waters. The diameter of the basin or crater, in one direction, is fifty-six feet, and forty-six in another.

No. 32.

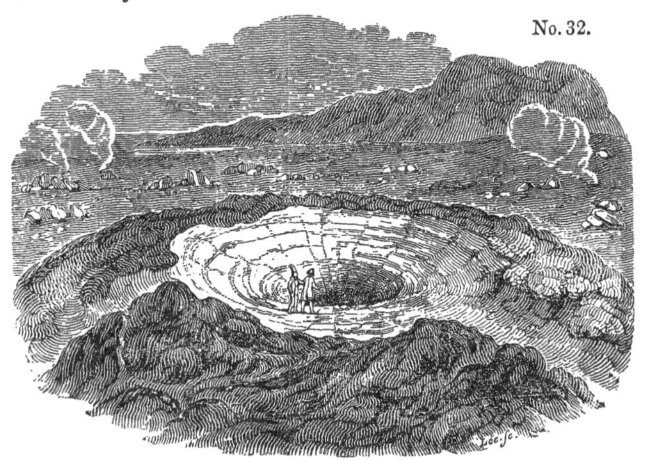

*View of the Crater of the great Geyser in Iceland**.

In the centre is a pipe seventy-eight feet in perpendicular depth, and from eight to ten feet in diameter, but gradually widening as it opens into the basin. The inside of the basin

* Reduced from a sketch given by W. J. Hooker, M.D., in his "Tour in Iceland," vol. i., p. 149.

is whitish, consisting of a siliceous incrustation, and perfectly smooth, as are two small channels on the sides of the mound, down which the water makes its escape when filled to the margin. The circular basin is sometimes empty, as represented in the above sketch, but is usually filled with beautifully transparent water in a state of ebullition. During the rise of the boiling water up the pipe, especially when the ebullition is most violent, and when the water flows over or is thrown up in jets, subterranean noises are heard, like the distant firing of cannon, and the earth is slightly shaken. The sound then increases and the motion becomes more violent, until at length a column of water is thrown up perpendicularly with loud explosions, to the height of one or two hundred feet. After playing for a time like an artificial fountain, and giving off great clouds of vapour, the pipe is evacuated, and a column of steam then rushes up with amazing force and a thundering noise, after which the eruption terminates. If stones are thrown into the crater they are instantly ejected, and such is the explosive force, that very hard rocks are sometimes shivered into small pieces. Henderson found that by throwing a great quantity of large stones into the pipe of Strockr, one of the Geysers, he could bring on an eruption in a few minutes*. The fragments of stone as well as the boiling water were thrown in that case to a much greater height than usual. After the water had been ejected, a column of steam continued to rush up with a deafening roar for nearly an hour; but the Geyser, as if exhausted by this effort, did not give symptoms of a fresh eruption when its usual interval of rest had elapsed.

In the different explanations offered of this singular phenomenon, all writers agree in supposing a subterranean cavity where water and steam collect, and where the free escape of the steam is intercepted at intervals, until it acquires sufficient force to discharge the water. Suppose water percolating from the surface of the earth to penetrate into the subterranean cavity A D by the fissures F F, while at the same time, steam, at an extremely high temperature, such as is commonly given out from the rents of lava-currents during congelation, emanates from the fissures CC. A portion of the steam is at first condensed into

* Journal of a Residence in Iceland, p. 74.

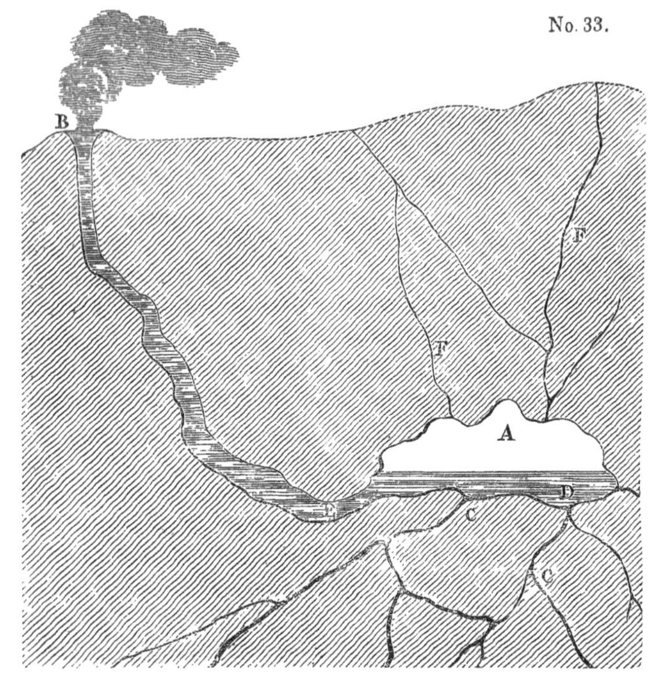

Supposed section of the subterranean reservoir and pipe of a Geyser in Iceland.

water, and the temperature of the water is raised by the latent heat evolved, until, at last, the lower part of the cavity is filled with boiling water and the upper with steam under high pressure. The expansive force of the steam becomes, at length, so great, that the boiling water is forced up the fissure or pipe E B, and a considerable quantity runs over the rim of the basin. When the pressure is thus diminished, the steam in the upper part of the cavity A expands until all the water D is driven to E, when this happens, the steam, being the lighter of the two fluids, rushes up with great velocity, as on the opening of the valve of a steam-boiler. If the pipe be choked up artificially with stones, even for a few minutes, a great increase of heat must take place, for it is prevented from escaping in a latent form in steam, so that the water is made to boil up in a few minutes, and this brings on an eruption.

Now if we suppose a great number of large subterranean cavities at the depth of several miles below the surface of the earth, wherein melted lava accumulates, and that water penetrating to these is converted into steam, this steam, together with other gases generated by the decomposition of melted rocks, may press upon the lava and force it up the duct of a volcano, in the same manner as it drives a column of water up the pipe of a Geyser. But the weight of the lava being immense, the hydrostatic pressure exerted on the sides and roofs of such large cavities and fissures may well be supposed to occasion not merely slight tremors, such as agitate the ground before an eruption of the Geyser, but violent earthquakes. Sometimes the lateral pressure of the lower extremity of the high column of lava may cause the more yielding strata to give way, and to fold themselves in numerous convolutions, so as to occupy less space, and thereby give relief, for a time, to the fused and dilated matter. Sometimes, on the contrary, a weight equal to that of the vertical column of lava, pressing on every part of the roof, may heave up the superincumbent mass, and force lava into every fissure which, on consolidation, may support the arch, and cause the land above to be permanently elevated. On the other hand, subsidences may follow the condensation of vapour when cold water descends through fissures, or when heat is lost by the cooling down of lava.

That lava should often break out from the side or base, rather than from the summit of a lofty cone like Etna, has always been attributed to the immense hydrostatic pressure which the sides of the mountain undergo, before the lava can rise to the crater. This conclusion is too obvious not to have met with a general reception; yet how trifling must this pressure be when compared to that which the same column imparts to the reservoirs of aëriform fluids and melted rock, at the depth of many miles or leagues below the surface!

If earthquakes be derived from the expansion by heat of elastic fluids and melted rock, it is perfectly natural that they should terminate, either when a volcanic vent permits a portion of the pent up vapours or lava to escape, or when the earth has been so fissured that the vapour is condensed by its admission into cooler regions, or by its coming in contact with water. Or relief may be obtained when lava and gaseous fluids have, by distending the strata, made more room for themselves, so that the weight of the superincumbent mass is sufficient to repress them. If we regard earthquakes as abortive volcanic eruptions at a great depth, we must expect them to succeed each other for an indefinite number of times in the same place, for the same reason that eruptions do; and it is easy to conceive that, if the matter has failed several times to reach the surface, the consolidation of the lava first raised and congealed will strengthen the earth's crust, and become an additional obstacle to the protrusion of other fused matter during subsequent convulsions.

As most volcanos are in islands or maritime tracts, the neighbourhood of the sea seems one of the conditions necessary for the ascent of lava to great heights. Even those volcanos which lie inland form part of a chain of volcanic hills, and may be supposed to have a subterranean communication with the extremities of the chain which are in the neighbourhood of large masses of salt-water. Thus Jorullo, in Mexico, though itself no less than forty leagues from the nearest ocean, seems, nevertheless, connected with the volcano of Tuxtla on the one hand, and that of Colima on the other, the one bordering on the Atlantic, the other on the Pacific ocean. This communication is rendered the more probable by the parallelism that exists between these and several volcanic hills intermediate*. Perhaps the quantity of water which percolates from the surface of the land is sufficient to contribute to the violence of earthquakes, without producing so much steam as is required to bring on a volcanic eruption. But when the sea overlies a mass of incandescent lava, and the intermediate crust of the globe is shaken and fissured by earthquakes, it may well be supposed that a convulsion of a different kind will ensue. If an open fissure be caused like that which traversed the plain of S. Lio, on Etna, in 1669, so that the water descends at once upon a mass of melted lava, eruptions will probably burst forth along the line of this aperture, the steam rushing up, together with gaseous emanations from the lava, and carrying up scoriæ with it. But from what we know of the wave-like motion of the ground during earthquakes, there is good reason to conclude that a continuous communication will rarely be formed between the sea and a bed of lava at great depth below, because the alternate

* See Daubeny's remarks on this subject,—"Volcanos," p. 368.

rising and falling of the earth causes chasms to open and again to close in violently. In the same manner, therefore, as yawning fissures shut again after engulphing trees and houses, so great masses of water may be swallowed up, and the sea may immediately afterwards be excluded. Suppose then a volcanic vent to be once formed by a submarine eruption, all the water engulphed will, on penetrating to subterranean reservoirs of heated lava, be converted into steam, and this steam making its way through the same channels by which elastic fluids escape in the intervals between eruptions, will drive melted lava before it. Successive eruptions will have a tendency to seek the same vent, especially if the peak of a cone is raised above the water; for then there will probably be no more than the pressure of the atmosphere in a great part of the duct leading to the crater.

Volcanos exhale, during eruptions, besides aqueous vapour, the following gases: muriatic acid, sulphur combined with hydrogen or oxygen, carbonic acid and nitrogen, the greater part of which would result from the decomposition of salt-water, a fact which, when taken in conjunction with the proximity of nearly two hundred active vents to the sea, and their absence in the interior of large continents, is almost conclusive as to the co-operation of water and fire in the raising of lava to the surface.

We have before suggested the great probability that, in existing volcanic regions, there are enormous masses of matter in a constant state of fusion far below the surface: this opinion is confirmed by numerous phenomena. Perennial supplies of hot vapour and aëriform fluids rise to certain craters, as in Stromboli for example, and Nicaragua, which are in a state of ceaseless eruption. Sangay in Quito, Popocatepetl in Mexico, and the volcano of the isle of Bourbon, have continued in incessant activity for periods of sixty or one hundred and fifty years. Numerous solfataras, evolving the same gases as volcanos, serve as permanent vents of heat generated in the subterranean regions. The plentiful evolution, also, of carbonic acid, from springs and fissures throughout hundreds of square leagues, is another regular source of communication between the interior and the surface. Steam, often above the boiling temperature, is emitted for ages without intermission from "stufas," as the Italians term them. Hot springs in great numbers, especially in tracts where earthquakes are frequent, serve also as regular conductors of heat from the interior upwards. Silex, carbonate of lime, muriate of soda, and many earths, alkalies and metals are poured out in a state of solution by springs, and the solid matter which is tranquilly removed in this manner may, perhaps, exceed that which issues in the shape of lava.

It is to the efficacy of this ceaseless discharge of heat, and of solid as well as gaseous matter, that we probably owe the general tranquillity of our globe; for were it not that some kind of equilibrium is established between fresh accessions of heat and its discharge, we might expect perpetual convulsions, if we conceive the land and the ocean itself to be incumbent in many extensive districts on subterranean reservoirs of lava. If there be reason for wonder, it is, as Pliny observed, that a single day should pass without some dreadful explosion. "Excedit profectò omnia miracula, ullum diem fuisse quo non cuncta conflagrarent *." But the circulation of heat from the interior to the surface, is probably regulated like that of water from the continents to the sea, in such a manner that it is only when some obstruction occurs to the regular discharge, that the usual repose of Nature is broken. Any interruption to the regular drainage of a country causes a flood, and, if there be any obstruction in the passages by which volcanic matter continually rises, an earthquake or a paroxysmal eruption is the consequence.

Michell has observed, that the wave-like motion of the ground during earthquakes, appears less extraordinary if we call to mind the extreme elasticity of the earth, and that even the most solid materials are easily compressible. If we suppose large districts to rest upon the surface of subterranean lakes of melted matter, through which violent motions are propagated, it is easy to conceive that superincumbent solid masses may be made to vibrate or undulate. The following ingenious speculations are suggested by the above mentioned writer. "As a small quantity of vapour almost instantly generated at some considerable depth below the surface of the earth will

* Hist. Mundi, Lib. ii., c. 107.

produce a vibratory motion, so a very large quantity (whether it be generated almost instantly, or in any small portion of time) will produce a wave-like motion. The manner in which this wave-like motion will be propagated may in some measure be represented by the following experiment. Suppose a large cloth, or carpet (spread upon a floor) to be raised at one edge, and then suddenly brought down again to the floor, the air under it being by this means propelled, will pass along, till it escapes at the opposite side, raising the cloth in a wave all the way as it goes. In like manner, a large quantity of vapour may be conceived to raise the earth in a wave, as it passes along between the strata which it may easily separate in an horizontal direction, there being little or no cohesion between one stratum and another. The part of the earth that is first raised, being bent from its natural form, will endeavour to restore itself by its elasticity, and the parts next to it being to have their weight supported by the vapour, which will insinuate itself under them, will be raised in their turn, till it either finds some vent, or is again condensed by the cold into water, and by that means prevented from proceeding any farther *."

In order to account for the retreat of the ocean from the shores before or during an earthquake, the same author imagines a subsidence at the bottom of the sea, from the giving way of the roof of some cavity in consequence of a vacuum produced by the condensation of steam. For such condensation, he observes, might be the first effect of the introduction of a large body of water into fissures and cavities already filled with steam, before there has been sufficient time for the heat of the incandescent lava to turn so large a supply of water into steam, which being soon accomplished causes a greater explosion. Sometimes the rising of the coast must give rise to the retreat of the sea, and the subsequent wave may be occasioned by the subsiding of the shore to its former level; but this will not always account for the phenomena. During the Lisbon earthquake, for example, the retreat preceded the wave not only on the coast of Portugal, but also at

* On the Cause and Phenomena of Earthquakes, Phil. Trans., vol. li., § 58—1760.

the island of Madeira and several other places. If the upheaving of the coast of Portugal had caused the retreat, the motion of the waters when propagated to Madeira would have produced a wave previous to the retreat. Nor could the motion of the waters at Madeira have been caused by a different local earthquake, for the shock travelled from Lisbon to Madeira in two hours, which agrees with the time which it required to reach other places equally distant *.

We shall not indulge at present in further speculations on the mode whereby subterranean heat may give rise to the phenomena of earthquakes and volcanos. No one, however, can fail to be convinced, if he turns his thoughts to the subject, that a great part of the reasoning of the most profound natural philosophers and chemists can be regarded as little more than mere conjecture on matters where the circumstances are so far removed from those which fall under actual observation. Many processes must be carried on in situations where the pressure exceeds as much that produced by the weight of the loftiest mountains, as the weight of the unfathomed ocean surpasses that of the atmosphere. The mechanical effects, therefore, of earthquakes at vast depths, may be such as can never be paralleled on the surface. The intensity of heat must often be so far removed from that which we can imitate by experiments, that the elements of solid rocks or fluids may enter into combinations such as can never take place within the limited range of our observations. Water at a certain depth may, as Michell boldly suggested, become incandescent without expanding, and remain at rest without any tendency to produce an earthquake. Air, if it ever penetrate to such depths, may become a fluid. Sir James Hall's experiments prove, that, under a pressure of about one thousand seven hundred feet of sea, corresponding to that of only six hundred feet of liquid lava, limestone melts without giving off its carbonic acid, so that it is only when calcareous lavas are forced up to within a slight distance of the surface, or into a sea of moderate depth, that the carbonic acid begins to assume a gaseous form, and to assist in bringing on a volcanic eruption.

But let us now turn our attention to those superficial changes

* Michell, Phil. Trans., vol. li. p. 614.

brought about by so many of the earthquakes within the last century and a half, before described. Besides the undulatory movements, and the opening of fissures, it was shewn that certain parts of the earth's crust often of considerable area, both above and below the level of the sea, have been permanently elevated or depressed; examples of elevation by single earthquakes having occurred, to the amount of from one to about twenty-five feet, and of subsidence from a few inches to about fifty feet, exclusively of those limited tracts, as the forest of Aripao, where a sinking down to the amount of three hundred feet took place. It is evident, that the force of subterranean movement does not operate at random, but the same continuous tracts are agitated again and again; and however inconsiderable may be the alterations produced during a period sufficient only for the production of ten or fifteen eruptions of an active volcano, it is obvious that, in the time required for the formation of a lofty cone, composed of thousands of lava-currents, shallow seas may be converted into lofty mountains, and low lands into deep seas. We need, therefore, cherish none of the apprehensions entertained by Buffon, that the inequalities of the earth's surface, or the height and area of our continents, will be reduced by the action of running water; nor need we participate in the wonder of Ray, that the dry land should not lose ground more rapidly. Neither need we anticipate with Hutton the waste of successive continents followed by the creation of others by paroxysmal convulsions. The renovating as well as the destroying causes are unceasingly at work, the repair of land being as constant as its decay, and the deepening of seas keeping pace with the formation of shoals. If, in the course of a century, the Ganges and other great rivers have carried down to the sea a mass of matter equal to many lofty mountains, we also find that a district in Chili, one hundred thousand square miles in area, has been uplifted to the average height of a foot or more, and the cubic contents of the granitic mass thus added in a few hours to the land, may have counterbalanced the loss effected by the aqueous action of many rivers in a century. On the other hand, if the water displaced by fluviatile sediment cause the mean level of the ocean to rise in a slight degree, such subsidences of its bed, as that of Cutch in 1819, or St. Domingo in 1751, or Jamaica in 1692, may have compensated by increasing the capacity of the great oceanic basin. No river can push forward its delta without raising the level of the whole ocean, although in an infinitesimal degree; and no lowering can take place in the bed of any part of the ocean, without a general sinking of the water, even to the antipodes.

If the separate effects of different agents, whether aqueous or igneous, are insensible, it is because they are continually counteracted by each other, and a perfect adjustment takes place before any appreciable disturbance is occasioned. How many considerable earthquakes there may be upon an average in the course of one year, throughout the whole globe, is a question that we cannot decide at present; but as we have calculated that there are about twenty volcanic eruptions annually, we shall, perhaps, not overrate the earthquakes, if we estimate their number to be equal. A large number of eruptions are attended by local earthquakes of sufficient violence to modify the surface in some slight degree, and there are many earthquakes, on the other hand, not followed by eruptions. Even if we do not assume, as many have done, that the submarine convulsions exceed in number and violence those on the land, in spaces of equal area, we must, nevertheless, reckon about three shocks exclusively submarine, for one exclusively confined to the continents.

We have said in a former chapter * that the aqueous and igneous agents may be regarded as antagonist forces, the aqueous labouring incessantly to reduce the inequalities of the earth's surface to a level, while the igneous are equally active in restoring the unevenness of the crust of the globe. But an erroneous theory appears to have been entertained by many geologists, and is indeed as old as the time of Lazzoro Moro, that the levelling power of running water was opposed rather to the *elevating* force of earthquakes than to their action generally. To such an opinion the numerous well-attested facts of subsidences must always have appeared a serious objection, but the same hypothesis would lead to other assumptions of a very arbitrary and improbable kind, inasmuch as it would be necessary to imagine the magnitude of our

* Chap. x. p. 167.

planet to be always on the increase if the elevation of the earth's surface by subterranean movements exceeded the depression. The sediment carried into the depths of the sea by rivers, tides, and currents, tends to diminish the height of the land; but, on the other hand, it tends, in a degree, to augment the height of the ocean, since water, equal in volume to the matter carried in, is displaced. The mean distance, therefore, of the surface, whether occupied by land or water from the centre of the earth, remains unchanged by the action of rivers, tides, and currents. Now suppose that while these agents are destroying islands and continents, the restoration of land should take place solely by the forcing out of the earth's envelope—it will be seen that this would imply a continual distension of the whole mass of the earth. For the greater number of earthquakes would be submarine, and they would cause the sea to rise and submerge the low lands even in a greater degree than would the influx of sediment. Two causes would, therefore, tend to destroy the land; submarine earthquakes, and the destroying and transporting power of water; and in order to counterbalance these effects, shallow seas must be upraised into continents, and low lands into mountains.

If we first consider the question simply, in regard to the manner whereby earthquakes may prevent running water from altering the relative proportion of land and sea, or the height of the land and depth of the ocean, we shall find that if the rising and sinking be equal, things would remain upon the whole in the same state: because rivers, tides, and currents, add as much to the height of lands which are rising, as they take from those which have risen.

Suppose a large river to carry down sediment into a certain part of the ocean where there is a depth of two thousand feet, and that the whole space is reduced by the fluviatile depositions to a shoal only covered by water at high tide: then let a series of two hundred earthquakes strike the shoal, each raising the ground ten feet; the result will be a mountain two thousand feet high. But suppose the same earthquakes had visited the same hollow in the bottom of the sea before the sediment of the river had filled it up, their whole force would then have been expended in converting a deep sea into a shoal, instead of changing a shoal into a mountain two thousand feet high. The superior altitude, then, of a district may often be due to the transportation of matter at a former period *to lower levels.* It would probably be more consistent with the natural course of events, if, instead of a succession of elevatory movements, we were to suppose considerable oscillations before the district attained its full height. Let there be, for example, three hundred instead of two hundred shocks, each separated from the other by intervals of about fifty years. Let the mean alteration of level produced by each earthquake be ten feet, two hundred and fifty shocks causing a rise, and the other fifty a sinking in of the ground; although more time will have been consumed by this operation than by the former, we shall still have the same result, for a tract will be raised to the height of about two thousand feet. The chief difference will consist in the superior breadth and depth of the valleys, which will be greater nearly in the proportion of one-third, in consequence of the number of landslips, floods, opening of chasms, and other effects produced by one hundred additional earthquakes. It should be borne in mind, moreover, that some of the lowering movements, happening towards the close of the period of disturbance, may have given rise to strange anomalies, should an attempt be made to reconcile the whole excavation in various hydrographical basins to the levels finally retained. Perhaps, for example, the middle portion of a valley may have sunk down, so that a deep lake may intervene between mountains and certain low plains, to which their debris had been previously carried.

But to return to the consideration of the proportion between the elevation and depression of the earth's crust, which may be necessary to preserve the uniformity of the general relations of land and sea, on the surface. The circumstances are in truth more complicated than those before stated, for, independently of the transfer of matter by running water from the continents to the ocean, there is a constant transportation of mineral ingredients from below upwards, by mineral springs and volcanic vents. As mountain masses are in the course of ages created by the pouring forth of successive streams of lava, so others originate from the carbonate of lime and other mineral ingredients with which springs are impregnated. The surface of the land, and parts of the bottom of the sea are thus raised, and if we conceive the dimensions of the planet to remain uniform, we must suppose these external accessions to be counteracted by some action of an opposite kind. A considerable quantity of earthy matter may sink down into fissures caused by earthquakes, but this cannot be deemed sufficient to counterbalance the addition of mountain masses by the causes before adverted to, and we must therefore suppose, that the subsidences of the earth's crust exceed the elevations caused by subterranean movements. It is to be expected, on mechanical principles, that the constant subtraction of matter from the interior will cause vacuities, so that the surface undermined will fall in during convulsions which shake the earth's crust even to great depths, and the sinking down will be occasioned partly by the hollows left when portions of the solid crust are heaved up, and partly when they are undermined by the subtraction of lava and the ingredients of decomposed rocks. The geological consequences which will follow if we embrace the theory now proposed are very important, for if there be upon the whole more subsidence than elevation, then we must consider the depth to which former surfaces have sunk down beneath their original level, to exceed the height which ancient marine strata have attained above the sea. If, for example, marine strata about the age of our chalk and green-sand have been lifted up in Europe to an extreme elevation of more than eleven thousand feet, and to a mean height of some hundreds above the level of the sea, we may conclude that certain parts of the earth's surface, which existed whether above or below the waters when those strata were deposited, have subsequently sunk down to an extreme depth of *more than* eleven thousand feet below their original level, and to a mean depth of *more than* a few hundreds.

In regard to faults, also, we must, according to the hypothesis now proposed, infer that a greater number have arisen from the sinking down than from the elevation of rocks. If we find, therefore, ancient deposits full of fresh-water remains which evidently originated in a delta or shallow estuary, covered subsequently by purely marine formations of vast thickness, we shall not be surprised; for we must expect that a greater number of existing deltas and estuary formations will sink below, than those which will rise above their present level.

Although it would be rash to attempt to confirm these speculations by reference to the scanty observations hitherto made on the effects of earthquakes, yet we cannot but remark, that the instances of subsidence on record are far more numerous than are those of elevation.

Those writers who have most strenuously contended for the analogy of the effects of earthquakes in ancient and modern times, have nevertheless declared that the energy of the force has considerably abated. But they do not appear to have been aware that, in order to adduce plausible grounds for such an hypothesis, they must possess a most extensive knowledge of the economy of the whole terrestrial system. We can only estimate the relative amount of change produced at two distinct periods, by a particular cause in a given lapse of time, when we have obtained some common standard for the measurement of equal portions of time at both periods. We have shown that, within the last one hundred and forty years, some hundred thousand square miles of territory have been upheaved to the height of several feet, and that an area of equal, if not greater extent, has been depressed. Now, they who contend, that formerly more movement was accomplished by earthquakes in the space of one hundred and forty years, must first explain the measure of time referred to, for it is obvious that they cannot in geology avail themselves of the annual revolution of our planet round the sun. Suppose they assume that the power of volcanos to emit lava, and of running water to transport sediment from one part of the globe to the other, has remained uniform from the earliest periods, they might then attempt to compare the effects of subterranean movements in ancient and modern times by reference to one common standard, and to show that, while a certain number of lava-currents were produced, or so many cubic yards of sediment accumulated, the elevation and depression of the earth's crust were once much greater than they are now. Or, if they should declare that the progressive rate of change of species in the animal and vegetable kingdoms had always been uniform, they might then endeavour to disparage the degree of energy now exerted by earthquakes, by showing that, in relation to the mutations of assemblages of organic species, earthquakes had become comparatively feeble. But our present scanty acquaintance, both with the animate and inanimate world, can by no means warrant such generalizations; nor have they who contend for the gradual decline of the activity of natural agents, attempted to support such a line of argument. That it would be most premature, in the present state of natural history, to reason on the comparative rate of fluctuation in the species of organic beings in ancient and modern times, will be more fully demonstrated when we proceed, in the next division of our subject, to consider the intimate connexion between geology, and the study of the present condition of the animal and vegetable kingdoms.

To conclude: it appears, from the views above explained, respecting the agency of subterranean movements, that the constant repair of the dry land, and the subserviency of our planet to the support of terrestrial as well as aquatic species, are secured by the elevating and depressing power of earthquakes. This cause, so often the source of death and terror to the inhabitants of the globe, which visits, in succession, every zone, and fills the earth with monuments of ruin and disorder, is, nevertheless, a conservative principle in the highest degree, and, above all others, essential to the stability of the system.

INDEX.

London: Printed by W. Clowes, Stamford-Street.

Printed in the United States
By Bookmasters